The Patrick Moore Practical Astronomy Series

More information about this series at http://www.springer.com/series/3192

Building and Using Binoscopes

Norman Butler

Second Edition

Norman Butler
Tamuning, Guam

ISSN 1431-9756 ISSN 2197-6562 (electronic)
The Patrick Moore Practical Astronomy Series
ISBN 978-3-319-46788-7 ISBN 978-3-319-46789-4 (eBook)
DOI 10.1007/978-3-319-46789-4

Library of Congress Control Number: 2017930814

Printed on acid-free paper

This Springer imprint is published by Springer Nature
The registered company is Springer International Publishing AG
The registered company address is: Gewerbestrasse 11, 6330 Cham, Switzerland

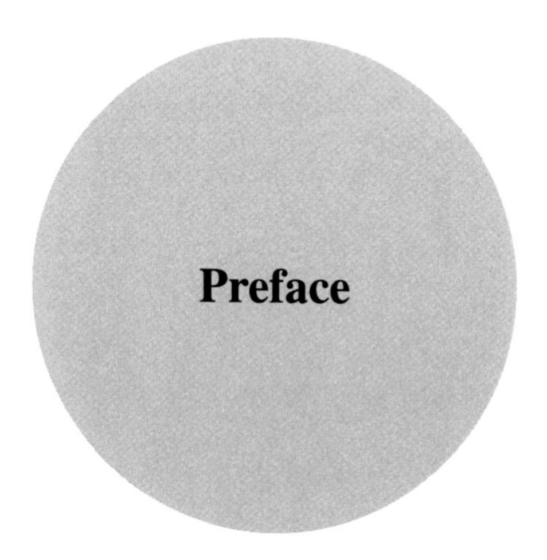

Preface

As an amateur astronomer living in Topeka, Kansas, in the 1970s, on clear nights, away from the city, one could easily see our spectacular Milky Way galaxy arching across the beautiful night sky. Seeing this wonderful sight night after night through my trusty telescope really got me thinking about how spectacular these wonderful celestial objects would look through a large binocular telescope. To realize my dream, in the late 1970s, I started on my quest to build my first large binocular telescope. By November of 1980, I had completed a dual 6 in. f/15 Cassegrain Dall-Kirkham binocular telescope (10 mirrors) on a clock-driven equatorial mount complete with a 360° steel ring OTA rotation system.

In terms of this book, this is not a "How To" book. It's a book that is designed to give the interested readers, amateur astronomers and amateur telescope builders (experienced or otherwise) ideas, suggestions and important information about "building" and "using" a refractor binoscope or binocular telescope. If you can already build and use a telescope and/or interested in doing so, then after you read this book, you should be able to take some of the more interesting and clever ideas that are presented in the following 10 chapters and incorporate them into building and using your own telescope, binoscope or binocular telescope someday. A binoscope and a binocular telescope are one and the same. They both do basically the same thing, which is allowing the observer to view celestial objects with two telescopes, using both eyes. In this book, I wanted to show as many photos as possible different kinds of homemade (some commercial) refractor and reflector binoscopes, binocular telescopes, and standard Dobsonian and Cassegrain telescopes to demonstrate how resourceful and creative today's amateur telescope makers really are. The homemade binoscopes pictured in this book should provide the interested reader with some good ideas on building their own binoscope someday.

It does not take a whole lot of imagination to understand the simple optics and mechanics behind building a binoscope or a large binocular telescope. What it does take is a little more expense for the cost of optics (times 2) and twice as much materials, time, and ambition to build, for example, a Dobsonian-style binocular telescope or binoscope compared to constructing a single telescope. Beyond that, once you have made the commitment to start your binoscope project by creating a robust design and doing all the advance planning and homework, then you should really look forward to your project. And after your binoscope project is finally completed, and you get your first views through it, then that's when the fun begins. Welcome to the binoscope club!

Please note: All of the author's own proceeds from the sale of this new Springer second edition book *Building and Using Binoscopes* and the first edition of the same book goes to help fund a junior college astronomy scholarship fund.

Northern Marianas Islands, USA Norman Butler
June 2016

Acknowledgments

I wanted to take this opportunity to acknowledge some individuals who encouraged me along the way to make this literary endeavor about binoscopes into a successful reality. I wish to extend my sincere thanks and deep appreciation to John Bauer, my late, great, dear friend, mentor, and astronomy professor during my early college years at San Diego City College. Without his inspiration, support, and guidance, I would have probably never made it this far in astronomy, especially as a career, and worked at the observatory level in Hawaii. Likewise, I wish to thank astronomers Fred Pilcher, Richard Wilds, Rick Schmidt, and Ray Lewis. All four of them, who in my distant past, encouraged me at some point to pursue some of my lofty goals in astronomy. Likewise, I wish to thank those individuals who contributed important images, information, website stories and illustrations for inclusion in my new book. And finally, I thank my dear mother Gladys and my brother Mike who were both always my main source of encouragement for all of my astronomy endeavors, no matter how outrageous they were.

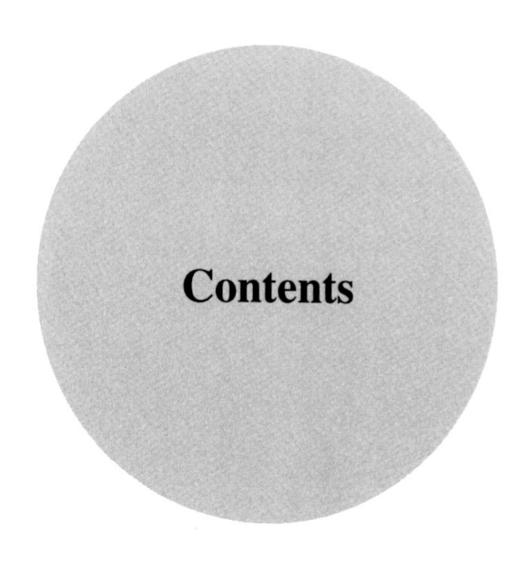

Contents

About the Author

Norman Butler is a noted, award-winning telescope maker who has made some very unique, one-of-a-kind binocular telescopes. He has also worked in the field of astronomy and electro-optical engineering for AVCO Everett Research Laboratory at the Haleakala Observatory on the island of Maui starting in the early 1980s, building electro-optical equipment for use on both 1.6 M and dual 1.2 M telescopes. A graduate of San Diego City College, Norman holds advanced degrees in physics and astronomy including a Ph.D. He also served in the US Navy as an opticalman on submarine tenders repairing submarine periscopes and optical navigational equipment throughout the entire 1960s. After a 20-year career in electro-optical engineering, starting in 1994, Norman relocated to Hong Kong and started working in nearby Shenzhen, China as a joint-venture manager in the electronics industry. In 2004, he became a permanent resident of Hong Kong and started teaching at Shenzhen Polytechnic College and Harbin Institute of Technology Shenzhen Graduate School in nearby Shenzhen, China. Norman retired in 2012 and now lives in the Northern Marianas Islands and enjoys searching for comets as well as other strange and mysterious cosmic interlopers under the beautiful dark tropical skies of Saipan and Guam.

Chapter 1

Why Binoscopes?

Just about all of us enjoy learning about the universe and looking at all of its celestial wonders through a telescope. It can be a lot of fun and educational at the same time. But when it comes to wanting to see more in the night sky than just observing with a single telescope, then that's when you start to think about getting a bigger telescope or even a big pair of binoculars. Using two eyes to view the universe with is perhaps one of the most satisfying ways to enjoy doing visual astronomy. Using a pair of binoculars can certainly make your observing experience a lot more satisfying. But what about using a binoscope or binocular telescope to observe the heavens with?

Binoscopes are a wonderful way to explore the night sky. This new book on binoscopes gives the reader a good opportunity to find out why a binoscope or binocular telescope is fun and exciting to use for astronomical observing. The book is filled with a wide variety of photos of homemade, commercial binoscopes and binocular telescopes and a few telescope cartoons thrown in for fun and levity. It also describes how binoscopes perform optically and what you can expect to see when you observe with a binoscope or a pair of big binoculars.

The book offers some practical advice, some suggestions and examples from other experienced binoscope makers on how to design and build a binoscope or a binocular telescope, as well as presenting some of my own homemade one-of-a-kind binoscope examples in Chap. 5 for doing so. In writing this book, it was important for the author to let the interested reader see a variety of interesting homemade and commercial binoscopes. There is also discussion and personal comments from actual binoscope makers and users about the advantages that a binoscope or a binocular telescope has compared to a single telescope. The book will help you decide to either try one, buy one, or make one of your own someday. If you want to build your own binoscope or binocular telescope, then

N. Butler, *Building and Using Binoscopes*, The Patrick Moore Practical Astronomy Series, DOI 10.1007/978-3-319-46789-4_1

you'll find out that making a binoscope or binocular telescope can be a lot of fun, especially if you intend to share the observing experience with others. A binoscope is also great for comet hunting. If you are an active comet hunter, then there is no better way to search the night skies looking for a new comet than using a binoscope. If anything is going to help you find a comet visually, then using a binoscope or binocular telescope is going to be your best bet. And don't forget to add a little bit of luck with that too.

This book is filled with lots of photos of homemade refractor binoscopes and binocular telescopes (some commercial) that were made by some very creative and resourceful amateur telescope makers from all over the world. Some of the new and clever ideas that amateur telescope makers are building into their homemade telescopes today may very well show up in commercial telescopes and binoscopes in the future. When you are trying to decide what you want and really need for your astronomical observing, a binoscope will provide you with some great visual observing using both eyes along with twice as much light gathering power compared to a single telescope of the same aperture. If you have the opportunity to view through a large binocular telescope or a big refractor binoscope, you won't forget the fantastic view, the detail, the vivid colors or the hues of the nebula and star clusters. You will always remember the wonderful stereo view of the planets and the large craters on the moon surrounded by a rugged lunar landscape. When you are observing with a big binocular telescope or binoscope, you are in for a truly memorable visual experience. Observing the heavens with a pair of big binoculars is a lot of fun too. However, we're talking about big binoculars that are in the 100–150 mm range. Not everyone can afford to go out and pay five or six thousand dollars for a pricey pair of 25 × 150 mm Fujinon binoculars. It's probably going to be a lot cheaper to just build a binocular telescope or a refractor binoscope and save a lot of time and money in doing it. One of the important things that most astronomers want to get out of their observing sessions is viewing these wonderful astronomical objects and seeing as much detail as their telescopes can provide. And that's why using two eyes to observe with will no doubt make their observing sessions a more memorable experience, especially if they're using a big binocular telescope or refractor binoscope to observe with. As a builder of binocular telescopes and refractor binoscopes, the author has always encouraged people at local star parties and national telescope makers conferences to look through his equipment so that they can enjoy the wondrous views that only a big binoscope or big binocular telescope can provide. Once they see the fantastic views these remarkable big binoculars offer, they too will become smitten with the big binoscope bug and want to observe the heavens with one of their own.

What Can One Expect to See in a Binoscope?

The first thing that will become apparent when you look into the eyepieces of a binoscope or binocular telescope, you'll begin to see fainter stars and brighter nebula with a wider field of view and greater depth, combined with a very noticeable stereo effect, especially with lunar and planetary objects. You'll see hints of more extended detail and greater resolution in most celestial objects. All of this combines to give the observer a very memorable viewing experience. Each observing session with your binoscope becomes special, where you will find new and interesting objects that you haven't seen before. Probably the best way to decide if you want a binoscope or binocular telescope is to try and find a local astronomy club in your area that has a monthly star party. Ask if someone in their astronomy group has a big binocular telescope or refracting binoscope. If they do, try to attend one of their star parties and then you'll have an excellent opportunity to look through a big binoscope and see why so many amateur astronomers enjoy using one during their observing sessions.

There have been many amateur astronomers who have taken either a homemade or commercial telescope and its Altazimuth or equatorial mounting and simply made a binoscope out of it. On the Internet are many telescope companies and Internet stores that sell a wide variety of accessories and that also carry a line of parts that can easily adapt a commercial (see Fig. 5.14) or even a homemade Altazimuth or commercial equatorial (see Figs. 5.28 and 5.29) telescope mount to handle a refractor binoscope. Sometimes it's just a little easier and perhaps cheaper to just modify your existing mount to carry two single telescopes and make a binoscope out of it. The actual cost of a refractor binoscope with a set of good ED optics (extra-low dispersion glass) is quite expensive and usually out of the price range of the average individual. A good achromatic binoscope on the other hand is far cheaper compared to a pair of expensive ED refractor optics and can still yield some very pleasing views when used in a refractor binoscope.

If you have made the decision to build your own binoscope, be it a binocular telescope or refractor binoscope, then that's when the real fun begins. Building your own binoscope becomes for most amateur telescope builders, a labor of love. Obviously, a binoscope is not for everyone that's interested in doing serious astronomical observing. And not everyone wants to build their own telescope or binoscope. But the author is sure there are many serious amateur astronomy minded individuals who would probably love to build their own telescope or binoscope, but finding the time and having the skills and necessary tools do it can be just another item on their wish list. However, for those who do build their own binoscope or binocular telescope, then for them, it becomes more than a project, it becomes a passion. It may take several months to build your binoscope or telescope project, but taking your time and building the binoscope or telescope the way you dreamed you could build it will make it turn out to even better than you ever expected. The key in building a good reliable binoscope or telescope, is taking your time without being in a hurry to complete it and building it with a good robust design and using quality materials that are long lasting and that can take some wear and tear without having to replace or repair things that normally shouldn't have failed in the first place. But things happen, and, over time, you may want to give your binoscope or binocular telescope a new coat of paint or replace a focuser or even remake something that can make it more efficient optically or mechanically. That's the fun of being an amateur telescope maker and building your own telescope or binoscope and the best part is…you did it yourself.

Going to Build a Monster Binocular Telescope?

From a historical perspective, the modern binocular telescope age really got started in the late 1920s when Mr. Hilmer Hanson of Nebraska built his first 6-in. Newtonian binocular telescope, a historic telescope (see Fig. 7.1). Since then, the evolution of binoscopes had been slow, until Lee Cain (see Fig. 5.2) in the early 1980s made a remarkably big 17½-in. Dobsonian binocular telescope (commonly called a “Dob” for short). Thirty-four years later, Dobsonian style binocular telescopes are a lot more common now than they were in the past. In the last 20 years, they have been getting bigger and bigger (see Figs. 1.36 and 1.4), and with the availability of larger and less costly thinner mirrors, it won’t be too long before we’ll see a real big “Monster” binocular telescope in the works. For example, building a big “Monster” 36-in. Dobsonian binocular telescope will be quite a job and will take a lot more time to construct and twice as much materials and expense compared to making a single 36-in. Dobsonian telescope. But after it is finished, it will undoubtedly provide the observer with celestial views that will be nothing short of spectacular. It too will become a historic telescope!

So what makes a telescope “Historical”? There are perhaps a few criteria that would possibly make a telescope have some historical interest. One good example would be if it was the “first” of its kind and the actual telescope maker who made it. A good example would be; Sir Issac Newton and his original Newtonian reflecting telescope. So how does the word “historical” relate to Hilmer Hanson and his unique homemade 1920s 6-in f/8 Newtonian binocular telescope? Answer is; Find another Newtonian binocular telescope that was built before the 1920. If in your search you happen to find one that was built before the 1920s and confirm who the actual builder was and the date it was built, then it most likely would be considered as “historical” binocular telescope. The source of this information would likely be found and documented in an old scientific magazine, astronomy book or periodical. Most likely, one would have to find and sort through quit a few different sources inorder to find a binocular telescope that actually predates Hilmer Hanson’s 1920s original Newtonian binocular telescope.

Before you start building a “Monster” Dobsonian binocular telescope, let’s take a look back in telescope making history and see how the big Dobsonian revolution was started. To begin with, let’s give some “kudos” to the late John Dobson (who the Dobsonian telescope is named after). In the 1960s and 1970s, John Dobson helped popularize the Dobsonian telescope (via The San Francisco Sidewalk Astronomers) with his use of cheap materials, cardboard Sonotubes, and thinner primary mirrors that he fashioned himself and placed in a simple Altazimuth mount. Dobson also let the public have an opportunity to view the heavens through his simple but nevertheless, humble Dobsonian sidewalk telescopes. The Dobsonian became the standard design for just about all Altazimuth mounted Newtonian telescopes during that period of time. The Dobsonian telescope became so popular that Coulter Optical Co. in 1980 decided to introduce a “Blue” cardboard Sonotube 13.1-in. f/4.5 Dobsonian. It later was changed to a “Red” sonotube and eventually offered in an f/7 version with a simple rectangular wooden "rocker box" at the rear to house the primary mirror and pinion bearings.

Coulter's Dobsonian telescopes eventually increased in size to 17.5 in. and in the late 1980s were considered the "Monster" Dobs of their time. Even though these big "rocker box" Dobsonians were extremely heavy and very bulky to transport, once set up, the views that these big "Dobs" provided were totally awesome and set the standard for the Dobsonian telescopes we see today (see Fig. 1.1).

Even today, a 1980s Coulter style "Red" or "Blue" Sonotube Dobsonian is still a crowd pleaser at local star parties. At national telescope maker's conferences, they are considered as a classic Dobsonian telescope to be admired as a standard even for modern Dobsonian telescopes that amateur telescope makers and commercial telescope companies are building today (Fig. 1.2).

Even though the lighter weight "open truss" (see Fig. 5.8) or "tube truss" (see Fig. 5.56) Dobsonian telescopes are probably the most common types in use today, Sonotube style Dobsonians are still around and probably will be for a long time. And don't we wish we could buy a big 13.1-in. Dobsonian today for the same price we bought one for in the 1980s (US$395.50)?

It's interesting to note (at least from the author's perspective) that after 1984, it took another 10 years or so to start seeing homemade binocular telescopes and refractor binoscopes beginning to "pop-up" at local star parties, amateur telescope maker events, and astronomy conferences. The idea of combining two big Dobsonian Newtonian telescopes together in a single Altazimuth mount for using

Fig. 1.1 A 1980s Dobsonian style Newtonian "rocker box" housed in an Altazimuth mount (Image credit: Halfblue-Wikimedia Commons)

Fig. 1.2 Greg Boles of Topeka, Kansas, is using a 1980s Coulter Optical "Blue" tube "rocker box" style 13.1-in. f/4.5 Dobsonian to observe the heavens with (Image credit: Rick Schmidt)

two eyes for viewing (in this author's mind) should have happened earlier, but for some reason was slow to catch on. Even more interesting is the fact that there was about a 60-year gap since Hilmer Hanson (see Fig. 7.1) built his first Newtonian binocular telescope back in the 1920s before we started to see any binocular tele-scopes appear in any real frequency.

It's possible perhaps it was just a matter of convenience for the 60-year gap, and amateur astronomers were perfectly happy just using a big Newtonian telescope for their astronomy observing and obviously, there were other reasons too. Since that time, hardly anything new appeared until 1984, when Lee Cain (see Fig. 5.2) built his remarkable 17.5-in. binocular telescope and Altazimuth mount. After that, the rest is history. Binocular telescopes and refractor binoscopes started to "pop-up" on occasion in the 1990s, and then amateur astronomers and amateur telescope makers began to realize that two eyes are better than one to view the heavens with, especially with a binocular telescope or refractor binoscope.

Since 2004, at least two 22-in. binocular telescopes have already been built (see Fig. 5.8). Even now, a new homemade 28-in. f/4.8 binocular telescope (mirrors and all) has been built by Mr. Joerg Peters of Germany. There is no doubt that it is the current "king" of the amateur-made binocular telescopes. With today's ATMs wanting to build bigger and bigger binoscopes, it's just a matter of time someday before we see a really big "Monster" 36-in. binocular telescope being unloaded off of a truck at a local astronomy club's star party or at a national telescope maker's conference. When that happens, then the sky's the limit when it comes to building a big binocular telescope. One can only imagine how large they will become in the next 25 years or so. So what will be some of the tasks you will need to do after you build a "Monster" binocular telescope? Below is a preview of what you can expect to do and the "estimated" amount of time it will take to do it if you plan to transport it to different observing sites on occasion.

1. Transporting it to and from your favorite observing sites will take a big concerted effort. Probably at least two people will be needed to help load everything into a small truck, and that may take more than 1 or 2 h to load. Don't forget to bring the coffee!
2. The estimated amount of setup time needed for assembly of the big 36-in. Dobsonian binocular telescope and its Altazimuth mount will probably require a minimum of two people at least 2 h of setup time after reaching your observing spot early in the day.
3. Once it has been setup, then comes the collimation and alignment. After the two big Dobsonian Newtonian optical systems have been collimated, someone in the early evening will get up on a big ladder (see Fig. 1.3), point the big binocular telescope at some bright star, center it in the field of view (FOV) in one of the telescope's two eyepieces, and do the final alignment "tweaking" by yelling out commands to someone on the ground to twist and tighten some knobs and bolts to

Fig. 1.3 A Dobsonian Bino (Image credit: Public Domain—Wikimedia Commons)

lock in both big Dobsonian binocular telescopes on the same object (two people and possibly 1 h>A total est. time for collimation and alignment 2 h minimum).

After the collimation and alignment has been completed, the "First Light" celestial object we want to check out with the big 36-in. binocular telescope would be M-42, the Great Nebula in Orion. M-42 is made up of wonderful vivid colors of flowing clouds of hydrogen gas, dust, and young bright stars that are in their infancy (see Fig. 1.6). Of course, there is a big list of objects that everyone wants to see, and we already have a line of anxious people waiting to get a look through our "Monster" Dobsonian. The line will get longer as the night wears on. The big "Monster" binocular telescope once it's setup and ready for observing would not only draw a long line of anxious individuals waiting for their turn to take a look but would be a big crowd pleaser too. As fun as it may be, there are some concerns taking a big "Monster" binocular telescope to local star parties or even popular telescope makers conference.

1. Requires the use of a big ladder (see Fig. 1.4) that people have to climb up and down on.
2. If a big "Monster" binocular telescope has "no" GOTO or motorized drive capabilities, it would take time and some effort to locate each object and position (push and pull) it until the object is brought into the field of view (FOV). It would be a lot more difficult to maneuver a big Dobsonian by hand at the zenith. (A motorized dual-axis drive system is a must.)
3. Because of the overall height and maneuverability of a big "Monster" binocular telescope, there probably would be very little observing at the zenith. It would be a little "shaky" for an individual standing near the top of a tall ladder trying to

Fig. 1.4 A Texas-size 48.9-in. Dobsonian called "Barbarella" (Image credit: Astronomy Technology Today)

maneuver the big dual "Dob" into position and observe objects at the zenith and keeping them in the field of view at the same time. It can be done, but be careful when you do.

Besides having to climb a "tall" ladder and trying to maneuver the big dual Dobsonians into position by hand will be challenging enough, but viewing with a big "Monster" binocular telescope will be the ultimate visual observing experience an observer can have.

Now, consider standing on a ladder nearly 20 ft off the ground to do some observing through a big 48.9-in. Dobsonian nicknamed "Barbarella"? This is probably one of the biggest amateur Dobsonian made to date. (On a sidenote, Utah amateur Mike Clements has since made a 70-in. Dob!) Even its Altazimuth mounting is huge, dwarfing a small car. The mirror alone weighs approximately 715 lb and is 4.8-in. thick. The big "Monster" open truss Dobsonian is approximately 20 ft. long. From looking at the photo above, another 5 ft. or 6 ft. of ladder might be needed in order to reach the zenith. Try and picture a big "Monster" Dobsonian binocular telescope this size at a local star party, in Texas maybe (Figs. 1.5 and 1.6).

To coin a phrase, "one picture is worth a thousand words," and the photo of the Great Nebula in Orion taken by the Hubble Space Telescope pictured below proves it. The imagined view of M-42 in a big 36-in. binocular telescope will no doubt be absolutely "stunning" and one that will be remembered for a long time or at least until the next observing session with the big binocular telescope.

Fig. 1.5 (Cartoon credit: Jack Kramer)

Fig. 1.6 The Great Nebula in Orion (Image Credit: Hubble-Wikimedia Commons)

How Does Surface Flatness Relate to Optical Quality?

One of the important things to remember when you have a binocular telescope or you're planning to build one is that no matter how good your primary mirror is, the quality, or surface flatness, of your secondary mirror and star diagonal is equally important. The secondary mirror and star diagonal are critical optical components of your binocular telescope or refractor binoscope and your telescope's optical performance can depend on their true optical quality.

For example, if you have a primary mirror that is certified to be 1/10 wave and your secondary mirror is rated at a 1/4 wave, then what does it really mean in terms of its optical quality? The majority of optical manufacturers specify the optical surface flatness of their optics in a "peak-to-valley" wavelength configuration. Keep in mind that a manufacturer's "peak-to-valley" surface flatness specification can sometimes be a bit imperfect and even slightly misleading. The table below is a very brief description adapted from Optical-technologies.info of Peak-to-Valley vs. RMS of what one should expect to see in the advertised optical quality of optics. The same information can also apply to secondary mirrors, optical flats and mirror star diagonal too.

Surface flatness (peak-to-valley)	Optical Quality	Type of Application
Less than λ/2 (1/2 wave)	Very low (poor)	Recommended for noncritical diverging applications only.
λ/4 (1/4 wave)	Low	One of the highest test standards for cube beam splitters. Not considered to be acceptable for high power applications if wave-front error control is the primary element.
λ/10 (1/10 wave)	Good	Considered the accepted standard for a quality manufacturer. Suitable for most laser and scientific applications.
λ/20 (1/20 wave)	Very good	Preferred surface flatness for critical wave-front error control application, such as interferometry and/or the use of intense femtosecond laser applications and optical laboratory experiments.

According to Jean Texereau, the famous French optician, the minimal acceptable surface accuracy for a secondary mirror is 1/8 wavelength. It is generally believed that the cheaper commercial telescopes do not meet this requirement or get close to it (Fig. 1.7).

Some manufacturers of telescope optics often refer to the optical standard called "Rayleigh's criterion." This is the criterion for the resolving power or angular resolution of an optical instrument. The formal Rayleigh criterion states that the images of two-point objects are resolved when the principal maximum of the diffraction pattern of one image falls exactly on the first minimum of the diffraction pattern of the other image. Basically, if you have a telescope, it's normal that you

Fig. 1.7 (Cartoon credit: Jack Kramer)

would want to know if its optical system is "diffraction limited" or not. If it isn't, and you want to do some serious astronomy with it, then obviously you will never be happy with its overall performance.

In a typical telescope optical system in which the resolution is no longer limited by imperfections in the lenses, but only by diffraction itself, it is said to be diffraction limited.

For example, point-like stellar sources that are separated by an angle smaller than the angular resolution of a telescope's optical system therefore cannot be resolved. A telescope that has a near-perfect optical system may have an angular resolution less than 1 arcsecond, but astronomical seeing and other atmospheric effects make reaching this angular resolution limit extremely difficult to achieve.

The angular resolution R of a telescope can usually be determined by:

$$R = \frac{I}{D}$$

Where:

λ is the wavelength of the observed radiation and

D is the diameter of the telescope's objective or aperture.

Resulting R is in radians. Sources larger than the angular resolution are more commonly called extended sources or diffuse sources (example: nebulae), and smaller sources are commonly called point sources. For example, in yellow light with a wavelength of 580 nm, for a resolution of 0.1 arcsecond, we need $D = 1.2$ m. This formula, for light with a wavelength of about 562 nm, is also called and known as Dawes' limit.

The formal Rayleigh criterion is close to the resolution limit discovered earlier by the noted English astronomer W. R. Dawes who recorded skilled celestial observers with tested experience observing close binary star systems (or double stars) that were the same or nearly equal in brightness. The result, with D inches θ in sub-arcseconds, was found to be slightly narrower than the original calculation compared to the Rayleigh criterion. Using airy disks as a point spread function, the math shows that at Dawes limit, there is a modest 5% dip between the two maximum peaks, whereas at Rayleigh's Criterion, there is an extremely large and quite noticeable 26.4% dip. Using the point spread function will typically allow resolution of close binary systems with an even smaller separation.

Calculations used to determine Dawes' limit using basic common units of measurement can present different elementary forms (see below) (Fig. 1.8):

$$R = \frac{4.56}{D} \quad D \text{ in inches}, R \text{ in arc seconds}$$

$$R = \frac{11.58}{D} \quad D \text{ in centimeters}, R \text{ in arc seconds}$$

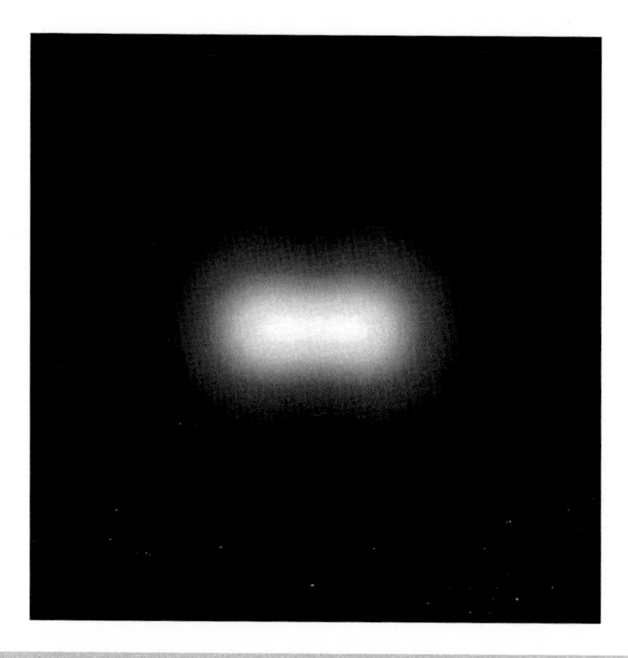

Fig. 1.8 Diffraction pattern matching Dawes' limit (Image credit: Geek3- Creative Commons)

where

D is the diameter of the main lens (aperture) and

R is the resolution power of the telescope—(Adapted from: Angular Resolution– Wikipedia).

It's important at this point to include some discussion about atmospheric effects on seeing and star images in general. Without an atmosphere, a small star would have an apparent size more commonly called an "Airy disk."

Interference patterns are created when light waves that pass through an aperture or opening of a telescope become disrupted and disturbed. A typical optical telescope normally has a round, circular, or even an annular aperture. An annular aperture is defined as a circular outer border that has a circular central obstruction that is basically considered round and or circular. A good example is a secondary mirror of a Newtonian reflector or a Cassegrain catadioptric telescope. The circular aperture produces a diffraction pattern that distributes the light from a point source of light, such as a stellar image or star, into a bull's-eye target-shaped image which is commonly referred to as an Airy disk (see Fig. 1.9). The Airy disk in the figure above is a simulated image of a point source that is monochromatic (single-wavelength) light and is an example where astronomical seeing is considered excellent, thus producing a stellar image that appears nearly perfect without the blurring and other initial effects. The term "gamma stretch," which is called a "stretch function," has been applied to enhance the overall contrast and brightness of the extremely faint outer rings of the point source image. Spider vanes that

Fig. 1.9 A computer generated image of an Airy Disk pattern (Image credit: Public Domain–Wikimedia Commons)

secure the telescope's secondary holder in a ridged fashion within the telescope's optical tube have various configurations and shapes and are sometimes jokingly referred to by some telescope makers as a "necessary evil" in a telescope. Spider vanes produce another effect commonly referred to as Newtonian Spikes that will be discussed in more detail later in the chapter.

Diffraction, in simple terms, is an optical interference effect caused by the bending of light around obstacles or structures in its path (e.g., the edges of a telescope tube or its secondary holder spider vanes) similar to the way ocean or lake waves are bent or deflected around dock pilings or a jetty's edge. All telescope optical systems show faint light and dark diffraction rings around a star's Airy disk at high magnification, as the diffracted light waves alternately cancel out and reinforce or strengthen each other. Diffraction rings are intrinsically faint and an inexperienced or novice observer's inability to see them wouldn't normally be considered to be a real or immediate concern. For example, in a perfect refractor telescope, approximately 84% of the light would typically be imaged in the Airy disk itself, with half of the remaining light falling in the first diffraction ring and the left over balance would be distributed among the second, third, fourth diffraction rings, etc. Since the first diffraction ring is approximately six times the area of the Airy disk itself, its fainter light is spread out and dispersed over a much greater area.

So as a result, the brightness of the first diffraction ring is actually less than 2% than that of the Airy disk. The other diffraction rings are even fainter. It is easy to

Fig. 1.10 Diffraction spikes from various stars seen in an image taken by the Hubble Space Telescope (Image credit: Public Domain—Wikimedia Commons)

see how a less experienced beginning or novice observer can struggle separating the intrinsically faint diffraction rings from the bright Airy disk. Catadioptric and reflecting telescope diffraction rings appear to be almost twice the overall brightness of those produced by a refracting telescope. The major reason is due to the additional diffraction caused by their secondary mirror, secondary holder and spider vane obstructions. But their overall brightness is still low compared to their Airy disk (only 4% as bright in comparison to the first ring). Optically speaking, a catadioptric telescope's diffraction with its increased ring brightness will appear as having lower contrast and with a noticeable (though slight) loss of sharpness on planets, binary stars, and star clusters when compared with that of a refractor telescope. The spider vanes holding a Newtonian reflector's secondary mirror create additional diffraction spikes that also lower contrast and radiate out from each star's image, an effect particularly visible on long-exposure photos. The image in Fig. 1.10 shows the diffraction spikes of various stars in a globular cluster radiating outward from the bright images of the stars taken by the Hubble telescope.

A catadioptric telescope also has a round secondary mirror shadow, as shown in Fig. 1.14, but does not have any visible bright diffraction spikes and spider vane shadows. A typical telescope star image that's a result of diffraction would normally be inversely proportional to the telescope's diameter. However, when light first enters the Earth's atmosphere, the different stratified layers of temperature combined with different extreme wind speeds and air currents have a tendency to warp and bend the light waves, leading to bloated star images that are often distorted and scintillated in appearance. The effects of the atmosphere can be shown as moving and turbulent cells of rotating air. At many of the well-known observatories around the world, the initial effects of air turbulence is often noted on scales that are greater than 10–20 cm and at visible wavelengths that are below the best seeing conditions, and this will frequently lower the resolution limit of some of the world's finest optical telescopes to be about the same as achieved by a space-based 10–20 cm telescope. The "seeing disk" diameter is most often defined as the "full width at half maximum" (FWHM), which is the common and accepted measure of the astronomical seeing conditions. Based on the definition, "seeing" is considered a constantly changing variable and will often differ from location to location, from night to night, hour to hour, and frequently changes on a scale of minutes or even seconds. Professional and amateur astronomers often talk about "good" nights with moments of clear conditions and low average seeing disk diameter, and "bad" nights where the atmospheric environment produces seeing disk diameters so large and bloated that not all of the observational data is considered useable. In some cases, the observational data is ignored. A medium to high power eyepiece can demonstrate a typical night of "bad seeing" when it is used to look at the moon. The FWHM of the "seeing disk" diameter is measured in arcseconds and it is normally abbreviated with the symbol of (″). For example, 1.0″ seeing is considered a good one for average astronomical sites. When compared to the cities, especially the larger ones, the seeing conditions are typically a lot worse. Good seeing nights are normally clear, especially those with cold nights without gusts of wind or turbulent air currents. For example, when warmer air moves upward via the convection currents, it will often degrade the seeing conditions much the same as wind and clouds do. At some of the best high-altitude observatories located on such mountaintop sites such as Mauna Kea and Haleakala in Hawaii, the wind brings in stable dry air which has not previously been in contact

Fig. 1.11 (Cartoon credit: Jack Kramer)

with the ground and sometimes provides seeing conditions as good as 0.4″ arcseconds (Fig. 1.11)—(Adapted from: Astronomical Seeing—Wikipedia).

Diffraction Pattern of Obstructed Optical Systems

By Rene Pascal

With the only exception of the "Schiefspiegler" (Oblique Telescope, like the design by Anton Kutter), every telescope with the main objective being a mirror is obstructed, and the quality of the image suffers more or less from the resulting disturbance of the diffraction pattern. As a rule of thumb, we have learned that a central obstruction of less than 20% (linear diameter) is negligible for the quality of the resulting image. But what is the effect of more complicated obstructions, like thick vanes carrying the secondary mirror or other constructive elements that protrude into the optical path? Since this may be an important question when planning and constructing telescopes and other optical instruments, I ran some simulations to get an idea of the influences on image quality.

Figure 1.12 (right) shows the simulated diffraction pattern that would result from a circular unobstructed opening (left) in monochromatic light. The diffraction pattern is displayed in a logarithmic gray scale as it would be seen by the human eye with a scaling from black to white of eight astronomical magnitudes (8 mag.) The field of the simulated views is 6″ (arcseconds) wide. This will be true for all of the following simulated diffraction patterns unless otherwise stated.

If we know the energy distribution in the diffraction pattern, we are able to simulate the image that would result from imaging an object with an instrument with exactly this entrance pupil but an otherwise perfect optical system. This gives us the opportunity to compare the influence of obstructions on the image quality

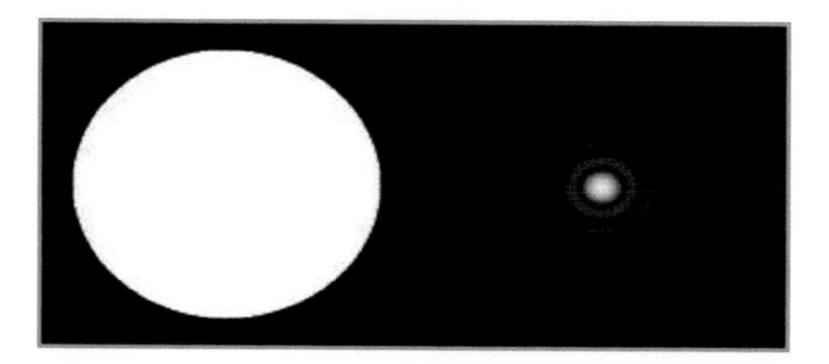

Fig. 1.12 Diffraction pattern Airy disk (*right*) of an unobstructed circular entrance pupil (Image credit: Rene Pascal)

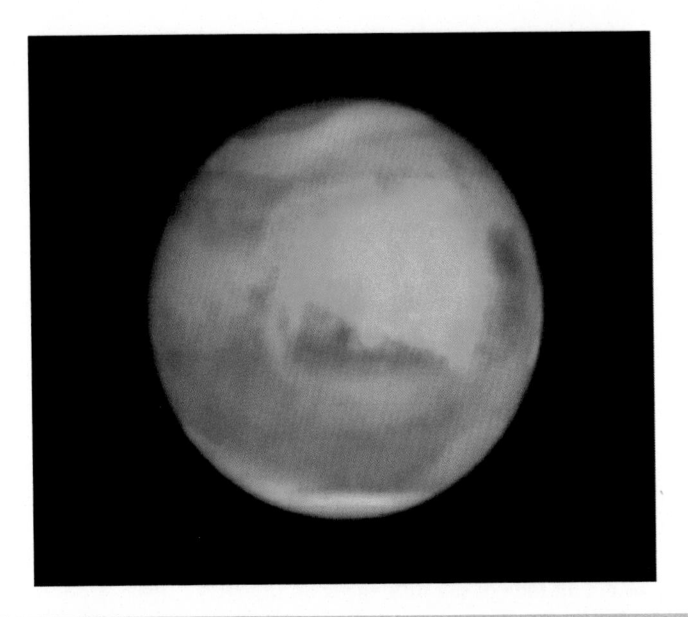

Fig. 1.13 Diffraction pattern Airy disk (*right*) of an unobstructed circular entrance pupil (Image credit: Rene Pascal)

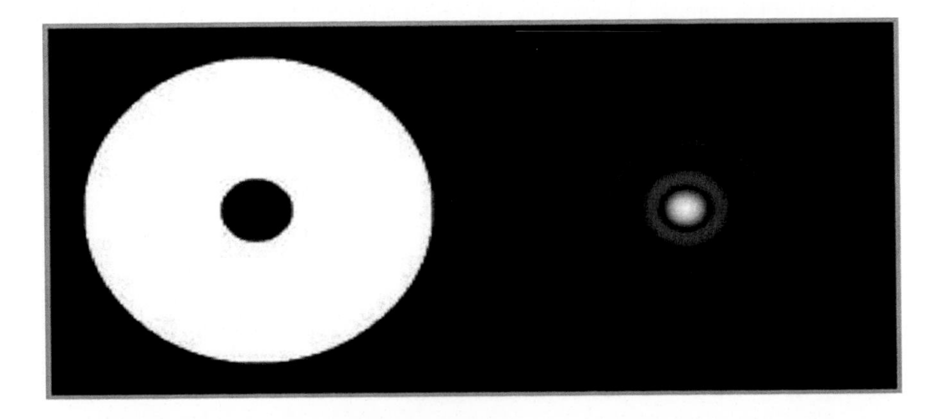

Fig. 1.14 Twenty percent central obstruction (*left*) and the resulting diffraction pattern (Image credit: Rene Pascal)

bare from different optical conditions of the instruments used normally for such side-by-side tests. I used Mars (an image from the Hubble Space Telescope) as test object, since its surface with the many low contrast features of different size is very sensitive to a decrease in image contrast of the optical instrument (Fig. 1.13).

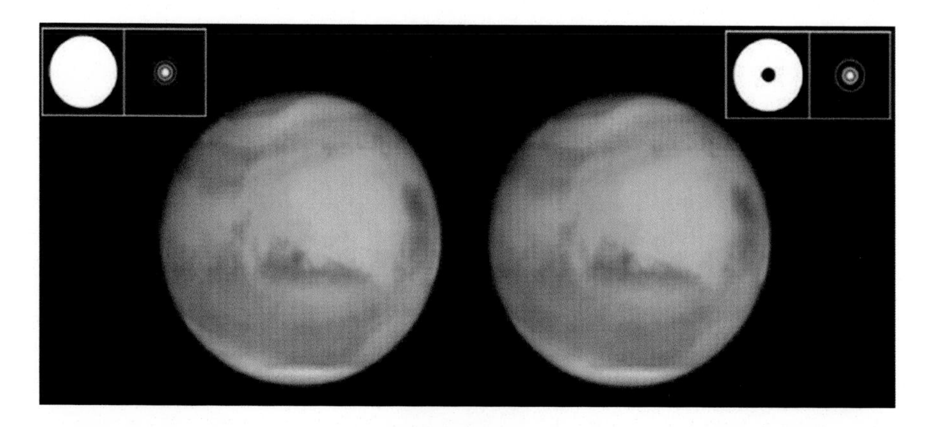

Fig. 1.15 Mars imaged with an obstructed instrument (*left*) and an instrument with 20% central obstruction (*right*), the resulting image contrast and visibility of surface structures is nearly identical. At the *top* the entrance, pupil and the resulting diffraction pattern is shown. The pattern is displayed at actual size with respect to the images of Mars (Image credit: Rene Pascal)

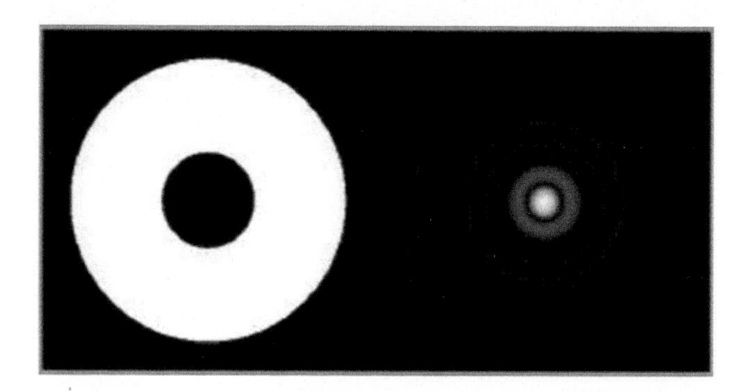

Fig. 1.16 Thirty-three percent central obstruction (*left*) and the resulting theoretical diffraction pattern (*right*) (Image credit: Rene Pascal)

As already mentioned, we have learned that a central obstruction of 20% (Fig. 1.15, left) is negligible for the image quality in most cases. The simulated view in Fig. 1.15 (right) tells the same: the left image is for an unobstructed instrument, the right one is with 20% central obstruction. Virtually no loss in contrast is noticeable (Figs. 1.14 and 1.15).

This clearly changes if we introduce 33% (Figs. 1.16 and 1.17) or 50% (Fig. 1.18) central obstruction. More light falls into the outer diffraction rings, lowering the contrast of the image. Sometimes you hear the statement that an instrument with 30% central obstruction is unusable for high magnifications and planetary observations. This is simply not true, as can be seen in the simulated

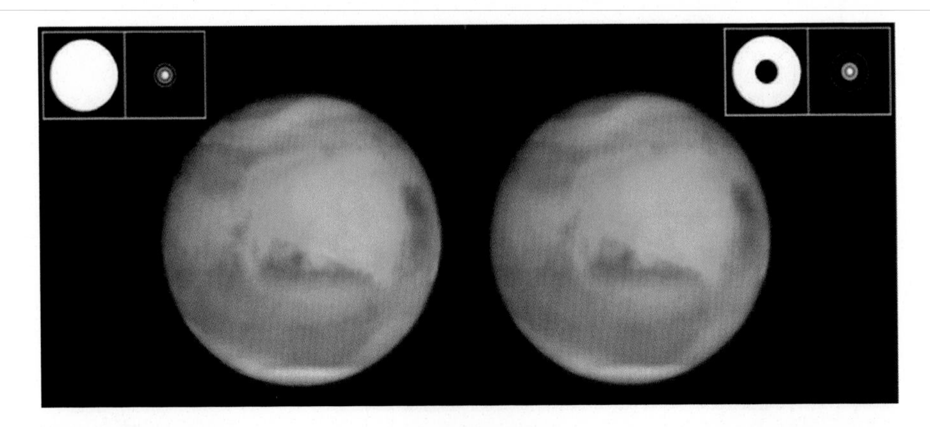

Fig. 1.17 Mars imaged with an obstructed instrument (*left*) and an instrument with a 33% central obstruction (*right*) (Image credit: Rene Pascal)

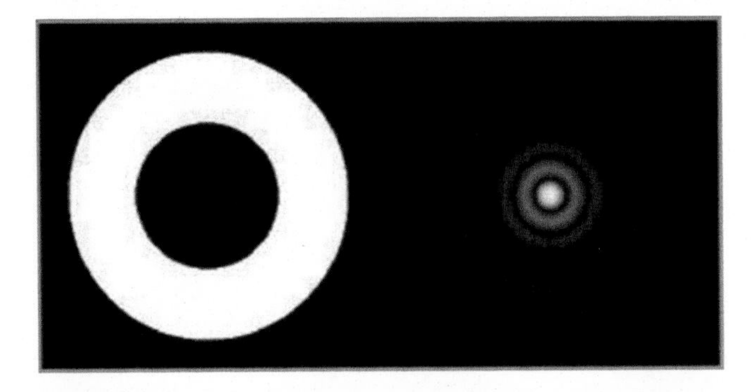

Fig. 1.18 Fifty percent central obstruction and the resulting diffraction pattern (Image credit; Rene Pascal)

views, and if a central obstruction of about 30% is needed for constructive requirements, I would not hesitate to build the instrument this way.

At 50% central obstruction (Fig. 1.19, right), the loss in contrast is distinct. On the other hand, one can imagine that even this difference should be difficult to figure out at a side-by-side test, if the conditions of the instruments under test (state of collimation, temperature acclimatization) are not identical.

More interesting than the central obstruction that has been discussed rather often is the influence of the vanes that carry the secondary mirror. In Figs. 1.20 and 1.21, the influence of vanes of 6 mm thickness on a 400-mm instrument (i.e., 1.5%) is simulated. Normally vanes are thinner than that.

Even really massive supporting "vanes" that reach into the optical path have very little influence on planetary contrast. The geometry of Fig. 1.23 (right) is often

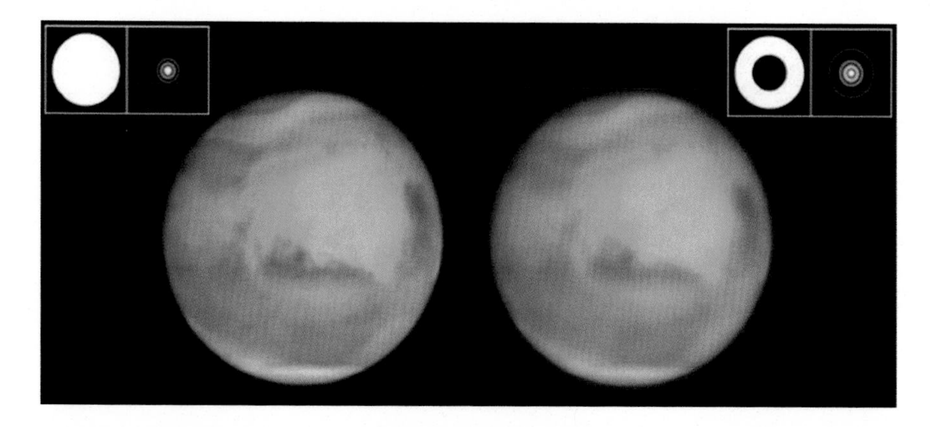

Fig. 1.19 An instrument with 50% central obstruction (*right*) again compared to the image with an unobstructed instrument (*left*). The difference in surface contrast is clearly visible (Image credit: Rene Pascal)

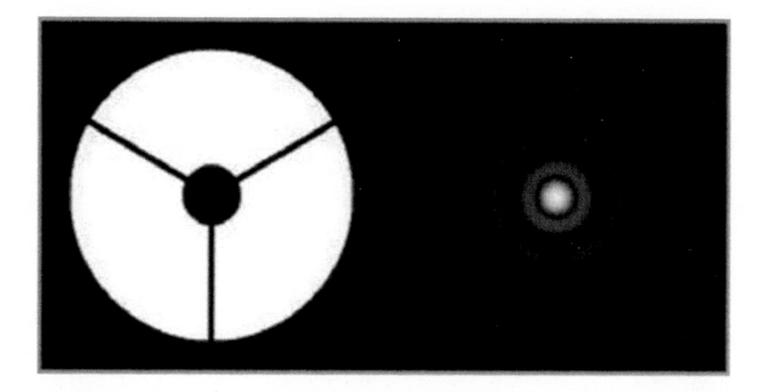

Fig. 1.20 A three-vane spider that is 1.55 of the diameter of the main mirror thick, combined with a central obstruction of 20% (Image credit: Rene Pascal)

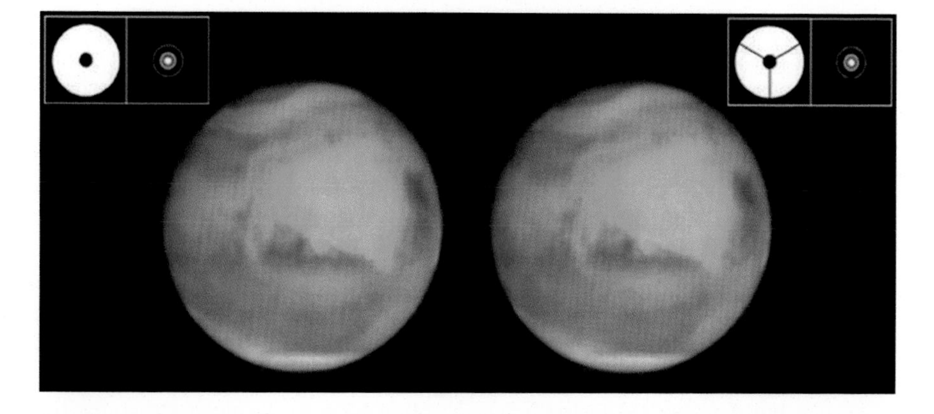

Fig. 1.21 An instrument with 20% central obstruction (*left*) compared with an instrument that additionally has a three-vane spider of 1.5% mirror diameter. The influence on planetary contrast is negligible (Image credit: Rene Pascal)

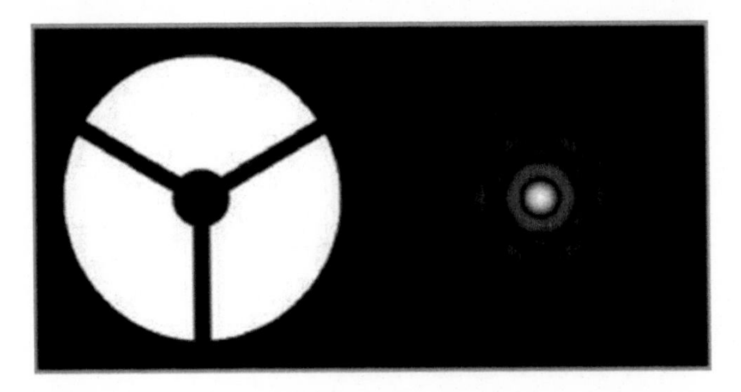

Fig. 1.22 Twenty percent central obstruction with vanes of 5% of the diameter of the main entrance pupil and the resulting diffraction pattern (Image credit: Rene Pascal)

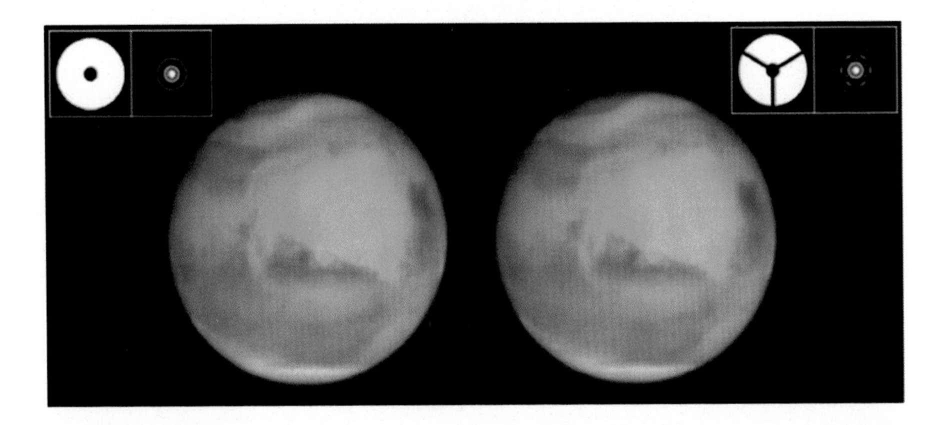

Fig. 1.23 The loss in contrast for such massive vanes is astonishingly low, even though the influence on the diffraction pattern (corresponding to the view of a star) is prominent (Image credit: Rene Pascal)

found in lightweight (robotic) telescopes that carry an electronic camera instead of a secondary mirror. The camera (or the secondary optical component) is supported by a tripod that reaches directly from the cell of the main mirror to the first focus. The geometry of Fig. 1.22 (left) would be for a tetrapod, respectively (Fig. 1.23).

Following the simulations, a rule of thumb for the vanes would be a three-vane spider with vanes with diameter under 5% of the linear diameter of the mirror is uncritical for the contrast in planetary observations—that is, a much higher value than normally assumed! For a four-vane spider, this value should be lower than 3% (Figs. 1.24 and 1.25).

When building optical instruments, we often have to decide if structures like mirror clamps or the focuser should be allowed to reach into the optical path or if

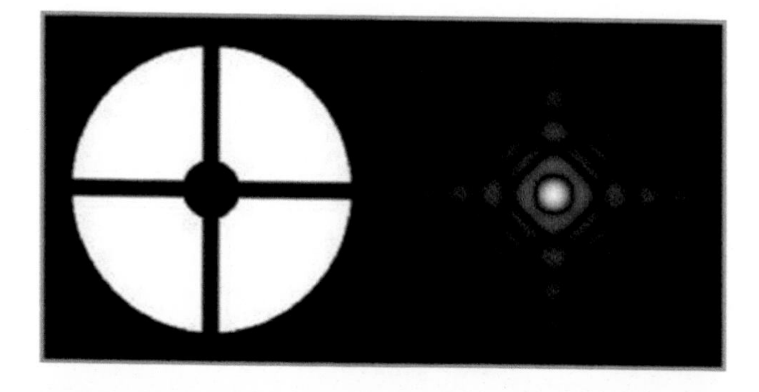

Fig. 1.24 A four-vane spider of 5% clearly has more influence on the diffraction pattern than a three-vane spider of 5% (Image credit: Rene Pascal)

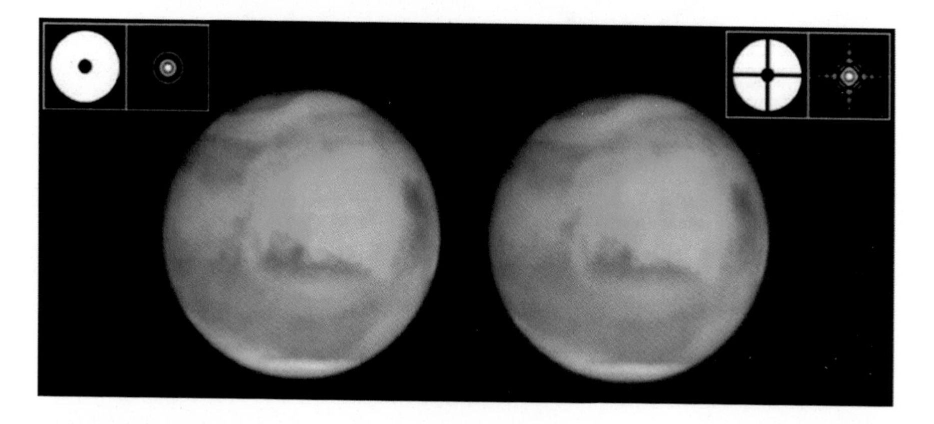

Fig. 1.25 Four supporting structures of the same size in Fig. 1.26 will result in a somewhat lower contrast image due to the higher amount of diffracted light outside of the Airy disk (Image credit: Rene Pascal)

we should spend some effort to prevent this, sometimes at the cost of a higher weight of the instrument. In Fig. 1.26 (bottom), a theoretical instrument with many constructive structures reaching into the optical path is presented, together with the influence on the diffraction pattern. When discussing the influence of vanes on the image, we have neglected the diffraction figures far away from bright objects like stars up to now. In practice, they have no influence on planetary contrast, but they are visible because of the immense dynamic of the human eye. This dynamic is not used when observing planetary surfaces (Fig. 1.27).

Figure 1.28 shows the diffraction pattern of several vane configurations. The field of view is 60″ (arcseconds) wide, again simulated for a 400-mm diameter objective. At left the geometry of the obstruction is shown, the three images show simulated views with a dynamic range from black to white of 8, 12, and 16

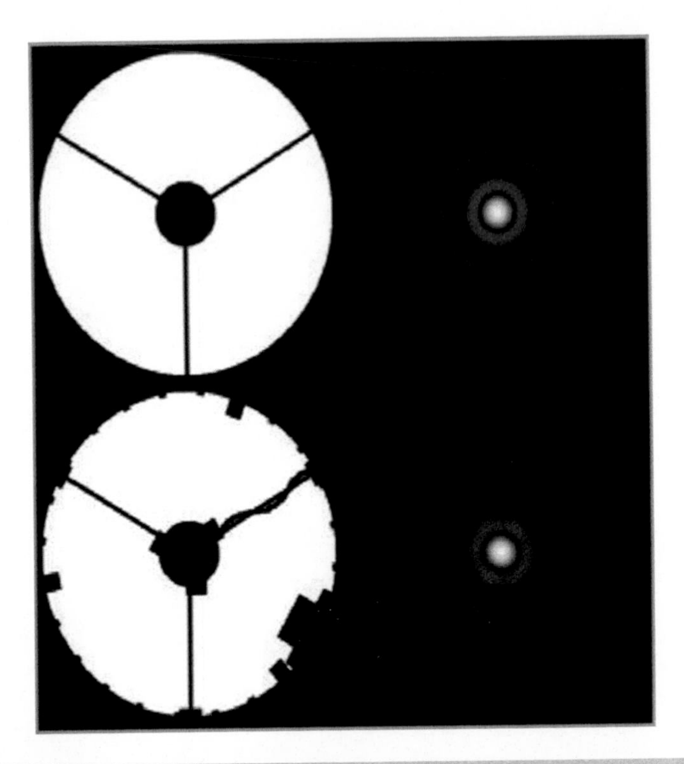

Fig. 1.26 An instrument with spider and secondary (*top*) compared with a theoretical instrument with several additional structures reaching into the optical path (*bottom*), like a focuser, vane mountings, mirror clamps, a heating cable for the secondary and some screws (Image credit: Rene Pascal)

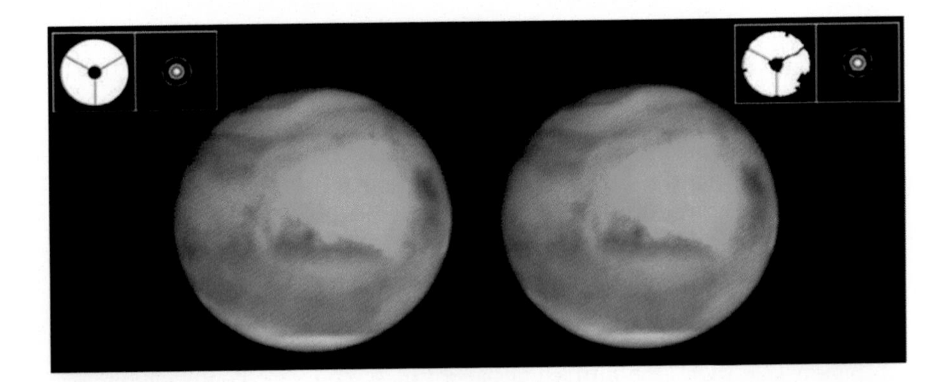

Fig. 1.27 The loss in surface contrast for the instrument with all the structures reaching heavily into the optical path is quite low (Image credit: Rene Pascal)

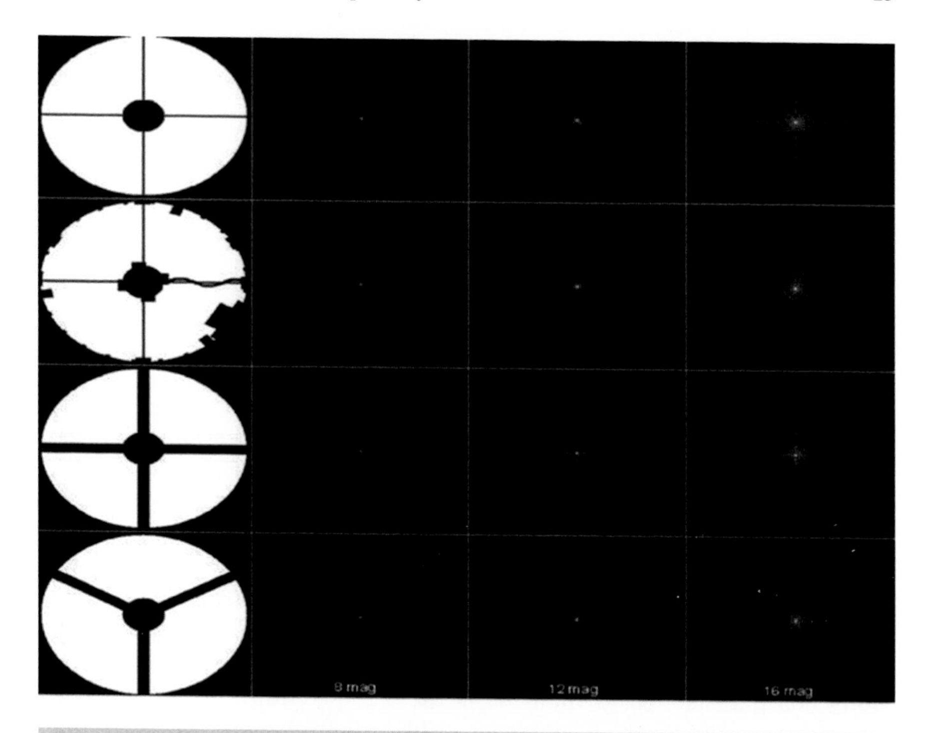

Fig. 1.28 A realistic view of diffraction figures in simulated wide field views in color (Image credit: Rene Pascal)

magnitudes, from left to right: Upper row: a four-vane spider of 0.4%, corresponding to the use of 1.6 mm steel sheets for the vanes on a 400 mm reflector. Second row: again the 0.4% spider, but additionally with many other mechanical parts that reach into the optical path. Third row: a four-vane spider of 5%, corresponding to 20 mm supporting tubes, for instance. Fourth row: a three-vane spider of 5%. We get six diffraction spikes, but they are less bright than for four vanes.

Testing the Program

For the simulation of the diffraction pattern, a program was used that was developed some time ago in Microsoft Visual C++. The software was critically tested resulting in simulations I am sure are accurate and correct.

One critical test among others was the simulation of a complicated apodizer disk from literature that was developed for a search for extraterrestrial planets close to stars on the proposed and meanwhile cancelled NASA space telescope "Terrestial Planet Finder – C". The complex apodizer diffracts most of the light that normally falls into the diffraction rings, to other directions. This results in a diffraction pattern that has the diffraction rings suppressed by several magnitudes within a certain zone around the central maximum, enabling an excellent contrast for a search for faint objects close to the star observed, on the expense of having strong diffraction structures in a zone around 90° to this orientation.

Rene Pascal
Germany

Adapted from "Diffraction Patterns Of Obstructed Optical systems" with permission from: Rene Pascal (http://www.beugungsbild.de)

Why Do Spider Vanes Cause Diffraction Spikes?

Diffraction is a phenomenon displayed by wave fronts as they pass along the edge of a telescope's secondary holder's spider vane and, as a result of the wave nature of light, it is scattered about at right angles to the spider's vane edge. For example, in a typical Newtonian telescope with a three-legged spider that has "straight" vanes, you get six diffraction spikes (see bottom Fig. 1.28). The central obstruction or secondary is round and/or circular in its cross section. Any given point on the telescope's primary mirror will be at right angles to a point around the edge of the central obstruction, resulting in the entire image is covered in diffraction spikes. Diffraction spikes are not as strong or intense as those created by the spider vane (because each spike is caused only at a particular point rather than the whole length of the line or edge), but they are in fact still there and they manifest themselves as a slightly lower contrast image. Some reflectors are fitted with curved spiders that nearly eliminate the visible spikes, but again the spikes as a result diffract over the entire image and make the noticeable contrast worse. You can see the effect of this if you compare the field of view through a refractor and a reflector. From an extremely dark sky location, the background appears almost jet black with a refractor. Through a reflector, it appears a dull gray. This is why most expert opinion equates a refractor to a reflector of larger aperture in terms of comparing their performance. A faint star is readily distinguishable on the black background of a refractor but can easily get lost in the scattered light of a reflector. It is also one of the main reasons why dedicated planetary observers, who are obsessed with contrast for whom contrast is everything, are willing to spend thousands on incredibly refined and well-corrected refracting telescopes instead of hundreds on reflectors that have significantly larger apertures.

Note that even an optical window with no support vanes will not completely solve this diffraction problem. The edge of the main imaging optic (lens or mirror) and the secondary mirror causes the same effect. This is seen as concentric rings of diffraction around stars at high power, a bulls-eye or target artifact. This is a "circular spike." Some Newtonian scopes have curved vanes that cause larger, fainter curved spikes that are not as noticeable in the overall picture. But they do lower the overall contrast a little. When it comes to telescope spider vanes, no matter how fine the spider vanes are, they will diffract light from a star, and this appears as diffraction

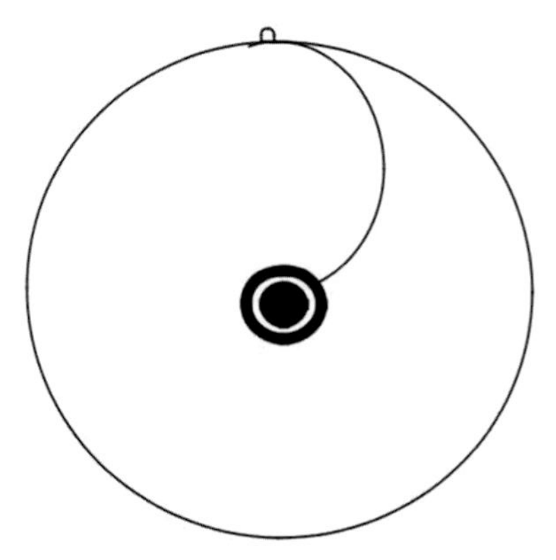

Fig. 1.29 A single half-curved spider vane (Image credit: The Author)

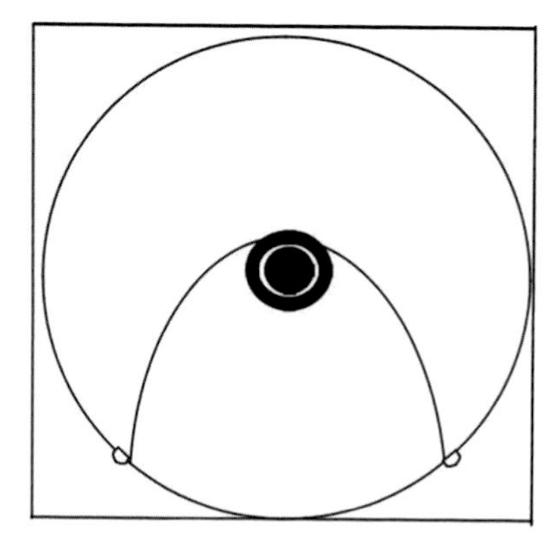

Fig. 1.30 A single-curved spider vane loop (Image credit: The Author)

spikes which are the Fourier transform of the support struts. In reality, diffraction spikes represent a loss of light that could have been used to help image a faint star.

The single half-curved spider vane shown in Fig. 1.29 has very few attributes when it comes to using it in a Newtonian telescope. It will vibrate a bit from side to side if the telescope is bumped, can be difficult to adjust or center in the exact middle of the telescope tube, and will not allow for out of center mirror cells or tube alignment, not a good choice for a spider vain support system.

The single-curved spider vane loop (Fig. 1.30) once installed in a Newtonian telescope tube has a tendency to shake a little from side to side. It is in fact a little easier to adjust than a single half-curved spider vane (Fig. 1.29), but it must be mounted precisely in telescope tube in order to properly center the secondary mirror. It does have a less noticeable diffraction pattern.

Even though there is a curved vane acting as another diffraction surface, the curved nature (a 180° curve) causes the light to be evenly distributed throughout the

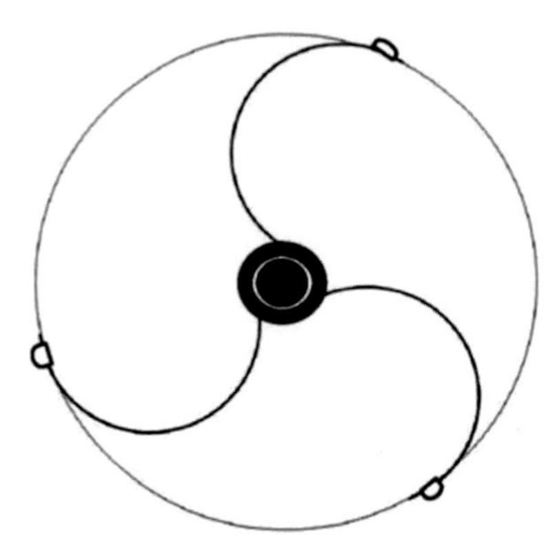

Fig. 1.31 A three-curved vane spider (Image credit: The Author)

diffraction rings. The result is the absence of the familiar diffraction spikes. In fact, a stellar image through such a similar equipped Newtonian telescope produces an almost identical star image as through a Cassegrain telescope. Actually, it's better, in terms of diffraction, in that the thin-curved vane produces less additional diffraction than the increased size of the Cassegrain secondary. It's almost like having your cake and eating it too. So what's the trade-off? In the case shown, none, really. The length of such a curved secondary is about equal to the sum of the lengths of the vanes in a 3-vane spider. So the total diffraction surface is about the same, but the spikes are almost absent. Compared to a 4-vane spider, this type of curved spider has less total diffraction. The only real concern is that the simple curve vane spider design is only adequate for Newtonian reflectors and Cassegrains with apertures up to 8 or 10 in. Beyond that the secondaries get so heavy that one needs to resort to 3 or 4 S-curved vanes to support the secondary. With the S-curve or similar designs, there is a trade-off. Clearly 3 S-curved vanes are longer than 3 straight vanes. So if one wishes to have multiple-curved supports around the aperture to support the secondary, eliminating the diffraction spikes comes at a cost of more total scattered light. If one were choosing which spider vane design to use for use in their new binocular telescope project, then the answer should be obvious. Choose the spider vane design that provides the most support and at the same time produces the least noticeable diffraction spikes. Obviously the curved spider vane designs seem to produce the least noticeable diffraction spikes, and that should be considered if you're doing more than just visual observing all the time. But remember, when building your binocular telescope or binoscope and you're going to be using a spider vane secondary support system, the spider vanes (curved or straight) need to be secured in exactly the same position and location in both optical tube assemblies in order to maintain what is called "diffraction spike harmony" when viewing (Fig. 1.31).

The three-curved vane spider once installed in the telescope tube is very stable, even when bumped or shaken. A three-curved vane spider is easy to install and center inside the Newtonian telescope tube. The design makes it easier to adjust, install, and remove a secondary mirror within the telescope tube (Fig. 1.32).

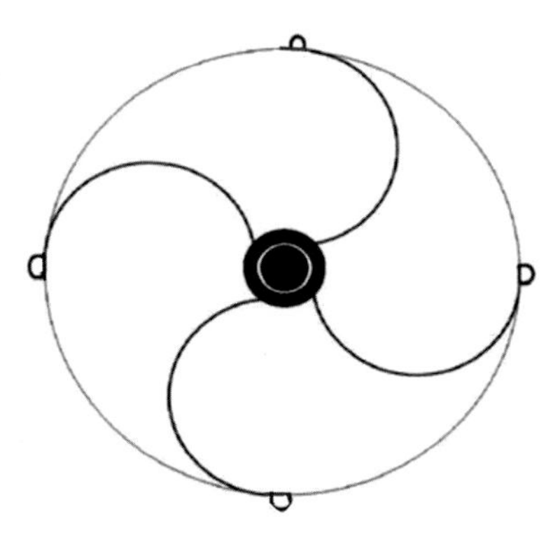

Fig. 1.32 A four-curved vane spider (Image credit: The Author)

The four-curved vane spider once installed within a Newtonian telescope tube is very stable, an excellent choice for truss tube Dobsonian telescopes that have four sides, is easy to install, and supports the secondary mirror holder in a very solid fashion.

Adapted from "Types of Curved Vane Spiders" with permission from Hans Wiest (http://www.1800Destiny.com).

There are other types of spider vane configurations to use in a Newtonian telescope that are equally well suited to support a secondary mirror and it's holder and present a minimum obstruction to the incoming light path. Depending on the amount of secondary mass the spider vane(s) has to support, a single spider vane with an appropriate thickness can also securely support a secondary mirror and holder assembly. One often sees a secondary mirror holder system supported with three straight vanes even in Newtonian telescopes as small as 4-1/4 in. with a small secondary. In reality, and in most cases, a single straight spider vane of sufficient thickness (example; 1/16 in. thick) would work just fine to support a small secondary and its holder assembly, especially in a small 4-1/4 in. f/6, f/7 or even f/8 and beyond Newtonian telescope that's carrying a small secondary. And there are other Newtonian telescope examples that present a similar consideration. One of the reasons why most of us choose to use a three straight spider vane configuration is the idea that three are "always" needed regardless of the size of the secondary. Although not as common, four straight spider vanes are sometimes found in place of three. If you happen to walk into a store that sells telescopes, don't be surprised if you see a 4-1/4 in. or 6 in. Newtonian telescope with four straight spider vanes supporting a small secondary. Keep in mind that if you're either buying or even consider making a small Newtonian telescope, that anything in the incoming light path is going to be an obstruction, even a thin spider vane or the head of a screw will cause less light to reach your primary mirror.

In some interesting configurations, individuals have chosen to use a series of thin metal wires to secure a secondary and it's holder within a Newtonian telescope tube in place of thin metal spider vane. However, is there an advantage to using a series of wires strung 120° apart from each other to secure and support a secondary and it's holder within a Newtonian telescope optical tube system? Not really. If a thin wire is, for example 1/32 in. thick, and a straight metal spider vane is of the same thickness, they will both have the "same" obstruction effect on any incoming light.

In the author's opinion, there is really nothing gained by using a thin wire verses a thin metal spider vane except maybe that it looks cool. In reality, there is no real advantage to this method, but it is another interesting and creative way to secure a secondary mirror and it's holder within a Newtonian telescope's optical tube system.

How Important Is My Secondary Mirror Size?

When it comes to selecting the proper secondary mirror size for your homemade Newtonian binocular telescopes, the concern is whether or not you want to achieve full illumination from your primary mirror. The term "full circle of illumination" means that your secondary mirror will be the largest optimal size that can divert 100% of the light reflected from your telescope's primary mirror. A telescope's secondary mirror size is based on its mirror's minor axis or width. But it also means that it will have a large central obstruction with less light making it to the primary mirror and a reduction of overall image contrast in your eyepiece's field of view. A smaller secondary mirror will in fact display a smaller central obstruction that results with more light making it to the primary mirror, but with less light diverted from your primary mirror to your eyepiece. Because secondary mirrors are an obstruction, they act to degrade your telescope's optical performance. The rule of the thumb for a telescope's secondary mirrors is: the larger the secondary mirror obstruction, the greater the degradation. A one-third secondary mirror obstruction, which is considered a larger ratio than what a visual Newtonian normally uses, will typically degrade a telescope optical system's resulting image quality by approximately one-sixth wave. Changes that are less than one-eighth of a wave are very difficult to perceive visually. If using a smaller secondary mirror is your choice, it will present better overall image contrast in your telescope's field of view. As a rule, as long as the overall size (area) of your secondary mirror is less than 20% than that of the primary mirror, its overall effect is extremely difficult to distinguish visually. The actual size (major and minor axis dia.) one chooses for the secondary mirror depends on the following factors: focal length, primary mirror diameter, distance of the point to the optical axis of the primary mirror, and the area around the focal point which should receive as much light as the focal point itself. The formula is given below:

$$\text{minor axis of secondary} = d + \frac{r * (D - d)}{f}$$

d=diameter of the fully illuminated field, r=distance from the central axis of the primary to the image field (radius of tube plus height of focuser), D=diameter of primary, and f=focal length of the primary (f=aperture*focal ratio).

Examples of secondary mirror size calculation are primary mirror diameter=200 mm, focal length=1200 (f/6), distance between primary mirror's optical axis and its focal point=150 mm, and radius of area which should receive maximum possible light=10 mm. (This area would be smaller if the telescope was for visual use only) The secondary thus needs a minor axis of approximately 42.5 mm.

Note:
The highest resolution can only be achieved by a telescope is by the "full circle of illumination".

It's important to mention at this point in the discussion about the secondary mirror "offset." If one wants to have an area of total illumination (sometimes called "full" illumination) around the focal point, it will be necessary to apply a small offset to the center of the secondary mirror. The reason is that the secondary mirror is positioned at a 45° angle in respect to the optical axis of the primary mirror. The actual point or part of the secondary mirror that is nearest to the primary mirror intersects with the light cone at a distance greater from the optical axis as the point furthest away from the primary mirror. The secondary mirror offset is in fact away from the focuser. Once the offset of the secondary is actually applied, it will no longer be centered above the primary mirror. Thus, it can be seen in the inside and outside of focused images. It is also possible that it may have a small impact (the key word here is "small," if at all) on in-focus images. Focal ratios of f/6 and slower, the offset is so small that its application should be ignored or as in most cases, not applied at all. It can be applied for focal ratios of f/4 and f/5, but it should be applied for mirrors with even faster focal ratios (when used for imaging). If the scope is used for visual imaging only, the offset application provides no advantage and does not need to be considered.

It's always beneficial in the beginning of your telescope's construction to calculate the entire optical system to scale in order to know exactly where the secondary mirror should be positioned in the optical tube and where it will fit optimally in the primary mirror's cone of light. You can follow these simple steps:

- For example, measure the distance of your focal plane from the center of the secondary to where the focus will be outside of the tube. If your telescope's focal ratio is 6, then divide the focal plane measurement by the focal ratio.
- If the measured focal plane is 9 in., your primary mirror's focal ratio is 6, and the minimum secondary size is 1.5 in., then choose a 2-in. secondary mirror. This will allow for full illumination.

Note:
Considering the fact that your choices of secondary mirror sizes may be limited, it's best to just get the one closest to the calculated size that you can find.

But if that particular secondary mirror size is not available, then try and get the next largest size to ensure a full circle of illumination.

What Is the Limiting Magnitude of a Binoscope?

The chart in Fig. 1.33 illustrates the limiting magnitude advantages that a binocular telescope has over a single-mirror telescope of the same aperture.

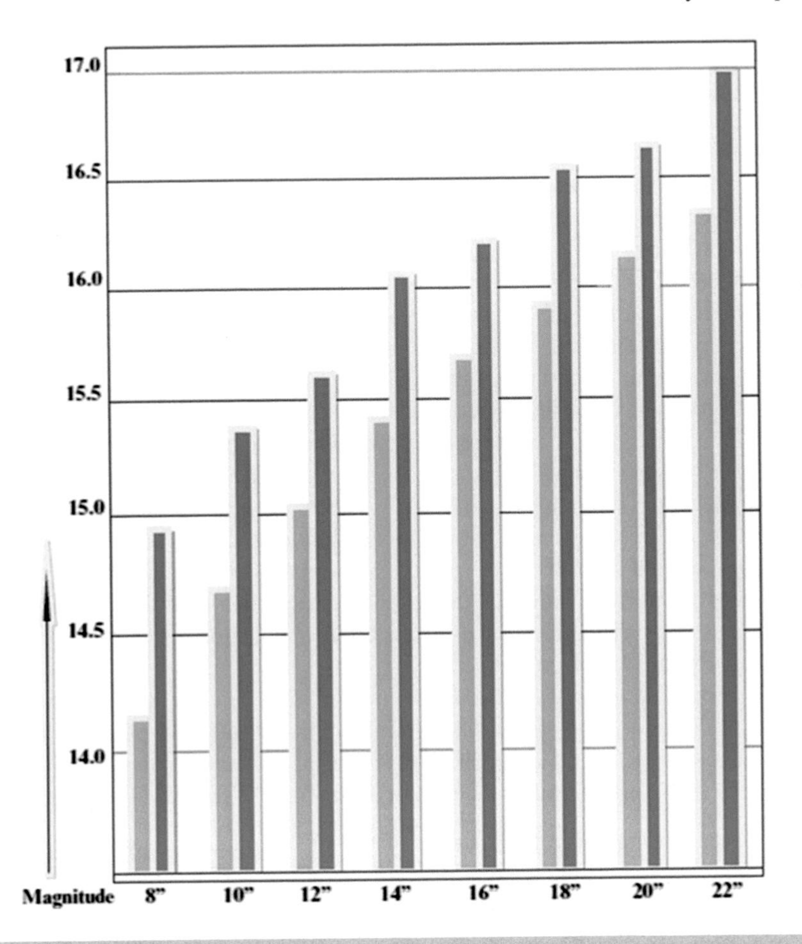

Fig. 1.33 Single mirror verses Double mirror (aperture in inches) (Image credit: the Author)

What Is the Light Gathering Power of a Binoscope?

With a refractor binoscope and a binocular telescope, you have either two separate mirrors or two objective lenses that are pouring light into both eyes from an object. For example, with a 16-in. Newtonian telescope, your primary mirror has a total surface area of 256 square inches of light gathering power and a 15.7 limiting magnitude under dark skies and good seeing conditions. With a 16-in. binoscope, the two big 16-in. mirrors produce a total of 512 square inches of light gathering power, which equate to a limiting magnitude under clear dark skies of approx. 16.3. The same equation applies to a refractor binoscope too; for example, if you have an 8-in. refracting binoscope that has two big objective lenses with a total surface area for gathering light equal to 128 square inches ($8\times8=64\times2=128$ square inches). Its limiting magnitude under clear dark skies is nearly 15th mag. compared to a single

8-in. refractor telescope that has a limiting magnitude of approximately 14.2 (see Fig. 1.33). When it comes to calculating the square surface area of an 8-in. mirror, we know that it has a calculated surface area of 64 square inches of light gathering power. The established approximate telescope limiting mag. formula is $7.5 \div 5$ log aperture (in./cm).

Note:
Telescope limiting magnification results can vary depending on seeing conditions, location, telescope optical quality, observer's age, experience, etc. However, when it comes to helping your telescope reach its maximum efficiency in terms of light gathering power, especially in reference to Newtonian telescopes, it would be extremely beneficial to have both the primary and secondary mirrors Silver enhanced. The same would apply for a mirror diagonal used with a refractor. Either way, it's always desirable to have about 96–98% of the incoming light being reflected off the surface of your primary and secondary mirrors than to have 92%. If you want to see how dramatic the term "reflectivity" is and how it relates to potential light loss in your own Newtonian telescope, simply cover your primary or secondary mirror with a piece of paper that covers approximately 25% of their reflective area and point it at the moon. You'll immediately have your answer. If your older Newtonian's primary and secondary mirrors are only "pumping out" up to 86% of their incoming light, without even having to ask, you'll know what you need to consider doing next. At 86% reflectivity for both mirrors, a simple math calculation will tell you your eye is only receiving about 74% of the original incoming light, and at that kind of optical output, those faint fuzzies out there are going to be even fainter and perhaps even fuzzier. When it comes to making up for unwanted light loss in a Newtonian telescope's optical system, then Silver enhancing your primary and secondary mirrors will certainly help.

In order to accurately determine the light gathering power of an 8-in. f/8 Newtonian binocular telescope, we need to subtract the two secondary mirror central obstruction diameters from both the primary mirrors' total. For example, if the secondary mirrors have an elliptical minor axis diameter of 1.33 in., then the total surface area of the secondary mirror's minor axis is $1.33 \times 1.33 = 1.77$ square inches. Subtract the secondary's minor axis total surface area 1.77 square inches $\times 2 = 3.54$ square inches from the combined total of the dual 8-in. Primaries ($8 \times 8 = 64 \times 2 = 128$ square inches) with a result of $128–3.54 = 124.46$ square inches of light gathering power. The same equation applies for any telescope that has a secondary or central obstruction. An example would be an 8-in. f/8 Newtonian telescope having three straight spider vanes. Each spider vane has an approximate obstruction length of 6.67 in. and obstruction thickness of approximately of 0.0625 in. To find the exact spider vane obstruction in square inches: > $6.67 \times 0.0625 \times 3 = 1.25 \times 2 = 2.50$ square inches. To find the total light gathering power of our 8 in. f/8 binoscope: > $124.46 - 2.50 = 121.96$ square inches.

Viewing through a binocular telescope or refractor binoscope of any aperture with both eyes will always be quite sensational compared to observing with a regular telescope. In reality, large binocular telescopes are just two Newtonians, or even two Cassegrain telescopes (a very rare combination) coupled together on the same

Fig. 1.34 (Cartoon credit: Jack Kramer)

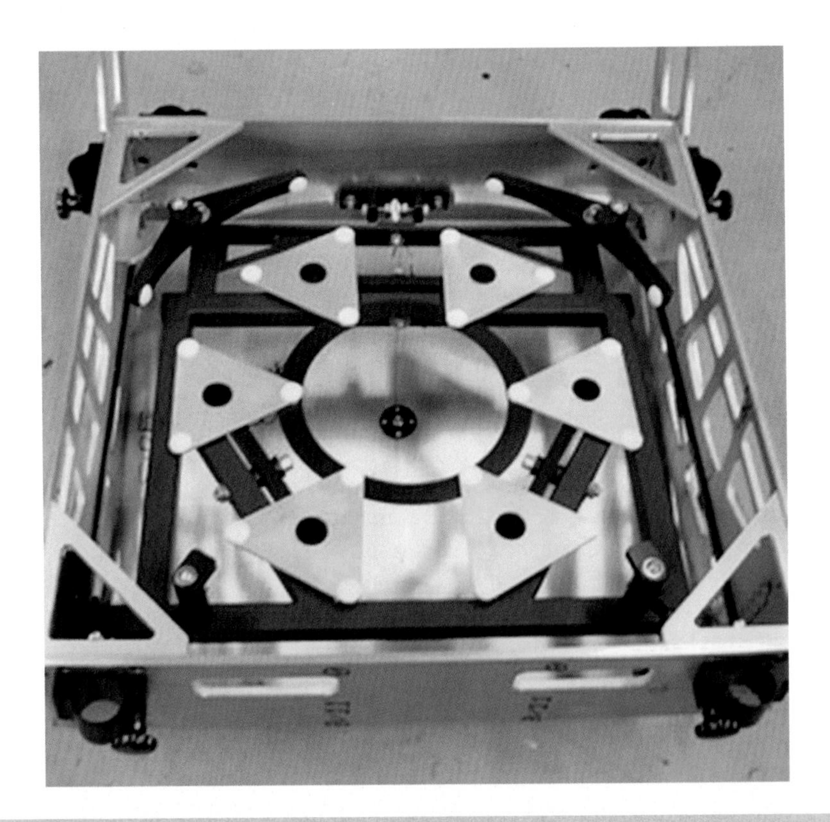

Fig. 1.35 A nicely designed 18-point suspension primary mirror cell that is looking for a home in a big Dobsonian telescope (Image credit: Aurora Precision)

Altazimuth or equatorial mount and optically aligned on the same object. The same goes for a refractor binoscope too. If you are going to buy a binoscope or binocular telescope, choosing which company to buy it from is not always easy. There are many good telescope companies and manufacturers (Meade, Celestron, JMI, Orion, Vixen, and Konus to name a few) as well as a host of Internet companies that offer a variety of telescopes to choose from. There are only a few companies (JMI, Avalon-Instruments and Vixen) that sell binocular telescopes and/or binoscopes. Trying to decide which is the best company to purchase your new binocular telescope, telescope or refractor binoscope from is sometimes more difficult than purchasing the telescope itself. Just be sure you get a good warranty with your purchase. If something goes wrong with your new telescope, you want to have it fixed as soon as possible without having a problem with the company who sold it to you. When it comes to buying a telescope, it's always desirable if you can "try before you buy" (Fig. 1.34).

The splendid looking 18-point primary mirror cell system in Fig. 1.35 was designed and built by Aurora Precision (aurorap.com) to hold a 20-in. primary mirror for a big Dobsonian Newtonian telescope. Enough can't be said about how important a primary mirror holder/cell is in terms of adequately supporting (without stress) a big primary mirror in a large Dobsonian telescope. Each equally spaced triangle point distributes support spread out over 18 equally spaced points of contact. This is a well-designed primary mirror holder for a very large Dobsonian telescope. A large Newtonian telescope needs a well-designed mirror cell to keep a heavy primary mirror secured firmly in place so it will stay collimated while

Fig. 1.36 If you have a 40-in. Webster Dobsonian like the one pictured above, get used to heights and standing on a tall ladder (Image credit: Eric Webster—Webster Telescopes)

Fig. 1.37 M-51 better known as the "Whirlpool galaxy" is a very bright spiral galaxy in the constellation Canes Venatici (Image credit: Sara Wager www.swagastro.com)

Fig. 1.38 (Cartoon credit: Jack Kramer)

moving the telescope about the sky for observing. For a good example, see the primary mirror cell shown in Fig. 1.35. It has what is called an 18 point suspension system that is designed to support a large primary mirror on a Newtonian Dobsonian or Cassegrain telescope. As I stated earlier, a well-designed primary mirror cell will keep your primary mirror properly supported and prevent it from shifting around once it is aligned and collimated. When shopping around for a good mirror cell, check out Aurora Precision for your primary mirror cell requirements.

Norm's Law states: "The bigger the Dobsonian…the taller the ladder" (Fig. 1.36).

When they get to be as big as the 40 in. Dobsonian in Fig. 1.35, then you may need a small truck to transport everything to the observing site and we're probably not talking about a one-man job. A monster Dobsonian this big demands attention; it obviously requires a little more setup time than a typical Dobsonian would. Judging by the height of the ladder, an observer would have to hang onto something for safety reasons to observe from that height. However, most of us amateur astronomers who love observing with a big telescope wouldn't mind literally having our heads in the clouds to look through a 40 in. Dobsonian telescope. The author would actually be the first in line to climb a ladder, regardless of how shaky it is, just to see the famous "Horsehead Nebula". If we had a big 40 in. Dobsonian sitting in our back yard, it would probably be mistaken for a two story apartment building because of its large size (Figs. 1.36, 1.37, and 1.38).

Further Reading

Aurora Precision, 20420 Boones Ferry Rd. N.E. Aurora, OR 97002–9401. http://aurora.com

Barbarella. (2008). 48 inch Dob. *Astronomy Technology Today*, *2*(6), 1374 North West Dr. Stafford, Mo 65757. http://www.astronomytechnologytoday.com

Chromey, F. R. (2010). *To measure the sky: An introduction to observational astronomy* (1st ed., p. 140). Cambridge: Cambridge University Press. ISBN 9780521763868.

English, N. (2011). *Choosing and using a Dobsonian telescope*. New York/London: Springer. http://www.springer.com

Kolmogorov, A. N. (1941). Dissipation of energy in the locally isotropic turbulence. *Comptes rendus (Doklady) de l'Académie des Sciences de l'U.R.S.S., 32,* 16–18. Bibcode: 1941DoSSR..32…16K.

Nightingale, N. S., & Buscher, D. F. (1991, July). Interferometric seeing measurements at the La Palma Observatory. *Monthly Notices of the Royal Astronomical Society*, *251*, 155–166. Bibcode:1991MNRAS.251..155N.

Pascal, R. Diffraction patterns of obstructed optical systems. http://www.beugungsbild.de

Steele, D. http://www.dobstuff.com

Tag Archives vs. RMS Optical Technologies. http://optical-technologies.info

Texereau, J. (1984). *How to make a telescope* (2nd ed.). Willmann-Bell, Inc. http://www.willbell.com

Tubbs, R. N. (2003, September). Lucky exposures: Diffraction limited astronomical imaging through the atmosphere. *Ph.D. thesis*.

Tubbs, R. N. (2006) The effect of temporal fluctuations in r0 on high-resolution observations. In *Proceedings of the SPIE*, Vol. 6272, p. 93T.

Types of curved vaned spiders, P.O. Box 5191 Pleasant, CA 94588. http://www.1800Destiny.com

Wager, Sara. http://www.swagastro.com

Webster Telescopes, 27843 Ford Road, Garden City, MI 49135. http://www.webstertelescopes.com

Wikimedia Commons. http://www.wikimedia.org

Chapter 2

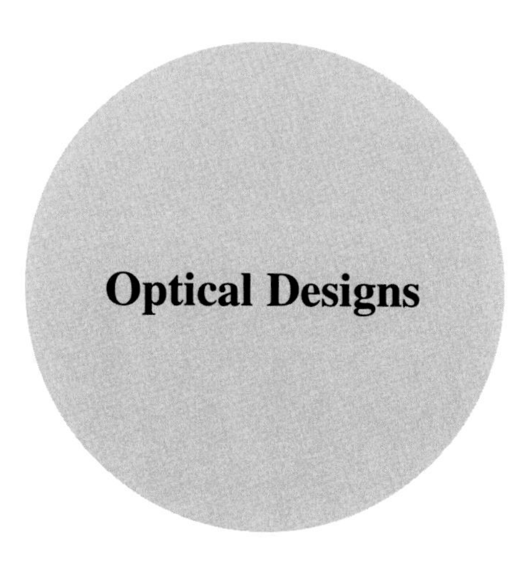

Optical Designs

For just about any telescope making project including a binoscope, or binocular telescope, it helps to become very familiar with the telescope's optical path that light will take through your telescope. This will help you in making the overall design of the binocular telescope or binoscope, especially in terms of the fabrication of the individual telescope parts. The entire overall design of the binocular telescope or binoscope is based on its overall optical focal length, focal ratio, and the size of the primary mirrors or objective lenses. A good idea before you start the construction of your binocular telescope or refractor binoscope is to draw the entire optical focal length diagram to scale on a computer. Using a computer with the appropriate software and a good optical ray trace program can show the position and angles of the mirrors (primary, secondary, and diagonals) including their focal lengths and actual position measurements in the optical system in either inches or millimeters. You can do this either to scale on the computer or draw it out by hand. Either way, you'll have an optical schematic accurately shown and drawn to scale on a computer to help you construct your binoscope.

The Newtonian optical design (see Fig. 2.25) for a Dobsonian binocular telescope is relatively simple to understand. There is nothing complex about where everything goes. Once you have established the correct dimensional position of each mirror and its focal arrangement in the optical system, then you can proceed with a high degree of confidence with the construction of your binocular telescope. Instead of trying to figure things out as you go and spending more time than necessary, create an optical path diagram to scale of your binocular telescope or binoscope on the computer and/or if need be, by drawing it out by hand (Fig. 2.1) as close to scale as possible. Ultimately this will help you from creating or eliminate the "bane" of all telescope builders, the dreaded "vignetting" (vignetting is a reduction of an image's brightness or saturation at the periphery compared to the image center - en.wikipedia.org) of

N. Butler, *Building and Using Binoscopes*, The Patrick Moore Practical Astronomy Series, DOI 10.1007/978-3-319-46789-4_2

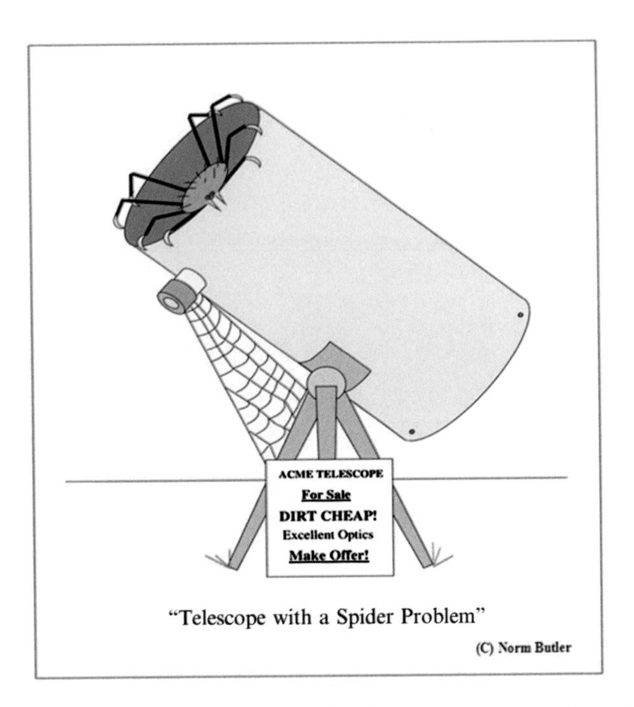

Fig. 2.1 (Cartoon credit: the Author)

the potential light path at the various aperture stops, field stops, baffles, eyepiece holders, eyepiece adapters, focuser draw tubes and also at or near the eyepiece focal plane (Fig. 2.10). This will make things easier and far more accurate in terms of the overall construction and fabrication of your binocular telescope or binoscope (Fig. 2.24, Fig. 2.25). Note: To avoid "vignetting" and cutting into your telescope's or binoscope's light path, the author recommends for example, leaving an estimated .002-inch to .004-inch (0.0508 mm to 0.1016 mm) clearance from the edge of a field or aperture stop...etc. to the periphery of the light path or light cone.

What Is Coma?

If you plan to make a binocular telescope with a fast focal ratio of f/4 or f/5 for example, then you'll have to put up with a lot of coma, an off-axis aberration. If you can't tolerate coma, then you'll need to use a couple of Coma Correctors. But before you go out and make that expensive purchase for a pair of high-quality Coma Correctors, let's first try and understand what coma really is. Coma is an optical defect in Newtonian telescopes with parabolic mirrors in which the focused star images appear more or less triangular, resembling a comet (comatic shaped) the closer the stars get to the edge of the field of view. The faster the telescope's focal ratio, the more pronounced the coma. The visual coma-free field of view (FOV) in a telescope in millimeters is roughly equal to the square of the telescope's focal

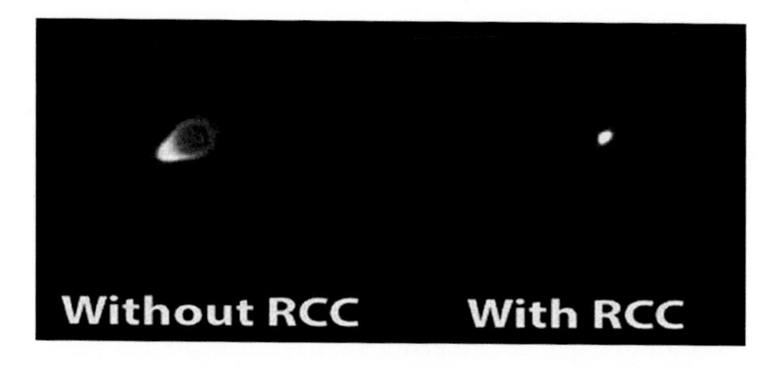

Fig. 2.2 A comatic star image at the edge of the FOV (Image credit: Rawastrodata—Wikimedia Commons)

ratio. For example, an f/5 Newtonian has a 25-mm field (5 squared = 25), while an f/6 Newtonian has a 36-mm field (6 squared = 36). With coma, the brightest portion of the star's image (coma shaped wedge) always points to the center of the field (see Fig. 2.2). This differs some-what from a telescope that is out of collimation, where the airy disks are all offset to the same side of the diffraction rings, no matter where in the telescope's field of view the star image is located or positioned.

How to Correct for Coma

If you plan to use Coma Correctors while you are observing with your binocular telescope, it would be beneficial to learn how they fit into your telescope's optical system and what their initial effect is optically and mechanically, especially when it comes to selecting the appropriate size and focal length eyepieces to use with it. A Coma Corrector can have from 1 to 3 optical elements, and some Coma Correctors, once installed, will increase your magnification approximately 15%. Plus, with up to three optical elements, you can expect some light loss with a Coma Corrector too. However, they do remove coma (see Fig. 2.2), and they flatten your field of view too. For example, a Coma Corrector can make an f/4 Newtonian telescope perform like an f/8. If you're going to make a binocular telescope or refractor binoscope that is f/5 or faster and you can't stand coma, then you're going to need two Coma Correctors. TeleVue Paracorr, Lumicon, Starzona, AstroTech, Baader Rowe, and ScopeCraft offer photovisual Coma Correctors.

How to Adjust the IPD in a Binoscope

Binoscopes can be refractors too. Some of the most creative amateur telescope makers in the country have made some very unique refractor binoscopes that prove to be very popular at local star parties and at amateur telescope making conferences. With so many people lined up to look through these splendid homemade binocular telescopes and refractor binoscopes, it was always an engineering challenge to come up with an effective way to adjust for so many different IPDs or more commonly called the "interpupillary distance."

One of the more clever and significant commercial designs to come along in the last few years is Avalon Instruments of Italy dual eyepiece (just a "twist" of a knob) IPD adjustment system. They also offer a splendid looking 107 mm binoscope that provides excellent lunar, planetary and deep sky images (Fig. 5.68). Note: When it comes to choosing what type of IPD system for use on their refractor binoscope commercial or otherwise, the author recommends using a prism IPD system verses mirrors. They're a lot cheaper, more efficient and easier to use.

The clever dual 2-inch focusing system red lever IPD adjustment system seen on the Bolton Group's 300 mm all-metal skeleton tube binocular telescope in Fig. 2.4 requires very little adjustment, especially for setting your interpupillary distance. Notice the dual individual threaded eyepiece focusers. This unique yet very simple design will in fact compliment the overall appeal and function of your binoscope, making others eager to have a look through it, especially at local star parties and astronomy conventions (Fig. 2.4).

Refractor binoscopes users have a few good choices for an IPD adjustment. One of the newer designs that seem to be popping up on some of the popular commercial binoscopes these days utilizes a pair of rotating rhomboid prisms and mirror diagonals to adjust for the IPD. When rotated at appropriate angles to each other, the dual rhomboid prism system can provide the observer with enough IPD rotational adjustment to view celestial objects comfortably using two eyes. Another possible option is to use two microscope eyepiece holders with eyepieces connected to your binoscope's focusing draw tube (see Fig. 2.21). For example, a pair of 1–1/4 in. eyepieces in a microscope eyepiece holder can be rotated in either direction without the image changing its original orientation. For use on refractor binoscopes, they would make an excellent dual eyepiece system complete with an IPD adjustment. It is even possible that a dual microscope eyepiece holder arrangement would work on a binocular telescope too, although I have yet to see one (see Fig. 2.10). The most commonly used binocular telescope dual eyepiece focuser arrangement is the use of two Crayford focusers (see Fig. 2.6) mounted directly opposite of each other. The two Crawford focusers are coupled to a pair of opposing focuser draw tubes that are optically and mechanically parallel to each other so that an adjustment for the IPD can be made as well as focusing each one individually on the same image.

The following photos and discussion will focus on some of the different kinds of mechanical focusers that are used on binocular telescopes and refractor binoscopes that can accommodate for the IPD adjustment. There are mechanical focusers that have an advantage over others and are more commonly used on binocular telescopes than other types. Keep in mind that there is no prefect mechanical focuser out there that can do it all but there are some that do more than others. The mechanical focuser is the one of the most-used instruments on a telescope when observing.

It allows the observer to focus the image, interchange eyepieces, star diagonals, and their accessories and therefore is in somewhat constant use. Not all mechanical focusers are of the same design even though they may appear similar. Mechanical focusers may be made out of machined aluminum, steel, while others may be made

Fig. 2.3 Baader Rowe Coma Corrector and field flattener (Image credit: First Light Optics)

Fig. 2.4 Dual 2-in. eyepiece focusing system red lever IPD adjustment system on a 300 mm all-metal skeleton tube binocular telescope (Image credit: The Bolton Group—www.deep-sky.co.uk)

out of a molded plastic material. Some require periodic maintenance and others do not. Some have a vernier scale or digital readouts for focusing purposes. Most do not. Some even have precision micro-focusing, while others use the standard rack and pinion for focusing. And there is the Crayford style focuser with its friction focusing. Crayford focusers (especially the "Reverse" Crayford style focuser) are becoming more popular on today's Dobsonian, Newtonian, and binocular telescopes. When choosing a good focuser for your binocular telescope, choose a precision-made focuser with all of the good mechanical elements in mind.

According to Wikipedia.org, the Crayford focuser was invented by John "Jack" Wall, who is a member of the Crayford Manor House Astronomical Society in Crayford, London, England. The original Crayford focuser, which is on display there, is a simple mechanical focuser designed for use on amateur telescopes and also found on commercial telescopes too. Crayford focusers are considered superior in design to standard rack and pinion focusers that are commonly found on today's telescopes. Instead of a typical gear rack and pinion, Crayford focusers have a simple mechanical design with a smooth spring-loaded shaft which holds the focusing tube against four opposing bearing surfaces that control its mechanical movement.

The Crayford focuser was originally demonstrated to the Crayford Manor House Astronomical Society, and then descriptions of his remarkable focuser were published in The Journal of the British Astronomical Association (February 1971), Model Engineer magazine (May 1972), and Sky & Telescope magazine (September 1972). Fortunately for amateur astronomers and amateur telescope builders (ATMs), Mr. Wall decided not to patent his clever and unique mechanical focuser; instead, he graciously donated it to the astronomical community.

The Reverse Crayford focuser pictured in Fig. 2.6 is a direct descendant of the original Crayford focuser described in Fig. 2.5. It is commonly used on binocular telescopes as well as Dobsonian and Newtonian telescopes and provides the user a backlash-free precision-focusing movement that is superior to typical rack and pinion focusers. Generally speaking, the Reverse Crayford focuser is a desired item to have on just about all homemade and commercial binocular telescopes. However, there are still a lot of homemade binocular telescopes, Dobsonians, and standard Newtonian telescopes that use traditional rack and pinion focusers that are adequate

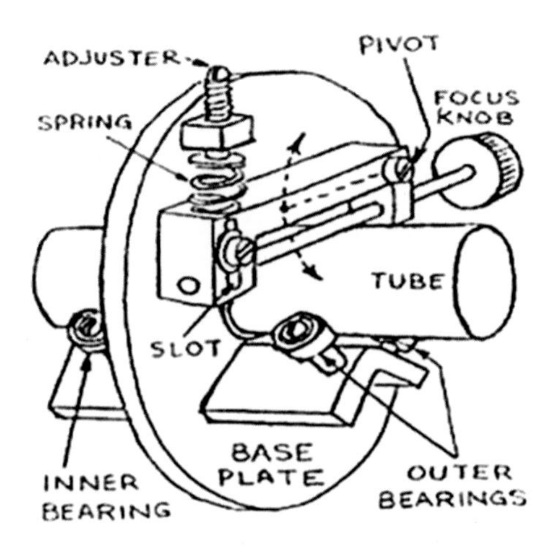

Fig. 2.5 The original Crayford focuser drawing by John "Jack" Wall (Image credit: Tamaflex—Wikimedia Commons)

Fig. 2.6 A Reverse Crayford focuser (Image credit: JMI)

Fig. 2.7 Bruce Sayre's elegant 14.5 portable binocular telescope has a simple IPD feature (Image credit: Bruce Sayre)

enough to provide the user with satisfactory focusing. If precision focusing is what you want on your binocular telescope, then the Reverse Crayford focuser is for you.

Bruce Sayre's portable 14.5 in. binocular telescope (Fig. 2.7) has two IPD carriages are attached to the altitude frame with linear bearings. To adjust IPD, the user rotates a knob just below the eyepieces. Counterclockwise rotation increases eyepiece separation; clockwise decreases it. This will cause the carriages—one supporting each side's OSS—to move laterally and in opposing directions through a right-angle drive. Counterweights attached to the bottom of the frame balance the telescope. The carriages also provide for convergence (or optical parallelism). Each OSS is connected to its underlying carriage through bolts and adjusting nuts. The IPD

Fig. 2.8 Arie Otte has figured out an efficient way to adjust for his IPD on his binocular telescope (Image credit: Arie Otte)

adjustment involves no change to the focal plane or upper end optics positions. Taking the IPD adjustment out of the optical path simplifies and lightens the upper end structure. This eliminates the errors in collimation or convergence that moving tertiary assemblies or rotating upper cage rings can introduce. Based on its mechanical description and how it functions, it certainly looks like Bruce has devised an excellent mechanical way to adjust the IPD on his portable 14.5 in. binocular telescope.

Mr. Aire Otte of the Netherlands has figured out a very efficient and uniquely clever way to adjust for his IPD on his 13-in. f/5 binocular telescope (see Fig. 2.8). A small section of round rack and pinion gear drive is mounted on both of upper tube assemblies (directly below the eyepieces) that can rotate or turn one or both of the upper optical tube ends. Arie calls them "turnable tops." The key word that he wants us to remember is "turnable." And it appears that Arie has indeed done his homework when it comes to designing and building an efficient way to adjust for the IPD. He has kept his IPD adjustment simple and very user friendly, and that's an important thing when it comes to being able to enjoy your own creation without having to make unnecessary adjustments in order to enjoy the problem-free use of it.

Of course there are other ways to adjust for the IPD on a binocular telescope and each way has its own advantages and disadvantages. At a big star party, for example, there are usually many people lined up waiting to look through your binocular telescope or binoscope and as a result, it's hard to accommodate for all of their different IPDs. Some designs that can accommodate for the IPD may take a little more adjustment time in order to set it close enough for someone to view comfortably through your binocular telescope. When you have 10 or 15 people waiting anxiously in line for their chance to look, that may be a disadvantage.

So what's the best design to use for adjusting the IPD on a binocular telescope? First of all, there is really no such thing as the "best design" for adjusting the IPD on a binocular telescope. There will be more discussion on that very subject later on in the chapter. To begin with, one would need to choose a good accommodating focuser first as the base system or platform to make the IPD adjustments from. There are some focusing designs that are more efficient to use on a binocular telescope than others. And there are some that shouldn't be used at all. What we're really looking for is an efficient and quick way to adjust for the IPD on a binocular telescope, and to do so we need a focuser design where you have the fewest moving parts. For starters, a single stalk horizontal focuser system (sometimes called a "sled" or "slide" focuser) could be considered a good alternative for use on a binocular telescope. JMI uses a similar focusing system. One of the main reasons why a single stalk horizontal sled focuser would be a good option to use on a binocular telescope is because it has very few moving parts. It also maintains a low profile on a telescope. The sled part of the focuser carries the secondary mirror on a single stalk along with an eyepiece holder as they travel along the optical axis in the horizontal plane of a Newtonian telescope. Any excess slop or movement that could potentially develop in the draw tube of standard rack and pinion focuser would not be apparent in a single stalk horizontal focuser, primarily because there is no draw tube. Perhaps the only drawback to using a sled-type focuser is that the focal plane will move in relation to the tube wall and the diameter of the illuminated field of view slightly changes. To control such a small parameter, you may wish to reduce the secondary's size to begin with when using various eyepieces. With a sled type of focuser (see Fig. 7.25), the eyepiece remains stationary in a holder and is attached to the sled part of the horizontal focuser. The sled of the horizontal focuser could be made out of wood, aluminum, or even plastic. The sled's secondary holder

Fig. 2.9 A commercial single stalk secondary holder sled focuser on a Newtonian telescope (Image credit: the Author)

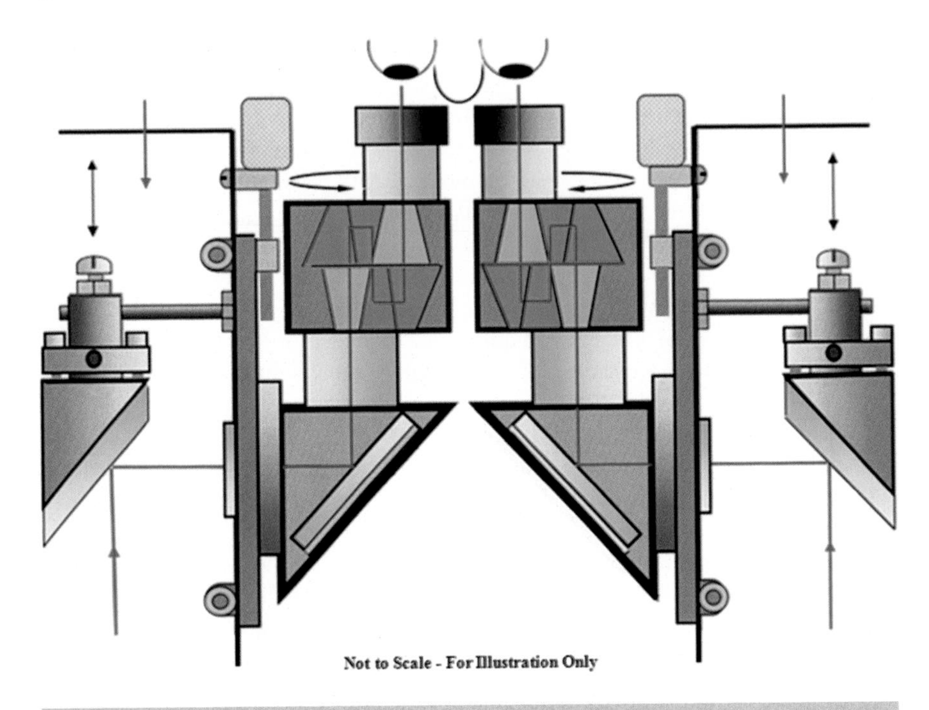

Fig. 2.10 Dual microscope binocular eyepiece holder for the IPD adjustment and a horizontal lead screw sled focuser (Image credit: the Author)

single stalk can be a rectangular, square, or round solid diameter rod (aluminum) that is connected to the bottom of horizontal slide of the sled focuser (see Fig. 2.9).

Using a single stalk horizontal sled focuser on a binocular telescope can have some important advantages over a typical standard rack and pinion or reverse focuser, and one of them is that the focus travel is not outward but parallel to the telescope tube and the optical axis. A "sled" or "slide" focuser also provides a lower profile, and this can make it more convenient for placement of diagonals for use in a binocular telescope. To simplify the IPD adjustment and make it even more user friendly, try using of a pair of 1¼-in. microscope prismatic eyepiece holders (see Fig. 2.10). They can make the IPD adjustment quite simple and will accommodate for an IPD for just about everyone. With today's optical coating technology, a prism manufacturer and/or an optical coating company can coat a pair of prisms with dielectric coatings that will allow them to transmit over 99.9% of the light. Using a pair of microscope prismatic eyepiece holders allows you to swing the eyepieces in or out for the IPD adjustment without changing the image's original orientation.

As shown in Fig. 2.9, the single stalk sled focuser is quite efficient and relatively simple in design with very few moving parts. There is no spider or vane assembly in its design, only a single solid stalk to hold the secondary mirror and it's holder in position (Fig. 2.11).

Fig. 2.11 (Cartoon Credit: Jack Kramer)

The design and construction of a sled focuser can be relatively simple, especially if one uses a rack and pinion from an old focuser assembly. The rack and pinion once attached to the sled will provide the linear track for focusing just like a standard focuser, except the focus travel is in the horizontal axis. The "sled" part can be made out of aluminum, wood, or plastic and attached securely to the top front section of a Newtonian telescope using a linear track guided by either bushings or bearings, so it can travel easily in a smooth linear fashion while focusing. A "sled" focusing system would be a good consideration for your next binocular telescope project (Figs. 2.9 and 2.10).

The aluminum tubes on Dave Trott's 6-in. binocular telescope are hinged and can be rotated slightly for the IPD adjustment, a very interesting and clever mechanical design that works! Below is a description on how his IPD adjustments are made using his dual tube hinge system (Figs. 2.12 and 2.13).

Each scope is movable for the interpupillary adjustment. With the 13-in. scopes, only one tube assembly is movable and that causes a slight "head twist" which is awkward. Comfort is the name of the game with these scopes! After experimenting with some heavy-duty door hinges, they can be made very stiff by replacing the hinge pin with a bolt and tightening the nut to a fairly high torque. The hinges are set at the greatest possible longitudinal separation to maintain parallelism. The alignment procedure is critical. Aligning this kind of scope is not a trivial process, so you don't want to do more than a touch-up very frequently. It is time consuming and frustrating, and having done it enough times, one will know the fastest and most effective techniques to use.

You start with a very rough field rotation adjustment made by looking carefully at the focusers to make sure they are parallel with the scopes. The scopes will of course already be pretty close to being perfectly parallel already. Remember, though, the scopes may look a bit off, while the optical axis is dead-on parallel.

Fig. 2.12 Dave Trott designed this clever hinged dual 6-in. binocular telescope that is similar to a basic pair of binoculars (Image credit: Dave Trott)

Fig. 2.13 Pictured are the push/pull bolts for vertical (on the *left*) and horizontal (on the *right*) IPD adjustment (Image credit: Dave Trott)

If the adjustment bolts tell you the two scopes are not aligned, do not worry. It's the optical axis of each scope that counts! Take the scope outside and have a look. You must select something very distant. By far the easiest time to do this is when there is a nice bright moon. A very distant mountain might do as well. Hopefully, there will be a nearly matching image in each scope. If not, it's trial and error till you get something similar in each eyepiece. Do the adjustment for horizontal first. Then adjust till the vertical is good. Always do the horizontal first because the eyes will compensate quickly for even a substantial horizontal misalignment. You will probably see it "snap in" at about this time as the eyes and brain work to make the images merge. But, you'll probably also feel some "strain." If you back the images out of adjustment in the vertical plane and look carefully at the moon, you will notice that one image may be slightly tilted with respect to the other. Check a few of the brighter craters and their relative position. A first-quarter moon is easiest because any tilt is a dead giveaway. Then rotate the tertiary mirrors until the images are identical. This will give you a headache. Take a short break and come back to it again in a few minutes. You will probably now have to readjust the horizontal and vertical all over again, though you should be pretty close. When you are right-on, you will know it. The image will, as Dave says, "sing" to you sweetly! You will wonder how you could have ever thought you enjoyed looking through one-eyed telescopes!

Dave Trott

Adapted from "Dave's 6-Inch Binocular Telescope" with permission from Dave Trott (http://davetrott.com)

David Moorhouse from New Zealand writes: Eye spacing varies greatly between different observers; therefore, there must be some mechanism by which the intraocular distance can easily be adjusted. The interpupillary distance ranges from roughly 53 to 73 mm with an average IPD for the adult population of approximately 63 mm. One of the restrictions this presents is that it is almost impossible to use 2-in. eyepieces. Therefore, my design only uses 1¼ in. eyepieces. There are two main systems that have been employed in order to alter the interocular distances (Figs. 2.15 and 2.16).

The first system involves focuser-type mechanisms to push the eyepieces closer together or farther apart; focus is then being achieved either by moving the eyepiece up and down or in the case of JMI moving the entire secondary and focus a mechanism vertically in unison. The main problem with this design is that every time the interocular distance is changed, the focus is also changed by a similar amount. The second approach, which is the approach that I have chosen, involves using a rotating secondary cage. This causes the eyepieces, focusers, and secondaries of the two different optics paths to rotate farther apart or closer together to each other. With this approach, the focus is consistent regardless of the interocular spacing (Fig. 2.14).

And focus is achieved with a standard focus mechanism. Looking at the photographs, you will see that I have used a reasonably complex system of slots in the upper ring with Teflon bushes, allowing the rotation to happen.

Fig. 2.14 Dave's dual eyepiece IPD adjustment system for his 16 binocular telescope (Images credit: David Moorhouse)

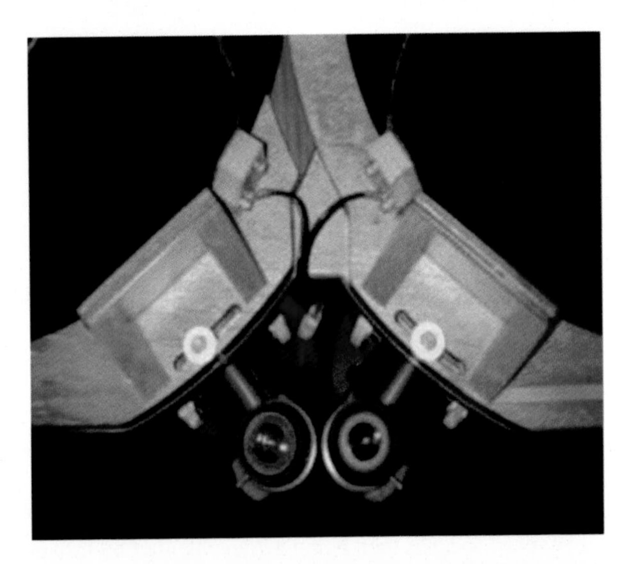

Fig. 2.15 Dave's dual eyepiece IPD adjustment system for his 16 binocular telescope showing the dual eyepieces adjusted "in" (Images credit: David Moorhouse)

Upon reflection this is probably far too complicated for the task at hand, and a much more simplistic design will be used to allow the tops to rotate in any future design. Firstly, the rings would have been much farther apart so as not to need the complex overlapping system. This was only done as I was trying to reduce the overall dimensions of the width of the telescope; however, I believe this was unnecessary. The slots could easily be replaced with simple blocks placed on either the inside or the outside of the ring system, again with Teflon bearings. In practice, this approach so far seems to work very elegantly. Only small horizontal and

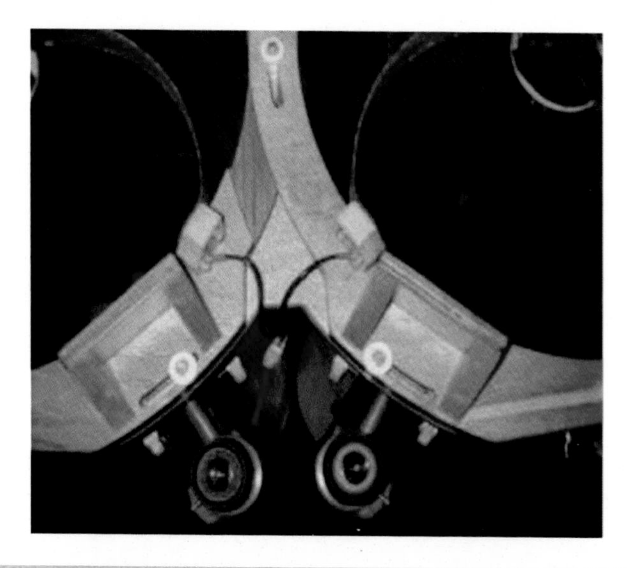

Fig. 2.16 Dave's dual eyepiece IPD adjustment system for his 16 binocular telescope showing the dual eyepieces adjusted "out" (Image credit: David Moorhouse)

vertical shifts have been noticed, while changing the interocular spacing and most people naturally tend to try and move the eyepieces apart or closer together as they would with a standard pair of binoculars. This is a very intuitive mechanism that is a plea-sure to use (see Figs. 2.15 and 2.16). The main reason why I chose curved spider vanes (Fig. 2.17) was for viewing believeability. I believe that visual astronomy needs to project a believable view to truly engage your mind. If you look through most standard telescopes, you will almost always see moderate to large size diffraction spikes caused by the straight spider vanes. Now imagine what would happen if you were to try and view this image through both eyes simultaneously. The unfortunate thing is that the diffraction spikes would be rotationally offset to one another.

You would see eight diffraction spikes, and your brain would not be able to make sense of the image it was seeing and ruin the reality effect (Figs. 2.17 and 2.18).

This would be a problem of course on bright sources. Using curved spider vanes, the diffraction energy is still present; however, it is a circular pattern. This sometimes makes the brighter stars to always appear out of focus or slightly bloated. However, the real appearance when seen with both eyes is perfectly acceptable, making such things as open clusters, often with red giant stars among them appear to hang silently just out of reach as if through a porthole window. Only on extremely bright stars is this bloating effect noticeable. The vanes are made from plate steel; these were then cut and rolled by a local engineering firm. The hubs are made up of two circular pieces of Bakelite held apart with three pieces of aluminum strap that we drilled and tapped to allow the assembly to be screwed together. Having three separate attachment points for the curved vanes also allowed me to feed power to some secondary heater units via the vanes. The secondary holder is a standard Novak

Fig. 2.17 Dave's curved spider vane holder system for his 16-in. binocular telescope (Image credit: David Moorhouse)

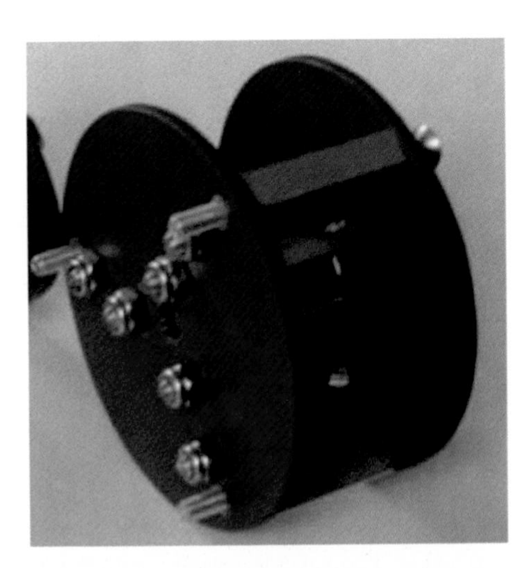

Fig. 2.18 Dave's three curve vain holder system for his 16-in. binocular telescope (Image credit: David Moorhouse)

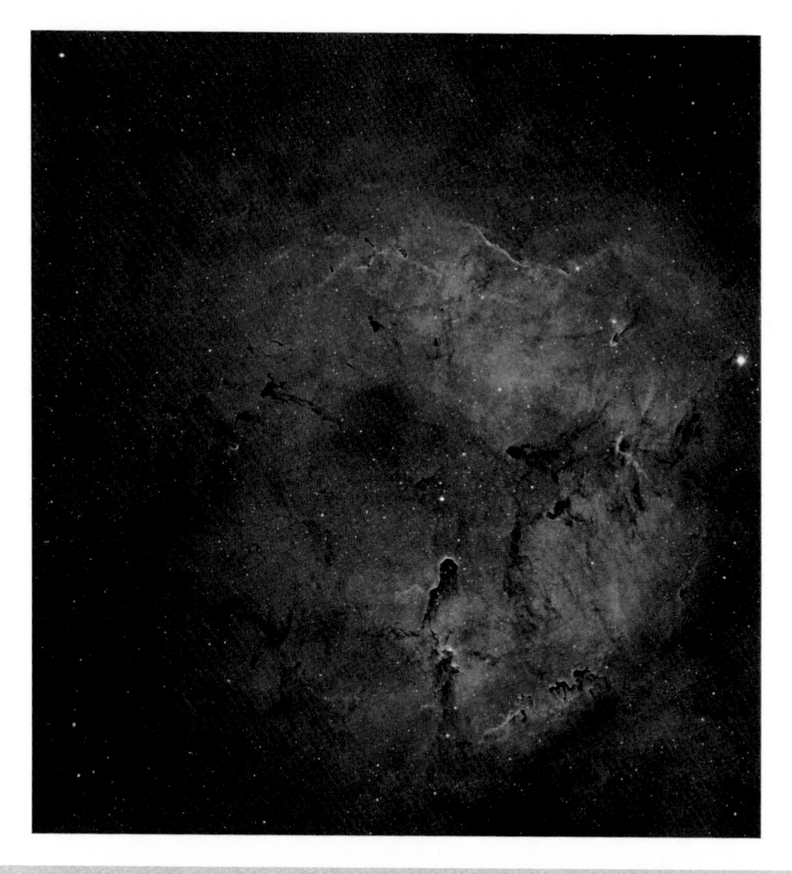

Fig. 2.19 IC1396 sometimes referred to as the "Elephant's Trunk" is located in the constellation Cepheus (Image credit: Sara Wager—www.swagastro.com)

mirror holder, although at 3.5 in. it is slightly larger than what would normally be considered the correct size for a 16-in. primary mirror. The main reason for this is that due to the tertiary mirrors, the secondary mirrors end up approximately 3 in. further away from the focus point. This means that the light cone at the point where it is intersected by the secondaries is slightly larger (Figs. 2.19 and 2.20).

David Moorhouse New Zealand

Adapted from "Dave's 16 Inch Binocular Telescope Page" with permission from David Moorhouse (http://www.binoscope.co.nz)

The author makes note of the fact that the use of a pair of microscope prismatic eyepiece holders on a refractor binoscope can easily accommodate for any IPD by simply rotating one or both of them. Using this method, the IPD adjustment can be made without changing the original orientation of the image. Just a small twist of the lock screw secures the eyepiece holder and IPD adjustment in place (Fig. 2.21).

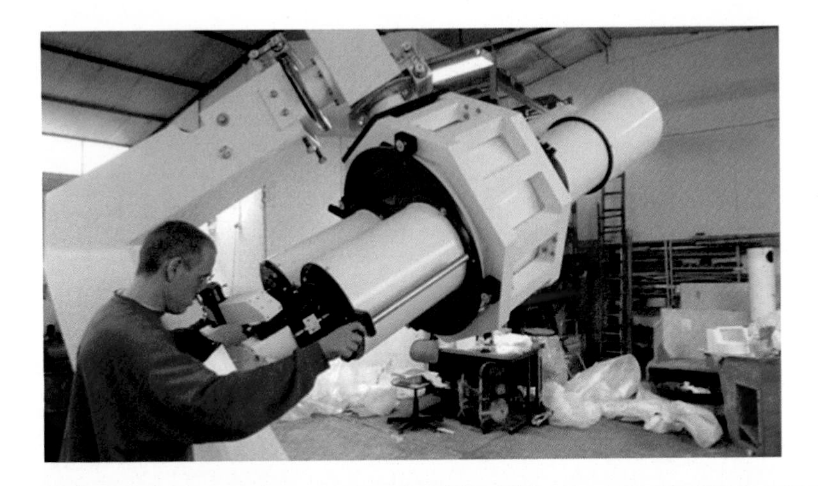

Fig. 2.20 Wouldn't it be great to have a big dual 12-in. F/7.5 APO refractor binoscope like this splendid computerized bino from APM setup in an observatory in your backyard? (Image credit: www.apm-telescopes.de)

Fig. 2.21 How about a pair of 1¼-in. microscope eyepiece holders to use for a refractor binoscope? (Image credit: the Author)

Sometimes you have to decide whether it is really worth it to spend the engineering time and resources to make an IPD adjustment on your binoscope, especially if you're going to be the "only" one using it most of the time. This dual 6-in. f/15 Cassegrain binocular telescope was made without an IPD adjustment for the single reason that only one person would be using this big binoscope majority

Fig. 2.22 A fixed IPD "zoom" binoscope eyepiece system on a pair of 6-in. F/15 Cassegrain binoculars (Image credit: Rick Schmidt)

Fig. 2.23 A nice example of a homemade Dobsonian binocular telescope (Image credit: ECeDee—Wikimedia Commons)

of the time. It was designed with an IPD of approximately 65 mm, about right for most people (see Figs. 2.22 and 2.23).

Note:
A mirror or a prism erects the image in the up-down direction, but the image remains reversed in a right-left sense.

If you have a refracting telescope, using a diagonal will turn the image left to right. In reality, in outer space there are no real "cardinal points" to distinguish direction, such as "up" or "down," "East" or "West." It's only relative to our position here on planet Earth. However, the confusion comes when you move the telescope and see the images in the eyepiece move in a different direction than you anticipated. Also, when observing with your telescope, you must try and understand the orientation of the image when checking it in reference to a star chart or sky atlas. Today, almost every refractor comes equipped with a star diagonal to redirect the light. Unless one prefers looking through a refractor without a diagonal, then

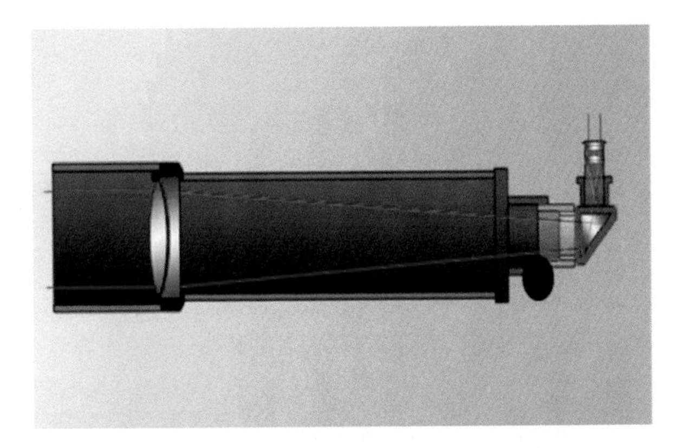

Fig. 2.24 Typical achromatic refracting telescope (Image credit: Starizona)

Fig. 2.25 Simple Newtonian optical path diagram (Image credit: Starizona)

Fig. 2.26 A binoviewer used in a reversed mode with a single eyepiece on a refractor binoscope or binocular telescope can provide spectacular views too (Image credit: www.celestron.com)

the light passing through the objective presents itself in a correct "normal, upside-down, right-left" image (Fig. 2.24).

A simple Newtonian optical path diagram shown below in Fig. 2.25 is the same optical path design used in Dobsonian binocular telescopes. An equal number of reflections as in a Newtonian reflector will present an "upside-down" image. The only difference is that the focal plane of a binocular telescope will normally extend a few inches further beyond the tube than the focal plane of a typical Newtonian telescope. Reverse Crawford focusers are commonly used in the construction of binocular telescopes. Each of the two focusers needs approximately 2 or 3 in. of vertical focuser travel to focus each eye independently as well as another inch or so of horizontal drawtube travel to adjust the observer's IPD accordingly. There is also the potential problem when using 2 in. eyepieces where the top outer diameter of their eyepiece plastic covers may be too thick and as a result, both eyepieces come in contact with each other and may "not" allow for a minimum inwards IPD adjustment. The average IPD for most individuals is 65 mm. It may be necessary to machine the top outer diameters of both eyepiece plastic or metal covers just a bit to allow for a greater inner IPD adjustment below 65 mm (Figs. 2.21–2.26).

An Interesting Binoviewer Concept

A high quality binoviewer like the one in Fig. 2.26 can provide the observer with some fantastic views of lunar and planetary objects. It's easy to reverse the intended function of a typical binoviewer, which basically consists of a simple rhomboid prism with a beam splitter (see Chap. 7—Binoviewers). When used in a reversed function, the binoviewer creates a beam-convergent system. It collects the optical light path of two binocular telescopes and combines them into one single image, increasing the effective light gathering power slightly more than that of a typical single telescope. So it's possible to use a typical commercial binoviewer on a binocular telescope or refractor binoscope (after some slight modification) in the reverse (opposite) "mono mode" to view lunar and planetary objects under high magnification using one eye. The advantage here is the potential for using a Barlow lens to study lunar and planetary features using a higher magnification because of the combined images from both binocular telescope mirrors or refractor binoscope objectives. Obviously, the use of a 2× Barlow lens along with the combined light grasp of the binocular telescope mirrors and/or refractor objectives has the potential for some splendid lunar and planetary views. However, this clever binoviewer concept is almost never seen being used at local star parties or national telescope makers conferences where there are binocular telescopes or refractor binoscopes.

One of the major reasons for this is that most people want to look through a big binocular telescope or refractor binoscopes with both eyes.

What kind of modifications would be needed if you really wanted to use a binoviewer for your observing pleasure on your binocular telescope? One of the first modifications you could make to your binoviewer would be to make or find a pair of 1–1/4 in. adapter inserts for each individual binoviewer eyepiece barrels. You can then insert them into both of the bino focuser's 1–1/4 in. eyepiece holders (Fig. 2.27). If you're using 2 in. eyepieces, then a 2 in. eyepiece adapter is needed inorder to insert it into the binoviewer, or you can alternatively use 1–1/4 in. eyepieces (recommended). Another good advantage in using a binoviewer in the "reverse mode" on a binocular telescope or refractor binoscope is that it is lot easier to detect if both optical tube assemblies are in accurate alignment with each other.

The greatest advantage with using the binoviewer in the "reverse mode" is the impressive visual images it can produce. This is especially true when observing lunar and planetary objects. (Thanks to Mr. Paul Van Slyke for the "reversed mode"

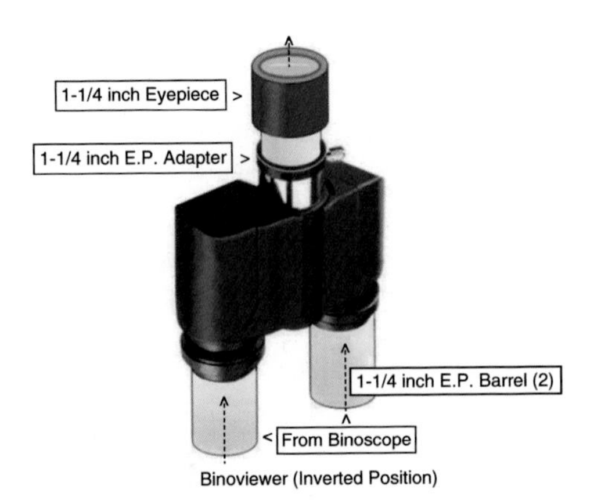

Fig. 2.27 Modified binoviewer in the "reversed" binoscope mode, sometimes called the "mono mode". (Image credit: the Author)

binoviewer idea). Springs for demonstrating the advantage of using a binoviewer in a reverse mode on his homemade 17.5-in. binocular telescope. In http://www.observatory.org.

Further Reading

APM Telescopes Goebenstrausse 35 66117 Saarbrueceken Germany. http://www.apm-telescopes.de
Celestron, 2835 Columbia St., Torrance, CA 90503. http://www.celestron.com
First Light Optics Ltd., Unit 53 Basepoint business Centre, Yeoford Way, Marsh Barton Trading Estate, Exeter, Devon EX2 BLB. http://www.firstlightoptics.com; http://www.binoscope.com
http://commons.wikipedia.org/wiki/Category:Dobsonian_Telescopes
Jim's Mobile Incorporated (JMI), 8550 West 14th Avenue, Lakewood, CO 80215. http://www.jimsmobile.com
Moorhouse, D. Dave's 16 inch binocular telescope page. http://www.binoscope.co.nz
Otte, A. Otte binoscopes. http://arieotto-binoscopes.nl
Starizona, 5757 N. Oracle Rd., suite 103, Tucson, Arizona 85704. http://www.starizona.com The Binoscope Company, J. Wyman Court, Corani, NY 11727.
The Bolton Group, http://www.deep-sky.co.uk
Trott, D. Dave's 6 inch binocular telescope. http://davetrott.com
Wager, Sara, http://www.swagastro.com

Chapter 3

Binoculars are considered binoscopes. We usually just think that binoculars are just binoculars. In reality, they are really in most cases a smaller version of a binoscope. Basically they are two small refractor telescopes that are optically aligned on the same object for viewing. Of course there are larger versions of the binocular. Some have objective lenses that are 150 mm in diameter and often 25 power or more. Being so large of a binocular, they have to be supported on a large tripod or pier. The view through a pair of 150-mm Fujinon binoculars is stunning to say the least. However, the expense that one has to pay for a pair of these fantastic large Japanese binoculars is often out of reach for the average individual.

There are so many different types, sizes, and brands of binoculars that range in just about every imaginable size and price range. An individual who is shopping around for a good deal would not have a problem finding a pair of good binoculars at a very reasonable price. And that's the key to finding a good deal on a quality pair of binoculars is to "shop" around and look for a good pair of brand name binoculars that are on sale or discount. Of course, the internet offers hundreds of binocular websites that sell just about every size and kind of binoculars. The best part about looking for a real deal on the internet is when you actually find a pair of binoculars and it only takes a credit card number to have them sent to your house within 5–10 days. However, returning them to the dealer for a refund or exchange may be difficult or impossible if they break. If you are buying binoculars on the internet, find out who the original manufacturer is and where they were made. You'll know more about what to expect once you have this information. Keep in mind that everything looks good on the surface until you do a little more research.

N. Butler, *Building and Using Binoscopes*, The Patrick Moore Practical Astronomy Series, DOI 10.1007/978-3-319-46789-4_3

A Condensed History About Binoculars

In order to better understand the evolution of binoculars, let's trace the origin of the first binoculars. The history goes far back in time and most of it is still a mystery, as it may remain for years to come. The historical evidence we do have starts with the discovery of glass around 3510 BCE. It then took about another 4950 years or so for it to be formed into something that was useful, such as a lens, which was ultimately used for the construction of the first telescope. But even the telescope's true origin is somewhat mysterious. It is generally accepted that binoculars came about as part of the natural evolution of the telescope. One of the big advantages that binoculars have compared to a single telescope, is that they provide a three dimensional view for the observer by producing a combined perspective. The result is a more realistic perceived field of view with greater depth.

History claims that the invention of modern binoculars is attributed to the Dutch spectacle maker named Hans Lippershey. In the winter of 1608, he discovered that a lens of a convex shape and a concave lens can be combined together to magnify an object of some distance. In simple terms, this discovery is the simplified "version" and origin for the telescope! After his initial discovery, Lippershey on October 2nd, 1608 decided to present his new telescope invention to the Netherland's States General. As a result of his newly invented telescope, it was requested that he build a telescope version for military purposes that could be used by an observer using both eyes. He complied with the request and as a result, created three sets of "two eye telescopes", which at the time, was not very impressive because of their poor optical performance. At the time, telescopes had low magnification and produced poor-quality images. Lippershey submitted a request for a patent on his new so-called telescopic invention but it was initially rejected. As a result, there is serious doubt as to whether or not Lippershey was in fact the first one to combine a convex lens with a concave lens. Nonetheless, by the early part of 1609, the early version of the "binoculars" or "spyglasses", which were referred to as telescopes, became extremely popular in Paris.

The first "actual" binoculars were the so-called "Galilean binoculars". Historical records tell the story of how the famous Italian physicist Galileo Galilei had heard about Lippershey's new optical invention and decided to build a telescope to experiment with. Over time, Galileo made around one hundred telescopes, with magnifications ranging from about three powers up to about thirty powers. Observing through his telescopes, Galileo was able to observe the craters on the Moon and Jupiter's four largest satellites, and that was the beginning of real astronomical research. Galileo's original telescopes consisted of only two lenses: a simple convex lens at the front end, called the objective lens, and a concave lens at the rear or eye end, which was called the eyepiece. With a Galilean telescope, the observer could see magnified, upright images. He could also use his telescopes to observe the night sky. Galileo during the early part of the 1600s was one of the few who could build telescopes that were good enough for observational astronomy. In Aug. of 1609, he demonstrated one of his early telescopes which magnified images up to about 8 or 9 times, to Italian lawmakers. Galileo's telescopes were also a profitable sideline business for him—he sold to merchants who found them useful both at sea and as items of commerce and trade. Galileo published his initial telescopic astronomical observations in his brief treatise entitled *Sidereus Nuncius* (*Starry Messenger*) which he published in March of 1610.

Galilean binoculars up until the 1850s were very popular accessories, used in theaters, operas and social events. They were often covered with ornaments such as pearls, silver, gold, ivory or bone and even colored leather (Fig. 3.46). Even their Galilean binocular shape was modified to resemble glasses of elegant beauty that held dignity and importance. The very simple Galilean design, which uses two simple lenses to produce an upright low magnification image, is still being used today in the most common "opera glasses".

A new and somewhat innovative type of binoculars was introduced in 1854 by the famous Italian optician Ignazio Porro. These popular Italian binoculars were named the Porro Prism binoculars. The Porro Prism binoculars were improved in the 1980s by manufacturers, such as the famous German optical company (now called Carl Zeiss AG) who specialized in quality of design in all of their products. The Porro Prism binoculars were designed to be wider and performed better than the Galilean binoculars, which as a result, made the once popular Galilean binoculars less sought after. However, starting with the late 1880s, the Roof Prism binoculars became the age's popular accessory item. The well-known French manufacturer Achille Victor Emile Daubresse is the creative designer of these binoculars and his incorporation of roof prisms may have appeared as early as the 1870s. Most roof prism binoculars use either the Abbe-Koenig prism design (named after Ernst Karl Abbe and Albert Koenig and patented by Carl Zeiss in 1905) or the Schmidt-Pechan prism design (invented in 1899) to erect the image and fold the optical path. They have objective lenses that are appropriately aligned with the eyepieces.

Roof Prism designs create a binocular that is narrower and a bit more compact than Porro prisms. There is also quite a difference in image brightness. Porro prism binoculars will inherently produce brighter images compared to Roof Prism binoculars of the same magnification, objective size, and optical quality. This is because the Roof Prism design employs silvered surfaces that reduce light transmission by approximate range from 12 to 15%. Roof Prisms designs also require tighter tolerances for proper alignment of their optical elements (collimation). This obviously adds to their overall cost since the design requires them to use fixed elements that need to be set at a high degree of collimation at the factory. Binoculars with a Porro prism design occasionally need their prism sets to be re-aligned to bring them into collimation. The Roof Prism fixed alignment design means the binoculars will normally not need to be recollimated.

There have been many other types of binoculars that have been created since the seventeenth century. If our old friend Galileo was still around, he would be amazed to see how far the evolution and technology of his original Galilean telescope has gone. Nevertheless, none of them have influenced or had such a profound impact on the market as did Lippershey's or Galileo's first optical glass creations.

Binocular Prisms

Today's new binocular profiles, styles, and configuration usually come in two general shapes, and these shapes or profiles are a function of the type of prism system that is used in the binocular's optical system. Porro prisms used in binoculars have that familiar "dog leg" shape with the eyepiece offset to the one side with respect to the objective lens. Because of their familiar wide bulky shape, they provide the observer with a solid binocular package for hand-held visual observing, especially if you enjoy doing some binocular astronomy and tracking down those wonderful celestial objects with your trusty pair of Porro prism binoculars. The other type we often see in binoculars is the roof prism binoculars with their familiar straight narrow shape with the eyepiece situated directly behind the objective lens. One of the advantages in using a pair of roof prism binoculars is the focusing wheel is usually placed directly under the observer's hands and fingers, which ultimately increases the binocular's focusing speed and overall user's comfort. A desirable attribute indeed (Fig. 3.1).

The primary function of prism in a binocular system is to provide the observer with a proper oriented image. Prisms are expensive, but they are needed to produce a correct image for the observer. Without prisms, everything would be upside-down with reversed images. Physically speaking, Porro prisms are in fact, wider than they

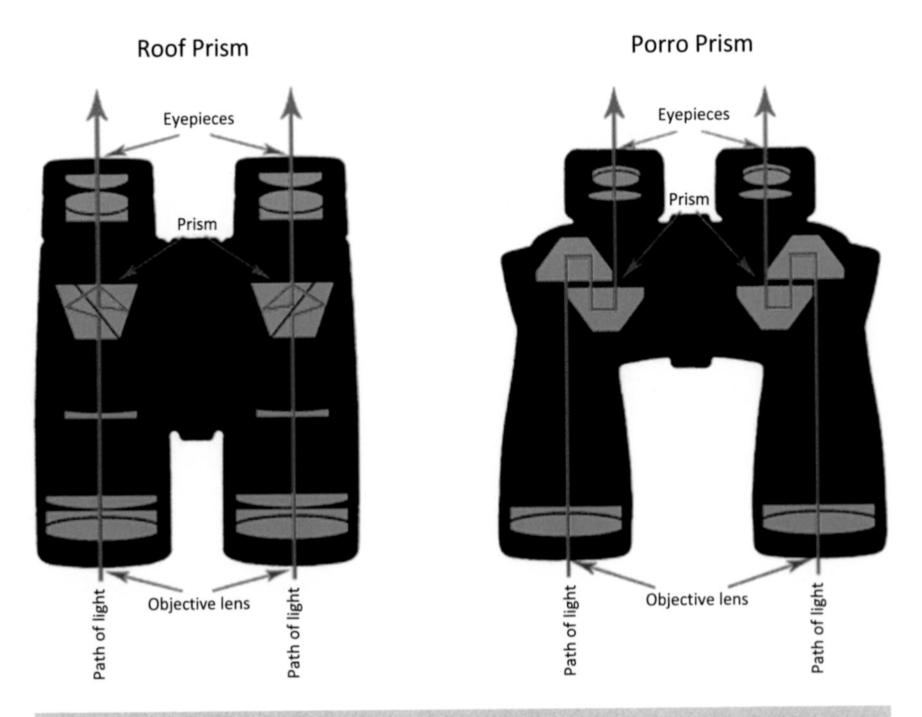

Fig. 3.1 In a typical binocular optical system, two types of prisms are commonly found (roof prisms and Porro prisms), joined together as in a set, which erect the image and shorten the overall length of the binoculars (Image credit: the Author)

are in terms of their length. Their physical profile shows them to have a shape for example, that is similar to a square-S design. If the prisms are made out of a quality optical glass and are properly aligned, then there is as a result, very little light loss or degrading of the overall image quality. An important disadvantage is using prisms is that they are rather large and somewhat bulky that require binocular housings that are somewhat relatively large for mounting purposes.

Binoculars with roof prisms are a relatively newer binocular design. Roof prism binoculars have a somewhat smaller house-shaped profile and are just a bit more compact compared to a pair of Porro prism binoculars. This allows them to be fitted into smaller and more compact binocular housings or spotting spots. However, standard roof prisms have a few inherent design elements that make them a little less desirable when it comes to choosing binoculars compared to a pair of Porro prism binoculars, especially when it concerns light loss. Generally speaking, a Roof prism produces images that are not as bright as a Porro prism because they have a % light loss at their mirror surfaces. Another reason why a roof prism is not as light efficient as a Porro prism is because the precise alignment and adjustment of a Roof prism is often more critical. Even a very small alignment error, less than the width of a human hair (0.002´´) will often degrade the image in a Roof prism just enough so as to make them visually uncomfortable for the observer to view with. As a result, it makes it more of a challenge to skillfully mount and secure them properly in a binocular housing. Another important reason is their images are split and then are rejoined again slightly out of phase when the light cone passes through a Roof prism. This results in an image that lacks less in visual resolution compared to a Porro prism of equal quality. Basically, the standard binocular Porro prism produces an image that is slightly sharper and brighter than a roof prism of the same standard optical quality. But thanks to today's coating technology, some Roof prisms can enjoy the benefit of enhanced coatings on their surfaces, which can reduce light loss to almost indiscernible levels. And even better yet, some of the newer roof prisms have special coatings that have a tendency to considerably reduce and even eliminate the phase problem. As a result of phase correction (or more commonly called PC), Roof prisms can produce images that are every bit the equal to some of the best Porro prisms used in the higher-quality binoculars. One of the draw-backs on this is that PC Roof prisms from some of the well-known higher quality binocular manufacturers such as Bausch and Lomb, Zeiss, Swarovski, and Leica are often very expensive and typically exceed a price tag of more than $2000. However, their instruments are considered some of the finest made and represent the highest standard in quality and excellence, especially when it comes to commercial binoculars.

When it comes to wanting the best in resolution and sharp images in a pair of binoculars, keep in mind that Roof prisms without phase correction and special coatings can only do so much when it comes to producing high quality images. If you want the best in resolution and sharpness, you'll have to pay the price for a pair of phase-corrected Roof prism binoculars or choose a good pair of binoculars with Porro prisms instead. If you compare Roof prism and Porro prism binoculars that are on sale at the same price, it shouldn't be too difficult to decide which one you want to spend your money on. But if you have a tight budget for your next pair of binoculars and want to spend less, then Porro prisms are the way to go. When you are out shopping around for a quality pair of binoculars for sale at a reasonable price, keep in mind that inexpensive Porro prism binoculars manufactured today are generally speaking, every bit the equal optically compared to the expensive phase-corrected Roof prisms binoculars. They should make your new binocular investment a much more satisfying experience when it comes to visual observing. Plus as an added benefit, Porro prism binoculars because of their slightly larger objective lens separation produce a more noticeable stereo effect than a pair of roof prism binoculars. And before you make a decision to buy a pair of binoculars, it's always good to know the type and kind of prism glass that may be inside of those coveted binoculars when you are out there shopping around

Adapted from "Binocular Prisms: Roof or Porro?" (http://www.shawcreekbirdsupply.com)

What Kind of Glass Are Binocular Prisms Made Of?

BaK-4 binocular prisms are also not-so-commonly referred to as Baritleichkron and are typically made out of Barium Crown glass. BaK-4 prisms are generally considered to be optically superior to BK-7 prisms as they come with a higher refractive index (1.569) and that have an inherently lower critical angle compared to other types of commonly-used prism glass. The result is that that less light is "lost" along the prism's periphery through what is commonly called "non-total internal reflection." Also, when light is lost through non-total internal reflection, it causes blue-gray segments or "cutoffs" in the exit pupil. It's interesting to note that many Chinese optical devices have been found labeled as having BAK4 prism glass, which in reality is not the same as the Schott BaK-4 prism glass. The Chinese BAK4, contrary to popular belief, is in fact, not Barium Crown glass, but instead, is made out of a cheaper phosphate crown glass that has a medium range refractive index of approximately 1.5525 and a dispersion factor of –0.0452 (Schott: –0.0523). It also has a higher permissible bubble count rate than Schott BaK-4 glass. Remember, bubbles are notorious for scattering light.

BK-7 prisms are usually found in the cheaper commercial binoculars. Even though they are said to be good enough for use as binocular prisms, they are for example, still considered to below the level of optical quality that are typically found in Bak-4 prisms. Prism optical glass quality will often vary widely between the various commercial binocular models. This is one of the many reasons why there is such a wide range in price that you will find in the stores that sell binoculars. The less expensive commercial binoculars usually come with the cheaper BK-7 prisms. When shopping around for a new pair of binoculars, if you look down the front objective lens of the binoculars, you can see the difference between BK-7 and BaK-4 (much better prism quality). If the binoculars are made with BK-7 prisms, if you look closely, you can actually see the BK-7 prisms with their squared-off sides. BaK-4 prisms present a more typical round somewhat circular shape, that ultimately results in more light that is transmitted with an image quality that is much sharper to the eye.

SK15 prisms are very high-quality prisms made from SK15 glass that minimize undesirable internal reflections and thus provide a very crystal clear high contrast image.

A Schott 2007 refractive index comparison of the different prism glasses that are described above:

	Refractive index	Abbe number
BK7	1.51680	64.17
BaK-4	1.56883	55.98
SK-15	1.62296	58.02

As the index table of refraction above illustrates, BaK-4 glass has a higher refractive index than BK7 as well as a steeper light cone, which in turn enables the manufacturer to make a shorter, and more smaller compact binocular design than those that use the BK7 prism glass.

Note:
BK7 however has a higher Abbe number and therefore has a lower light dispersion.

SK15 glass is a good compromise between the two with a higher refractive index than BK7 and BaK-4, permitting a more compact binocular, but still has a lower light dispersion than the BaK-4 prism glass.

Amici prisms are in the group commonly called "roof prisms" or otherwise known as "right-angle roof prism" designs that are often used in today's modern optics. An Amici roof prism was invented in 1860 by Italian astronomer Giovanni Amici and is similar to the Schmidt prism design. They revert and invert the image as well as bend the line of sight through a 90° angle. They bend a beam of light by 90° while at the same time invert the image. Roof prisms are excellent when used as prism diagonals in binocular and telescope optical systems, because they will erect an inverted image. They are also ideal for use in spotting scopes, and any optical instruments where it is desirable to take an inverted image from an objective lens, turn it right side up, and bend it through a 90° angle, to maintain the correct visual orientation. The various binocular body shapes that are typically offered commercially come in various somewhat different shapes that are often recognized and typically labeled by common three-letter codes: ZCF, BCF, BWCF, DCF, MCF, or UCF. Each of these three-letter codes typically represent an individual and even separate variation in the primary Roof prism or Porro prism designs.

Chromatic aberration which is also commonly called color fringing, achromatism, or chromatic distortion is a type of distortion or aberration in which an optical lens fails to focus all spectral colors where they converge. Chromatic aberration in binoculars and telescopes produces overall soft images, and color fringing adds edges with high contrast, similar to an edge between black and white. Described in another way, chromatic aberration is caused by the inability of a lens to focus different wave-lengths of light onto the exact same focal plane and/or by the lens magnifying various lights wavelengths differently. These particular types of chromatic aberration are most commonly referred to as "longitudinal chromatic aberration" and "lateral chromatic aberration" and can often occur at one and the same time or more commonly referred to as "concurrently". The amount of chromatic aberration in an optical system really depends on the inherent dispersion qualities of the glass. If you look at an image through a lens with chromatic aberration, color fringing may occur. To correct this, some higher-quality commercial binoculars use the more expensive, extra-low dispersion (ED) glass.

The Importance of the Exit Pupil in Binoculars

Exit pupils are basically a virtual aperture or opening in a binocular's optical system. In a typical pair of binoculars, the exit pupils can be seen in the center of the binocular eyepieces as two uniformly bright circles or discs of light while holding the binoculars at a short distance or at a normal arm's length and pointing the binocular objectives toward a reasonably bright area. Another easy way to see the exit pupil is to first totally pull back or fully retract the rubber eye cups and then focus the binoculars onto an area that's relatively bright. Then hold a white place card or piece of folded piece of paper up to the eyepiece, allowing the binocular eyepieces to project a small disk or relatively small circle of light onto the place card or white paper. Then move and position the place card closer to or further away from the front of the binocular eyepiece until you see a very small (the smallest) possible circle or disc of light. The overall diameter of the bright circle of light displaced by the center of the eyepieces is the actual diameter of the exit pupil. Note that the exit pupil should always appear to be round and circular and be of uniform brightness. If shadows are noticeably visible or present, this could be a possible indication of a binocular's overall quality level, which is considered poor or even worst. The size of exit pupil and its importance in viewing with a pair of binoculars cannot be understated. Note that the exit pupil is an important item, because the larger the exit pupil diameter, the greater amount of light that will be transmitted to each of your eyes. It is therefore an important aspect when comparing the theoretical brightness of two optical instruments and that's something important to consider when choosing your next pair of binoculars, especially for use in poor light conditions such as in the early dawn hours just before sunrise or dusk (sunset) and/or for astronomical observation. Another important thing to remember is that the larger the exit pupil diameter is, the brighter the field of view will be. This is an important consideration when using binoculars in dark and low light conditions and for astronomical viewing. The human pupil typically opens to about an average 2 mm or slightly larger during the daylight hours and about an average 7 mm in somewhat darken conditions or the night time. If you use a pair of binoculars with a 2-mm exit pupil in daylight, you will not be able to readily perceive the darker images. Also note there also will not be much noticeable overall variation or change in brightness whether you use binoculars with a 7-mm or 2-mm exit pupil.

Exit Pupil Math for Binoculars

$$Exit\ pupil = \text{aperture / mag.} = \text{exit pupil dia. (mm)}$$

8×50 *binoculars* — The equation is : $50 \div 8 = 6.25 >$ The diameter of the exit pupil is 6.25 mm

10×50 *binoculars* — The equation is : $50 \div 10 = 5.0 >$ The diameter of the exit pupilis 5.0 mm.

7×56 *binoculars* — The equation is : $56 \div 7 = 8.0 >$ The diameter of the exit pupil is 8.0 mm

The magnification and the diameter of the objective lens determine the size of the exit pupil. The diameter of the exit pupil determines how much light is transmitted to your eye. The actual diameter of the exit pupil is easily computed. Divide the diameter of the front objective lens (in millimeters) by the magnification of the binocular. For instance, take a pair of standard size 7×50 binoculars. Divide 50 (diameter of the objective) by 7 (the magnification) and you get approximately 7 mm (7.143 mm), which is the diameter of the exit pupil for 7×50 binoculars. When it comes to using a pair of smaller compact binoculars, the exit pupil of a pair of 8×21 binoculars is only 2.63 mm (21 divided by 8=2.63 mm). A lot less light reaches your eye from compact binoculars than it does from standard size binoculars. Light is what you will be sacrificing if you want to get a pair of compact size binoculars. An important question often arises: Does the size of the exit pupil matter? Generally speaking, as long as there is enough ambient light so that the pupils of your eyes are smaller than the exit pupils of your binoculars, but when the ambient light becomes dimmer, and the pupils of your eyes start to accommodate for the change of light by enlarging, the exit pupils of your binoculars may become the limiting factor. Referring to the 8×21 compacts in the example above, when it gets dim enough for the pupils of your eyes to exceed 2.63 mm in diameter, the binoculars are restricting the light available to your eyes. Ideally, human eyes in healthy condition can achieve about a 7-mm pupil opening. So a 2.63-mm exit pupil from your compact binoculars can be quite limiting in dim light. You can probably see more without your binoculars. But the 7×50 binoculars in the first example above have 7-mm exit pupils, as large as young, fully dark-adapted human eyes, so they never limit what you can see, even at night.

As one ages, the human eye loses its ability to adapt to dimmer light, as a result, for a person who is middle aged, their maximum pupil size is typically around 5 mm. As we get older, the exit pupil size we normally need becomes increasing smaller with age. The relative brightness or more commonly called the relative brightness index (RBI) in a pair of binoculars attempts to measure image brightness. It is calculated by squaring the exit pupil. For example, 7×32 binoculars have a 4.57-mm exit pupil (32 divided by 7=4.57). Their relative brightness index (RBI) is 22.88 (4.57 times 4.57=20.88). A relative brightness index (RBI) of 25 or greater is considered good for use in dim light. Since we already have discussed (above) how to calculate the actual exit pupil size, and what it means, the relative brightness index (RBI) becomes largely redundant.

The twilight factor is a somewhat subjective measurement that purportedly reveals how much detail one observer can supposedly see in twilight conditions. It tends to favor magnification somewhat, which is good for binocular sales. For instance, Celestron computes the twilight factor of 7×50 mm binoculars as 18.7 and the twilight factor of 10×50 mm binoculars as 22.4, even though the former has a 7.1-mm exit pupil and the latter only a 5.0-mm exit pupil. The increased magnification supposedly makes up for the decrease in brightness in "twilight conditions" (when the individual's eye is not yet fully dark-adapted). This rather somewhat artificial measurement can be useful for the average hunter and typical birdwatcher, since animals are often spotted just before sunrise and just after sunset

Astronomical binoculars often use higher magnification as well as larger aperture. For example, Celestron 25×100 mm binoculars have a 4-mm exit pupil. Vixen makes binoculars with replaceable eyepieces, meaning the exit pupil can be any thing. A good pair of quality prism binoculars is very handy for locating objects in the night sky. Once an object has been located with binoculars, it is easy to train a telescope on it for a more detailed view. The binocular astronomer needs very high quality, very bright binoculars. For general hand-held use, 7×50 m, 8×56 mm, and 9×63 mm binoculars will work very well. Choose the highest power that you can hold reasonably steady without a shaky image. Giant binoculars are in a class by themselves for binocular astronomy. These require a solid tripod mount, but reveal spectacular views of large objects like open star clusters. The 20×80 size is the most popular, and perhaps represents the best compromise between magnification, brightness, and field of view for general astronomical observing.

The larger the exit pupil, the less critical it becomes to align the binoculars to your eyes. When the light is dim, another effect of the size of the exit pupil becomes more important. Assuming two binoculars are the same power, the one with the bigger aperture will have the larger exit pupil. As the size of the exit pupil increases, the image appears brighter, up to the point at which the exit pupil is the same diameter as the pupil of the eye and the entire area of the pupil is illuminated. After that, further increases produce no real gain in brightness if any at all. As people age, the maximum opening of the pupil diminishes. The pupil of a healthy young person's eye has a diameter of about 2 mm in bright light, 5 mm in dim, and 7 mm in the dark. People in their 30s typically have a maximum pupil diameter of about 6 mm, which shrinks to 4.5–5 mm in their 40s. Before evening, at dusk, people in their twenties would see a brighter image through 7×50 (exit pupil dia. 7 mm) binoculars than through 7×35 (exit pupil dia. 5 mm) binoculars, but persons in their 60s would perceive no difference. Some manufacturers give a "relative brightness index" that is found by squaring the size of the exit pupil in millimeters. The highest number allowed in this index is 49, since human pupils open no more than about 7 mm. Binoculars eye relief is the distance between the eyepiece and the point of focus of the exit pupil (see Fig. 7.6). Currently binoculars that are commercially available have eye reliefs ranging from 1 to 23 mm. The former (1 mm) is overly small. Eyelashes coated with cosmetics, such as eyeliner, may dirty and contaminate an eyepiece lens. Binoculars with large eye reliefs around 15 mm are needed by

Fig. 3.2 Mr. Antony Kay and the 180 mm IJN binoculars, which have two-power eyepieces mounted on turrets that can be rotated to present the eyepieces for use. This binocular is 1.04 m (41 in.) long and weighs a hefty 64 kg (141 Ib). Once serviced, this giant binocular was filled with nitrogen. (Image credit: Antony Kay—OptRep—www.opticalrepairs.com)

Fig. 3.3 Here is an example of a Carl Zeiss 3×20 mm Teleplast (Circa: 1907–1911) It has Sprenger prisms and triple element field lens. (Image credit: Antony Kay—OptRep—www.opticalrepairs.com)

Fig. 3.4 A French inter-war binocular of unusual porro prism construction mounted on a flat plane. It has no IPD adjustment. (Image credit: Antony Kay—OptRep—www.opticalrepairs.com)

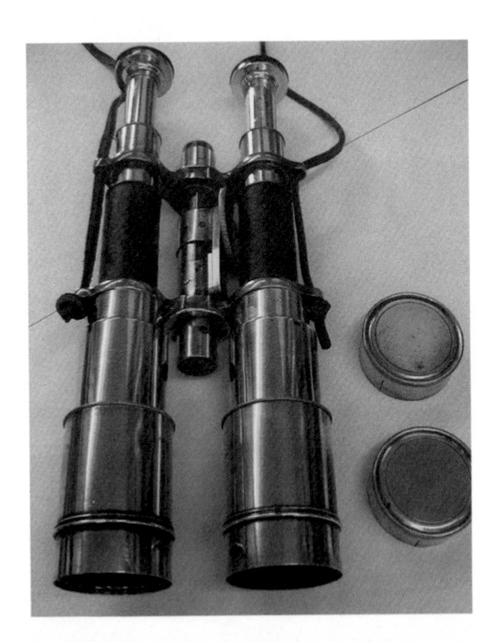

Fig. 3.5 A rare binocular telescope after restoration in the OptRep workshop. It was made by G&S Mertz (Munchen) circa 1850 and its specification is 10× (approx.) with 35 mm air spaced objective lens, a rack, pinion eyepieces and telescopic eyepiece tubes. Fully extended, 11 in. in length. (Image credit: Antony Kay—OptRep—www.opticalrepairs.com)

persons who wear eyeglasses while viewing. Even though the focusing adjustments of most binoculars can accommodate near and farsightedness, they cannot compensate for astigmatism. Some people prefer to wear their glasses while using binoculars, simply because they don't like taking them on and off while viewing (Figs. 3.2, 3.3, 3.4, 3.5, 3.6, 3.7, 3.8, 3.9, and 3.10).

Fig. 3.6 A Carl Zeiss 10 × 50 mm (*left*) "Dekar" (circa 1950) and a Hensoldt (*right*) 10 × 50 mm "Dialyt" of post war vintage. Both are collectables (Image credit: Antony Kay—OptRep—www.opticalrepairs.com)

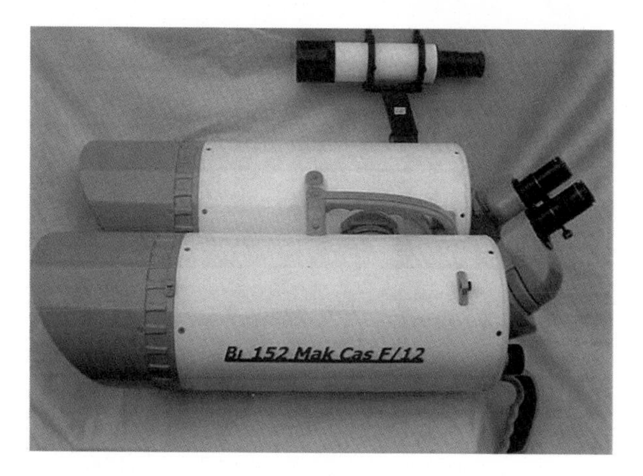

Fig. 3.7 These are Russian "Intes Bi 152mm Maksutov-Cassegrain f/12" binoculars with 1-1/4 in. astronomical eyepieces and an adjustable IPD. (Image credit: Antony Kay—OptRep—www.opticalrepairs.com)

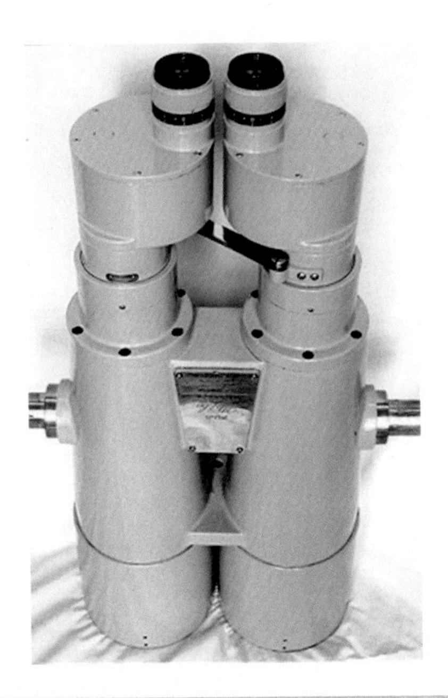

Fig. 3.8 A modern Fuji 40×150 mm "Meibo" binocular (Image credit: Antony Kay—OptRep—www.opticalrepairs.com)

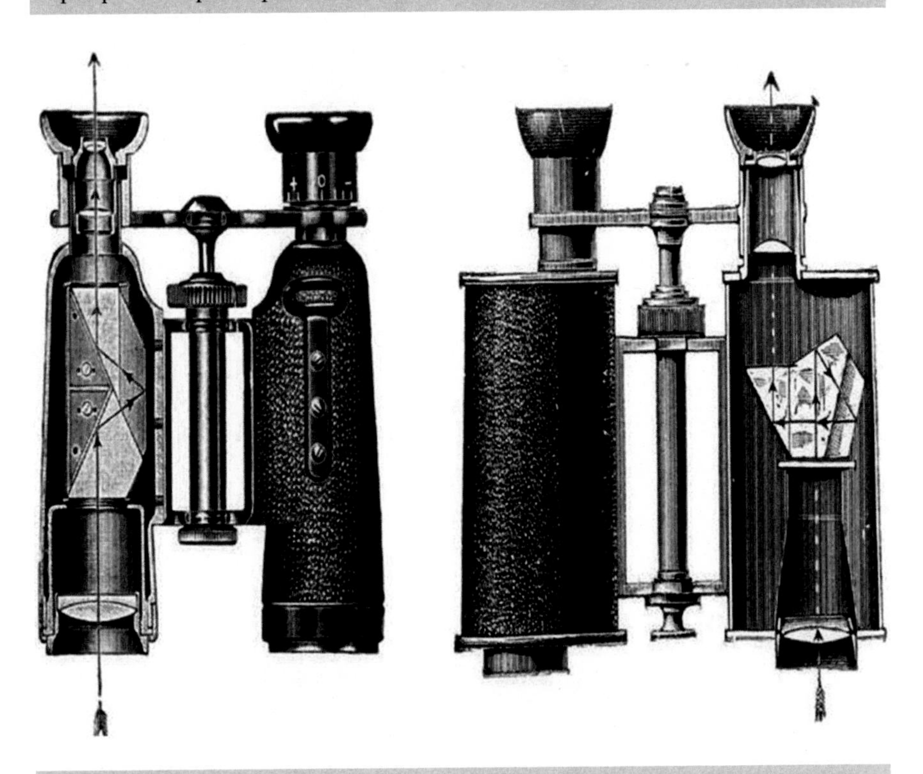

Fig. 3.9 Two pairs of the late-eighteenth century double prism binoculars designed by Montz Hensoldt (Image credit: Wikimedia Commons)

Fig. 3.10 (Cartoon credit: Jack Kramer)

The evolution in binocular design has come a long way since the early part of the nineteenth century. However, the use of double-prism binoculars to invert and revert the images in a pair of binoculars of that era was not totally indifferent compared to today's common binocular prism designs. It's interesting to see what a pair of the late-eighteenth-century double-prism binoculars that were first designed by Montz Hensoldt in 1984–1897 looked like as they were the common binocular design that was introduced during that era (Fig. 3.9).

Author's note: Having seen firsthand the excellent optical quality images from the Intes Bi 152mm f/12 Maksutov-Cassegrain, one wonders why this rare dual Maksutov-Cassegrain binoscope setup wasn't produced commercially. That's how good this Russian Maksutov-Cassegrain really is. I can only envy the owner of the one pictured in Fig. 3.7.

Binocular Mirror Mount

By Carl Vehmeyer

When I bought my 28 × 110 binocular in 2008, it was obvious that this instrument needed a big stable mount. The binocular weighs approximately 7 kg. Although I initially wanted to build a mirror mount for this binocular, I wasn't able to find a good enough mirror. In the end I build a parallelogram which worked fine but restricted observing to a maximum altitude of 60°. Also I found using a mount like this a bit uncomfortable, so a mirror mount remained on my wish list. When a friend started his own mirror-making company (Huygens Optics) last year, I asked him to make me a couple 160 × 120 mm flat mirrors. I had a lot of discussions with

Fig. 3.11 Carl Vehmeyer made this splendid example of a binocular mirror mount. Notice the "trap door" for his binocular access to the sky (Image credit: Carl Vehmeyer)

friends about the required quality of the mirrors. In the end, it seemed that a PV of ½ lambda is more than sufficient for a 28 × 110 binocular (Fig. 3.5).

The mirrors were installed at a 40° angle. This way one can easily see the zenith and can observe as low as 5°. Lower didn't seem useful and by setting the angle at 40°, one could get away with smaller mirrors. The mirrors are glued to four-point mirror cells. Small ball magnets are used to attach the mirrors to the cells; this way the mirrors can be easily removed for transport. Both mirrors are tilted on different axes to collimate the binocular. The mirror box was constructed with 9-mm plywood. A 4-mm thick aluminum plate was used to reinforce the box where the binocular is attached to it. The plywood used for the mount was 12-mm thick. Since I didn't see the need for high-quality wood, the mount needed a fair bit of paint (Fig. 3.11).

I initially planned on using the mirror mount on a table, but a tripod seemed more practical when observing at remote locations. Since the surveyor tripod that I was using was a little high for use with the mirror mount I shorted it. I recently updated the mirror mount with a lazy susan bearing, so I could dispense with the tripod when traveling. Both the tripod en tablet top configuration are used and are equally effective. For aiming the mount I use a quick finder or laser pointer. The first views with the new mount were fantastic, its like rediscovering the night sky. Merging the images is a matter of seconds and observing with binoculars has never been this comfortable. The only downside to the mirror mount is that some objects look a different because of the flipped image. This is especially apparent when viewing the North America nebula. Personally I find that an advantage to this is that I seem to see new details in objects that I have seen many times before (Figs. 3.12 and 3.13).

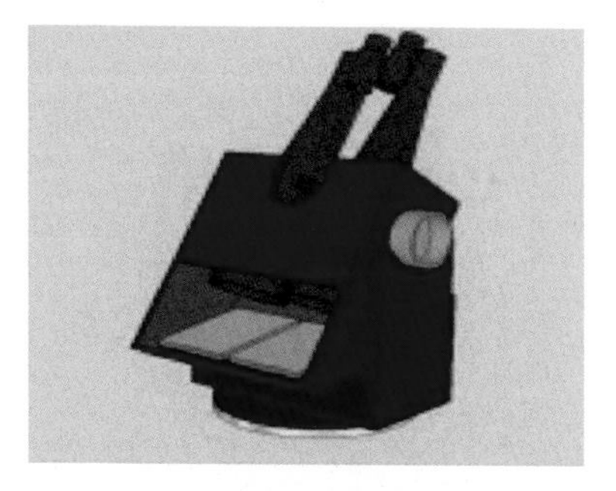

Fig. 3.12 Here is Carl's drawing of his binocular mirror mount showing the dual mirror system (Image credit: Carl Vehmeyer)

Fig. 3.13 (Cartoon credit: Jack Kramer)

Carl Vehmeyer
The Netherlands

Adapted from "Binocular Mirror Mount" with permission from Carl Vehmeyer (http://vehmeyer.net)

Downward Looking Binocular Mount

Why Build One?

By Rob Nabholz

I have found binocular astronomy to be a lot of fun. It is a great test of observing skills and can be a way to do some observing at times or places where a more involved setup of equipment is not possible (Figs. 3.6 and 3.7).

Fig. 3.14 A very nice, stable binocular mirror mount with a simple design (Image credit: Rob Nobholtz)

In observing with binoculars, I have found the experience much more rewarding if I can mount the binoculars. There are various methods of doing that, and I have played around with many of them. I came across the idea of a downward-looking mount using mirrors and liked the idea of comfort that looking down versus up would afford (Fig. 3.14).

I found a source for inexpensive first surface mirrors on the net and decided to build a mount.

There seem to be a couple of popular approaches to this type of mount. One uses a large mirror on a tilting platform. The advantage of that design is that the binocular eyepiece remains in a static position. The disadvantage is that it requires a rather large mirror to make it work, and "large" in this case also means expensive.

The second approach involves tilting the mount and binoculars together, allowing the use of the smaller, less expensive mirror. That is the path I took (Fig. 3.15).

Construction

Main beam at a 45° angle and is also three pieces, but the center piece, extends up through the main beam and is secured by glue and the construction is quite simple. Most of the unit is constructed from ½″ maple. The main beam is made from multiple pieces of maple glued and screwed together (mostly all of the screws are

Fig. 3.15 This simple wooden binocular mirror mount setup looks good from all angles (Image credit: Rob Nabholz)

from the backside and are not visible in these photos). The downstalk joins the screws. The binocular mounting stalk is a single piece glued and secured between the outer pieces.

The mirror rests on a platform of plywood and is secured with multiple dabs of silicone adhesive. The platform is attached to the beam at a 45° angle by a short piece of maple that, like the bino mount stalk, is sandwiched between the outer; pieces. It is secured in place by a ¼″ removable bolt that allows removal of the platform, making it easier to store and pack.

The entire unit attaches to the tripod using ¼″ 20 threaded insert in the bottom of the down stalk. It attaches to a pan and tilt head providing the movement necessary to scan the skies.

Using the Mount: First Light

Out under the stars for First Light, things did not go as smoothly as I had planned. My favorite astro binoculars are Fujinon 16×70s, so I mounted them up and anxiously moved into the eyepieces. My first view was a jumble of fuzzy elongated stars. Quick check of the focus maybe a little better, but not even close to the tight pinpoint stars I was used to seeing with these binoculars. After some experimentation using different methods, including checking the binocular's collimation, I began to suspect that the mirror might not be optically "up to snuff." To check that theory, I took off the 16× binoculars and mounted my Fuji 7×50s, and sure enough, the view was perfect. The mirror simply was not smooth enough to support the higher magnification of the 16×70s.

Since then I have mounted the 10× binoculars and enjoyed a very nice view, but I believe that magnification is about the limit for this mirror. A few months later, I noticed that the supplier (who was responsive and accommodating) has added some new products to his offering, mirrors rated to support higher magnifications (at a higher coast, as you would certainly reasonably expect).

Other Considerations

Using this mount, I have found two points regarding tripod features that I consider essential for maximum enjoyment. First is a nice damped tripod head. A damped head makes scanning with the mount a real pleasure. The second essential feature is a tripod with an adjustable center post. Tilting the mount changes the eyepiece height. Having it mounted on an adjustable center post makes adjusting to the new eyepiece much easier.

One other point: using this kind of mount where I live means having to deal with dew. As you might expect, the mirror can accumulate dew quickly on some of our Iowa summer nights. Be aware of this and cover the mirror when not in use, or the other favorite dew fighting tactics like blowers or maybe a heating system.

Aiming the mount is a bit of a trick and takes some getting used to. I have recently been experimenting with a green laser pointer as an aiming device. Mounted parallel to the binocular's optical axis, when adjusted, the laser hits the mirror and points out the current center of the view, much easier than trying to "eyeball" it.

Conclusions

I do enjoy using this mount. The downward look is quite a comfortable viewing. The mirror sometimes makes high magnifications impossible, so this is wide field, casual scanning—more of a "type b" behavior, just what the doctor ordered some nights when you feel like getting lost for a while (Figs. 3.8 and 3.16).

Rob Nobholz

Adapted from "Downward Looking Binocular Mount" with permission from Rob Nabholz (http://www.homebuiltastronomy.com)

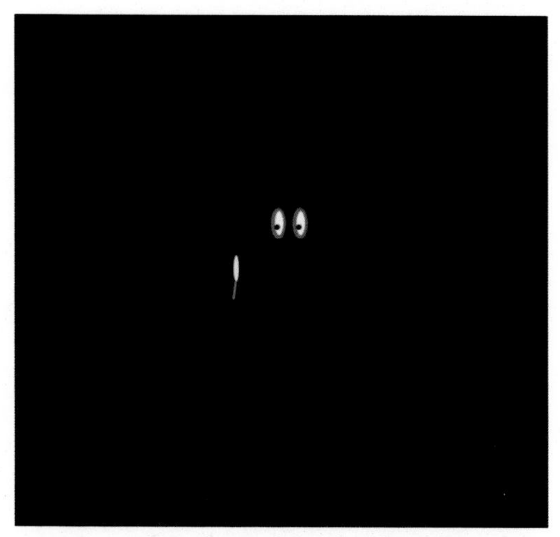

"The Universe Before the Big Bang"

Fig. 3.16 (Cartoon credit: the Author)

Making of *Voyager*

By Wayne Schmidt

The design and construction of Wayne Schmidt's motorized binocular chair for a pair of 8-in. diameter aperture binoculars.

Telescopes are great…as long as you don't have to use them. Flat, one-dimensional views, straining to reach the eyepiece and fighting the cold all conspire to challenge even the most stalwart astronomy enthusiast. My solution to these problems was to mount a pair of large binoculars on a motorized chair complete with electrical heaters. I named it Voyager (Fig. 3.17).

I used large binoculars because they reduce eyestrain, increase light sensitivity by 40%, color sensitivity 25%, and most importantly provide beautiful three-dimensional appearing images.

The following photo image of Wayne Schmidt's remarkable motorized binocular chair was taken at the 1995 Riverside Telescope Makers Conference (RTMC) at Camp Oakes. (Author's note: It should be mentioned that Wayne won a merit award at the 1991 RTMC for "Interesting Concepts In Motorized Observer's Chair").

I used f/8.0 mirrors because they provided optimum compromise between light capturing capability, ease of fabrication, and image quality. By using a longer focal length than the usual 4.0–6.0 seen on most reflective binoculars, I was able to use smaller secondary mirrors and thinner spider vanes to reduce diffraction effects. Here's what the optical path looks like: can look down into the observing area of the chair (Fig. 3.25).

Fig. 3.17 Wayne Schmidt standing next to his 8-in. f/8 binocular telescope (image credit: Wayne Schmidt)

Unfortunately, one of the negatives of this design is that the image rotates as the tubes rotate. This means that the simple method of tube rotation for adjusting the interocular distance can't be used. Instead, right-hand tube is mounted on a cradle equipped with sideways push-pull screw pairs front and back to adjust eyepiece separation. For merging the images in both scopes, the left-hand tube was mounted on a cradle whose rear end could be moved left and right and the right-hand tube with a cradle that moves its end up and down by turning adjustment knobs located in the front of the chair. This way the observer can set the alignment while using the binoculars in the actual viewing position. This is important because unavoidable flexure resulting from tilting the 350-lb chair with a 170-lb rider can throw the alignment off. The ability of adjusting on the fly makes tuning up the alignment easy at any time. The pivot points for the tubes are set at the eyepiece positions so alignment doesn't affect the interocular distance. By removing the opaque overhead light cover, you can look down into the observing area of the chair (Fig. 3.18).

The eyepieces in the top set are 50 mm Plossls that provide 32.5 power with a 1.4° field of view. They are mounted on sliding carriages that allow the focus to be quickly adjusted for higher powers using a pair of 2.0× Barlow lenses stretched to 2.8× to provide 90×. The eyepieces themselves lock into focusers made from flush valves. The smaller eyepieces are to a pair of very fine Adlerblick 7 × 50 binoculars. Their large field of view and erect image make them ideal finderscopes. Below that is a circular metal disk, which uses magnets to hold star charts in any desired

Fig. 3.18 Wayne Schmidt standing next to his motorized 8-in. f/8 binocular telescope at the 1995 RTMC (Image credit: Robert Stephens—RTMC)

position. In the middle of the star chart is a finder gauge consisting of two wire rings. The outer one corresponds to the field of view of the Adlerblicks, while the smaller inner circle matches the field of view of the 8-in. f8.0 binoculars. The wire hanging down on the left is a low power resistive heater for both pairs of eyepieces to prevent dew formation. (Speaking of heaters, the chair has an electrically heated vest and boots to make sure the rider stays comfortable on even the coldest nights.) (Figs. 3.19 and 3.20)

To move the binoculars forward and backward or up and down to use different eyepieces or reach the finder binoculars, the frame that holds the binoculars is equipped with two motors. One, mounted on the back of the rider's chair, moves the entire binocular assembly up and down using an electric motor driving a threaded rod through a captured nut. The 100+ lb weight of this assembly tended to grind down the nut and rod so it's important to keep both of them heavily greased and protected from dust. For moving the binoculars back and forth, they were mounted on a pair of linear bearings, and a second motor/captured nut drive pushed or pulled them as needed (Fig. 3.21).

Fig. 3.19 Wayne sitting at the controls of 8-in. f/8 motorized binocular telescope at the 1995 RTMC (Image credit: Robert Stephens—RTMC)

Fig. 3.20 Photo showing the dual eyepiece system of Wayne's 8-in. f/8 binocular telescope (Image credit: Wayne Schmidt)

Fig. 3.21 Photo showing the motor drive unit that powers Wayne's 8-in. f/8 binocular telescope (Image credit: Wayne Schmidt)

The motor drive that moves the binocular carriage forward and backward is a cheap rechargeable drill with its handle cutoff. Three similar motors drive the other chair motions.

At this point, it's necessary to make a comment about safety. When the chair is tilted back so that the observer is looking straight up, if the binoculars broke loose, all 100+ lb of them would come crushing down into the observer's eyes. For this reason, it's necessary to have multiple safety measures in place to insure that if this ever happens, fail-safe mechanical locks prevent the observer from being injured. The same thing applies to all aspects of such chairs. When being ridden, the observer is essentially imprisoned, and every aspect of safety must be carefully considered and addressed.

The chair's frame was made of 3/4-in. plywood strengthened where needed with 2×4-in. lumber. Straight-grained, kiln-dried wood was used through to avoid warping. The altitude bearings are 2-in. diameter metal casters. I found they produced less drag and ran smoother than traditional Teflon bearing. The altitude drive consists of a rubber roller pressing against a sector arc covered with 30-grit sand paper. The rubber wheel is driven by an electric motor. (Note that because this was a friction drive, it is essential to keep the center of weight the same or the drive could slip. I found this to be the chair's biggest weakness because if a 170-lb rider shifts his position by only 2 in., he's creating 340-in. pound of additional torque, enough to cause the altitude drive to slip. I tried a bent-bolt drive, but it slipped even easier.) For this reason, it's difficult for other people much different in size to myself to use the chair. I have counterbalancing weights that can be used to compensate for this problem, but they are awkward to use.

Seating area of the chair is padded with a 5-in. thick layer of contoured foam to provide uncompromised comfort. The rotating part of the chair rests on a 12-in. diameter Lazy Susan bearing stabilized with four casters mounted near the edge of the circular base. A similar drive system rotates the chair, except the entire disk of the base is used so the chair can turn 360°.

Ordinary toggle switches on the right-hand armrest turn the motors on, off and control the direction of slewing. Additionally, there's a switch that puts a 1-Ω, 20-W resistor into the power circuit to reduce the slew rate from 3° per second to 0.5° per second. Power is provided by a 6-V, 140-Ah deep discharge battery that has enough stored energy to run the chair for five nights, even in sub-zero weather.

I made the mirrors myself. They both pass six-point Millies-Lacroix diffraction limit test and have smooth surfaces and good edges. The final focal lengths were 64 in. and 64.37 in., well within the 1% agreement needed for good binocular image merging. It took an extra 5 h of work to get the second mirror to match the first, but then these were the first mirrors I'd ever made. A more experienced mirror making could certainly do it in less.

I used three 2-square-inch pads of foam mounting tape to secure each mirror to its cell. These mounts have held for years and never shown any sign of straining the mirrors or coming loose. The mirror cells themselves were simply two disks of 3/4-in. plywood with push-pull adjusting screws set at 90°, rather than the usual 60°, for easier collimation.

The secondary mirrors are 1.5 in. across and provide a 100% illumination field 20 min in diameter. Surveying 1866 astronomical objects indicated that this field would be large enough for 100% illumination of 95% of them. The illumination drop-off near the edge is unnoticeable. By using such a tight optical path, I was able to employ a better baffling system that increased the scope's contrast. In the end, using the smaller diagonals actually provided brighter, sharper images.

All three mirrors in each tube were coated with enhanced aluminum. This increased the overall reflection efficiency from 68% of normal coatings to 86%.

The tubes for each telescope were cut from 11-in. diameter concrete forming tubes purchased from Home Depot. Mounted on front of them are 15-in. diameter, 40-in. long baffling tubes used to reduce stray light from reaching the focal plane and reducing contrast (Fig. 3.18).

To decide how to coat the insides of the tubes, I conducted an experiment to see which surface treatment absorbed the most light. Above, from left to right, is flat black paint, black flocking paper, and sawdust sprayed with flat black paint, black cloth, and finally black velvet. The velvet was clearly the best so that's what I used (Fig. 3.22).

The four 0.02-in. thick spider vanes used in most telescopes produce very large diffraction spikes. To reduce this, I used a set of three vanes, which avoids spike reinforcement from in-line vanes, made of 0.008-in. music wire. The combined effect was to reduce the brightness of the diffraction spikes by 65%. While these

Fig. 3.22 Photo showing the different colors/materials that Wayne chose from to reduce reflections inside of the telescope tubes (Image credit: Wayne Schmidt)

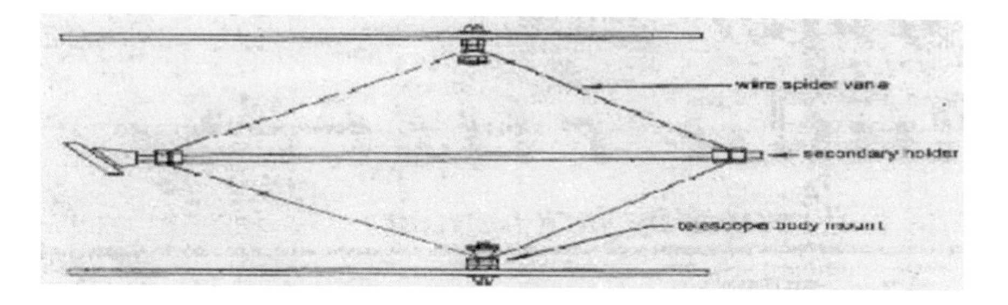

Fig. 3.23 Drawing showing Wayne's secondary holder design for his 8-in. f/8 binocular telescope (Image credit: Wayne Schmidt)

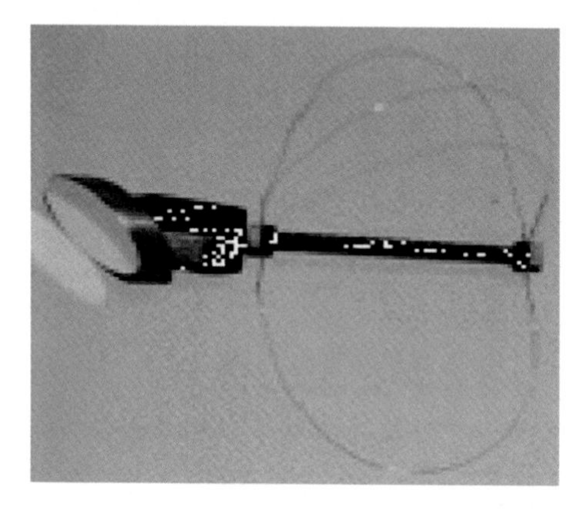

Fig. 3.24 Photo showing what the actual secondary mirror system looks like (Image credit: Wayne Schmidt)

spikes are easily seen in normal scopes when looking at second magnitude stars, *Voyager's* wire spiders don't show them until you're looking at first magnitude stars, and even then they are barely discernible. Here's a cross section of what they look like (Fig. 3.23).

A jig was used to make each wire loop (This is necessary so the nylon nut and nut cap that holds the wire firmly to the side of the telescope tube yet with enough slip so it can be adjusted and a close-up of the end of the central diagonal bar showing how two nuts tighten on the wire's end loops. The secondary position is adjusted closer or further from the primary mirror by sliding the wires through the nylon capture nuts mounted on the telescope's tube. The secondary itself is foam tape mounted on a wooden block and can be rotated by turning it on the threaded rod screwed into it that forms the central spar of the diagonal. Finally, the entire wire assembly can be angled by sliding one wire through its nylon nut further than the other two. Collimation using this system isn't easy, but the reduced diffraction losses are worth it, that they are all the same length (Fig. 3.24).

A word of warning: I found collimating a three-mirror optical path very difficult. There are so many rings within rings that have to all be concentric that I had a hard time keeping them straight and remembering what adjustment affected which ring. Fortunately, once the scopes were collimated, they tended to remain that way even after transporting them hundreds of miles in a car.

Originally the chair broke down into 10 manageable pieces. Assembly on-site took half an hour. However, it was a tight fit to pack it into my station wagon so after a year I rebuilt it onto a small trailer. Everything except the telescope cradle and telescope tubes was permanently fixed to it. This reduced setup time to less than 5 min. During transport I used wood wedges to raise the main chair off the Lazy Susan bearing so heavy bounces wouldn't flatten it. A white vinyl cover sewed to fit the chair protected it from weather. The only drawback to this was that it made the chair look like a construction site port-a-potty (Fig. 3.18).

So, how does it work? Outstanding! Besides the already mentioned advantages of binocular viewing, there is a very real psychological pleasure from physically looking in the same direction the telescopes are looking. One additional benefit I hadn't expected was the fact that when seated in the chair, the observer's head lays on a head rest. This immobilizes his head and therefore his eyes. By eliminating the small eye movements unavoidable when looking through a normal telescope, I found that images looked brighter, sharper, and much more concrete. I estimate that this increased the scope's limiting magnitude by at least half a magnitude (Fig. 3.19).

Large binoculars breathe life into astronomical views that must be seen to be appreciated, and riding a motorized chair in warmth and comfort enormously increases the joy of observing, creating a sense of identity with the objects inhabiting the night sky unattainable any other way. The Orion nebula becomes a huge glowing cloud ripped by valleys of stygian blackness. The Triffid's color can be clearly seen, and the swan looks so solid and three-dimensional it's hard to take your eyes off it. When riding *Voyager*, the universe comes alive, inviting exploration (Figs. 3.25, 3.26, and 3.27).

Wayne Schmidt

Adapted from "The Making of Voyager" with permission from Wayne Schmidt (http://www.waynesthisandthat.com)

A GOTO Binocular Chair

By Norm Butler

Made from an old discarded IKEA semi-reclining chair, the author's homemade GOTO binocular chair has a pair of servomotors located on each axis, which enables the observer, while holding a Celestron hand controller, to select each object in the 40 K object database via NexStar 4SE software. With just a touch of the Celestron hand controller's button, the servomotor-powered binocular chair

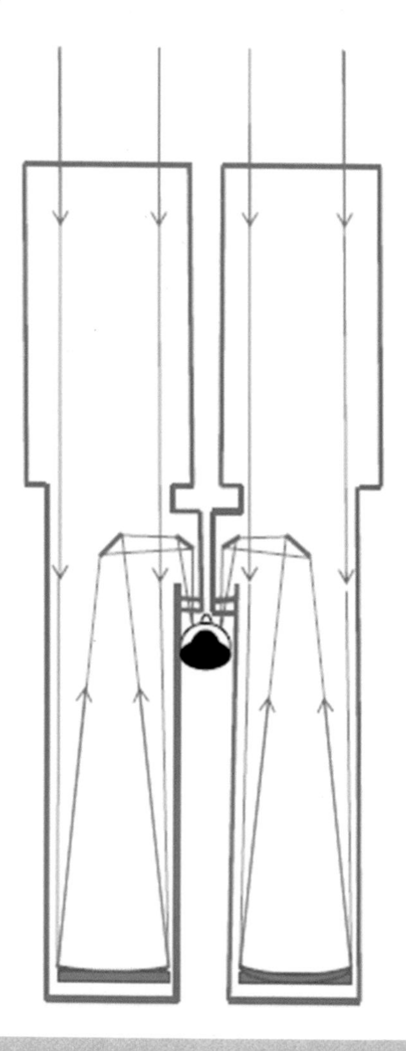

Fig. 3.25 A diagram illustrating the optical path that coming light takes through Wayne's dual 8 in. f/8 motorized binocular telescope (Image credit: the Author)

moves to each selected celestial object that the observer chooses to observe sitting in complete comfort (Fig. 3.28).

The idea of a GOTO binocular chair had never occurred to me, much less the prospect of actually building one, until a gentleman at the RTMC a couple of years ago who was a wheel-chair-bound binocular astronomy enthusiast mentioned to me during a brief conversation that it would be nice to sit in a comfortable computerized GOTO binocular chair and push some buttons on a hand controller to aim the chair, user, and binoculars to desired objects. As challenging as the project would be, it was a fascinating idea of comfortably observing the heavens with no more effort than pushing buttons on a hand controller.

Fig. 3.26 Photo showing how Wayne transports his motorized binocular telescope (Image credit: Wayne Schmidt)

Fig. 3.27 This is what the newer version of Wayne's original motorized binocular telescope looks like today (Image credit: Wayne Schmidt)

Having designed and built some pretty interesting telescopes in the past had pushed my electro-optical, and mechanical skills to the limit, but this one would probably be the most challenging of all of them. Still, once an idea takes hold, there's no going back. While working at Haleakala Observatory on Maui for AVCO Everett Research Laboratory in the early 1980s gave me some great engineering experience, especially with then the state-of-the-art computerized systems that ran

Fig. 3.28 The author's homemade GOTO servomotor-powered binocular chair (Image credit: the Author)

the big telescopes, at the same time I knew that even with today's advanced GOTO systems, it would be difficult to design and build a totally computerized GOTO bino chair. In considering the options, it was decided to not use big DC motors and chain or belt drives typical of motorized observing chairs, preferring instead to use small but powerful DC servomotors (torque monsters) wired in series to drive each axis, compact motors that can still produce enough torque to easily move a 200-lb person in both azimuth and altitude. The real challenge would be in finding the right kind of chair (reclining or otherwise) to do the job and thus eliminating first the typical "pitch-and-tilt" bino chair design with the observer's entire body being tilted forward and backward, often with feet raised overhead and head well below heart when the binoculars are pointed near zenith. This arrangement is familiar on some motorized bino chairs, and given the fact that the chair would be used to observe various assortment of celestial objects for extended periods of time, the feet-above-head arrangement just didn't appeal to me. I had searched for a while for a chair that could be easily modified and make it move with the same mechanical motion(s) as a typical Altazimuth GOTO telescope, where my body would be somewhat horizontal (at least leg and feet wise) while observing the heavens with my big binoculars (Fig. 3.28).

While dumping the trash one day, an old discarded IKEA chair was leaning next to the local trash bin. It's condition, even though a few years old, still appeared to be excellent. It was in good mechanical condition and worked splendidly as a semi-reclining chair. To make it into a fully reclining binocular chair, it would have to be

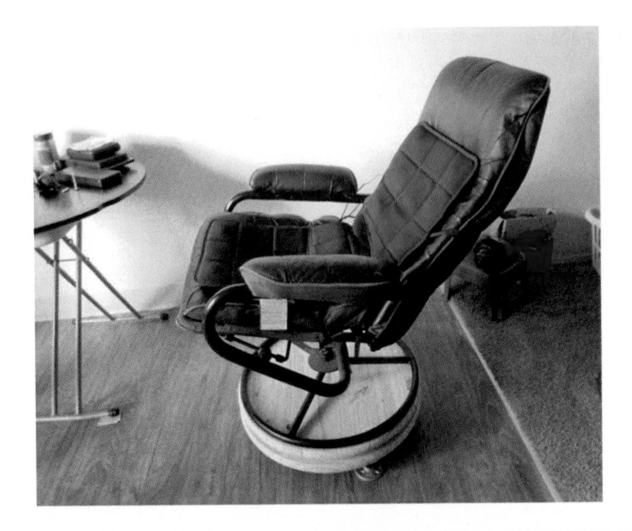

Fig. 3.29 The GOTO binocular chair in the early stages of construction (Image credit: Astronomy Technology Today)

stripped down to its bare frame and remove some of the existing mechanical components (the semi-reclining slide bars on the bottom sides of the seat, for starters) and modify others to allow the backrest to recline fully, without tilting the seat portion. The first build step was to mount the chair on a rotating platform. After searching the Internet for swivel bearings, I found a 17.5-in. diameter ball-bearing swivel (Lazy Susan bearing) rated for a 500-lb load on eBay, ordered it and then found two 23.5-in. round wood tops at Home Depot. After purchasing them, I mounted the IKEA chair on top of a round wooden top with the big swivel bearing centered, sandwiched, and secured in place between the two wooden tops and then tested the assembly against my 200 lb. body, and it rotated a full 360° with very little effort (see Fig. 3.29).

After performing the initial rotation tests, it was the right tie me to replace the original 23.5-in. bottom plate for one that had a greater diameter for additional stability. Fortunately, there was an old 30-in. diameter 5/8-in. thick wooden table top that was not in use at the time, so it became the new bottom plate. It was important to be able to level the base, so three 3-in. diameter pipe flanges were secured in place on the bottom of the 30-in. diameter wooden table top, bolting them in place approximately 120° apart, each combined with a ½-in. diameter 2.5-in. long threaded pipe with a 3-in. diameter pipe flange mounted on the bottom of each one. A simple turn or two of the pipe flange(s) is all that's needed to level the chair assembly at most viewing locations. There is a metal handle mounted on each side on top of the bottom wooden base for lifting the assembly (see Fig. 3.31). After it was mounted, the IKEA chair was stripped down to its bare frame and removed all of the mechanical components that prevented the back from reclining fully to start rebuilding it from "scratch." As in any project, it's easier do some of the more simple

Fig. 3.30 The 17.5-in. dia. swivel bearing for the rotational base for the GOTO binocular chair (image credit: the Author)

things first, like modifying a portion of an ironing board leg stand to serve as a footrest frame and then attaching it at the four pivot points of the original chair frame. The next modification to the original chair frame was to remove the side slide rails that were bolted to the bottom side sections of the seat, replacing them instead with nylon wheels salvaged from a skateboard found at a local swap-meet (Fig. 3.30).

The next important thing on my "to do" list was to make a strong stable system for supporting a pair of big binoculars securely in front of the observer. Ironically, a few days later, when once again emptying the trash in our local trash container, a square-tubed rectangular metal frame with the right proportions was spotted next to the trash container. Looking upward and thinking, the "Gods" are with me on this project and to be honest, the author wasn't even sure what the metal frame was originally designed for, but it looked like it could work just fine for the binocular support frame in terms of length and width. A few drilled holes and ¼–20 bolts were all needed to mount it securely to back of the chair in the position shown in Fig. 3.31. The use of a standard ¾-in. steel pipe and a four-way steel pipe connector for the rest of the binocular support system simplified the project without having to machine anything—to form a connector for the holder of the binocular support shaft, an 8-in. long 3/8-in. diameter threaded steel bolt and the use of a big plastic knob on the top end to mount the binocular holder (see Fig. 3.34). The resulting binocular support system can be adjusted to accommodate the observer's most comfortable head and eye position. The "U"-shaped support frame can be moved up or down by the observer no matter what position the back of the chair is in. It has a 3/4-in. spring-loaded pipe lever with a handle on the observer's right side that the observer can lift up or down to move the binoculars out of the way for entry into or out of the chair, aided by a counterbalance system on the rear section of the of

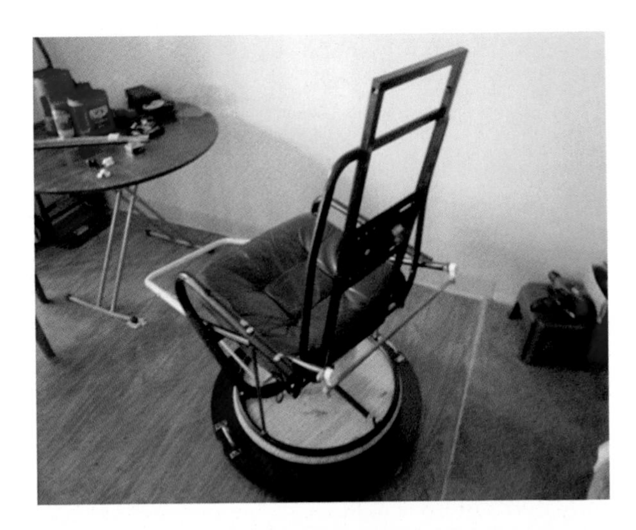

Fig. 3.31 The IKEA chair stripped down and ready to reassemble into a GOTO Binocular Chair (Image credit: Astronomy Technology Today)

the support frame. It can accommodate up to three round weights of 6 lb each; weight can be easily added or removed to accommodate for the weight of small or large binos. With the back of the chair tilting forward and backward during slewing or in the full go-to mode, with the observer's weight pressing the front of the chair, a counterbalance system was made to accommodate for this weight shift, to make it easier for the two altitude servomotors to move the back of the chair via a steel cable drive. For this, a set of barbell plates mounted on each side was used and secured at the end of two 20-in. long by 3/4-in. diameter support bars, totaling approximately 30 lb (more can be added or removed if needed). The weight system support bar is attached to the bottom rear of the of the bino chair and can be easily removed for transport via two 3/4-in. male/female pipe inserts.

For the altitude drive, two NexStar 4SE servomotors were used combined with the Celestron Sky Align software and NexStar hand controller (see Fig. 3.35). The two altitude servomotors are wired in series/tandem (see Fig. 3.32) and together provide plenty of torque to move the back of the bino chair and the observer it supports to just about any position. From the beginning, I wanted to make the altitude mechanical drive system as simple as possible. For me, simplicity makes a much smoother, problem-free system to operate and maintain, and after it was completed, I did not want to have to go back and change it later if it became a problem to operate or if it performed inconsistently (Figs. 3.33 and 3.34).

The altitude drive system consists of an 11-in. tangent arm (equivalent to a 22-in. diameter drive gear) that pulls the back of the bino chair forward and backward via a 1/8-in. diameter stainless-steel plastic-coated steel cable that wraps/unwraps approximately 3.6 times (11.938 in. total length) around a 1.0-in. diameter steel roller via 36-tooth gears located on each end of the steel roller that in turn mates

Fig. 3.32 The dual altitude servomotor steel cable roller drive system for the GOTO Binocular Chair (Image credit: Astronomy Technology Today)

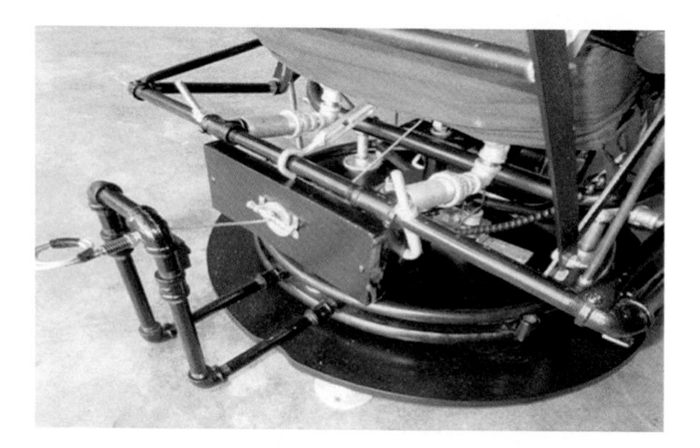

Fig. 3.33 The rear of the GOTO Binocular Chair with the altitude dual servomotor drive system in place (Image credit: Astronomy Technology Today)

Fig. 3.34 The binocular mounting and adjusting system for the GOTO Binocular Chair (Image credit: Astronomy Technology Today)

Fig. 3.35 NexStar 4SE electronics, servomotors, and controller for the GOTO Binocular Chair (Image credit: Astronomy Technology Today)

Fig. 3.36 The NexStar 4SE electronics package and azimuth drive for the GOTO Binocular Chair (Image credit: Astronomy Technology Today)

with the 16-tooth NexStar 4SE servomotor gear. The azimuth drive also uses two NexStar 4SE servomotors. Just like the altitude drive, the azimuth drive has both servomotors wired in series/tandem (see Fig. 3.35). The azimuth motors are coupled to 200-tooth 3.8-in. diameter gears (Fig. 3.36).

The NexStar 4SE Electronics Package. The entire GOTO electronics are housed in a small 6-in. diameter by 3/4-in. tall housing that fits neatly under the bino chair's rear support arm and is secured with two metal clips (see Fig. 3.35). The servomotor connectors, the hand controller, and the power cord plug into the PCB electronic board are located inside of this housing, and the housing is made water proof thanks to a rubber seal that surrounds its base (see Fig. 3.36). Shown is the NexStar hand controller, two NexStar 4SE azimuth servomotors wired in series (bottom), two NexStar 4SE altitude servomotors wired in series (top), NexStar 4SE electronics package (middle) and remote 12-V plug for the Celestron Power Tank (top). The housing design permits easy access to the sealed electronics package when needed.

To better handle the torque produced by the twin servomotors, the mounting brackets were reinforced by using a couple of steel 1/8-in. thick right-angle brackets and a 7-in. hose clamp that wraps around both servomotors. Adding the big hose clamp really kept the servomotors tightly engaged in the 3.8-in. diameter azimuth drive gears. The altitude drive cable is secured at the back of the bino chair with a ½-in. thick aluminum plate secured in place on a U-shaped ½-in. pipe bracket and a 3/8-in. bolt that has a lengthwise hole through which the steel cable passes. The cable is secured with a strong spring and lock ring. The spring takes up any slop in the cable while it wraps/unwraps during go-to and slewing.

Using the GOTO Bino Chair for the first time was great fun. It takes all the work out of searching for objects of interest to view with your big binoculars. With the touch of a button, the chair slews you and your binoculars to the desired object and then keeps it centered and tracks while you view it from a comfortable, natural sitting position. A motorized bino chair is great, but a GOTO bino chair is even more fun, especially if you build it yourself. Any creative, resourceful ATM can design and build a similar system. All parts used in mine are readily available and affordable.

To set up the GOTO Binocular Chair for observing, the observer levels the binocular chair via the three leveling pads and then levels the binoculars and making sure the back of the binocular chair's seat is in 90° position and then makes sure that front (foot rest) of the binocular chair is pointed in the north direction. Using the Celestron hand controller, one can input the time, date, and location and then select the appropriate Sky Align program (e.g., One Star Align). Then it's just a touch of a button on the NexStar hand controller, and the binocular chair will GOTO to the selected celestial object of interest that is within the 15 × 80-mm binoculars 3.5° field of view. The observer is always sitting in a comfortable position, while the binocular chair tracks the celestial object(s) at the appropriate tracking rate. Other times one can just enjoy scanning the skies while slewing or "slowing" in the "azimuth" mode using just the hand controller. Also, just hang a thermos of coffee on one of the arm rests, pausing briefly for a sip or two every now and then to keep you going through a night's observing session.

Any creative and resourceful ATM can build a GOTO Binocular Chair because the parts to build one are out there and you can find most if not all of the major pieces of hardware at your local Home Depot or a hardware store. A motorized binocular chair with a joystick is great, but a "GOTO" Binocular Chair can be even more fun, especially if you build one yourself. All parts used in mine are readily available and affordable. So, why not give it a try?

Norm Butler
Northern Marianas Islands

The StarChair3000

The StarChair3000 is a commercial motorized binocular chair that can move the observer about both in the vertical and horizontal axis in complete comfort via a joy stick controller. This kind of chair would provide the observer with the ultimate kind of observing experience, especially with a big pair of Fujinon 150 mm × 25 binoculars mounted in front of you (Fig. 3.37).

For the binocular astronomer who enjoys viewing the heavens in complete comfort in a motorized binocular chair, then the Sky Rover StarChair™ is for you. The author wanted to include the following information about this remarkable motorized binocular chair because it offers the binocular astronomer an excellent opportunity to enjoy sitting in a completely motorized binocular chair and moving to all parts of the binocular sky using a joystick controller sitting in complete comfort.

Being a binocular astronomer to some degree myself, I was fortunate enough to have seen and used the StarChair3000 on more than occasion, and each time I was more than impressed with its capabilities and operation. Consider the following

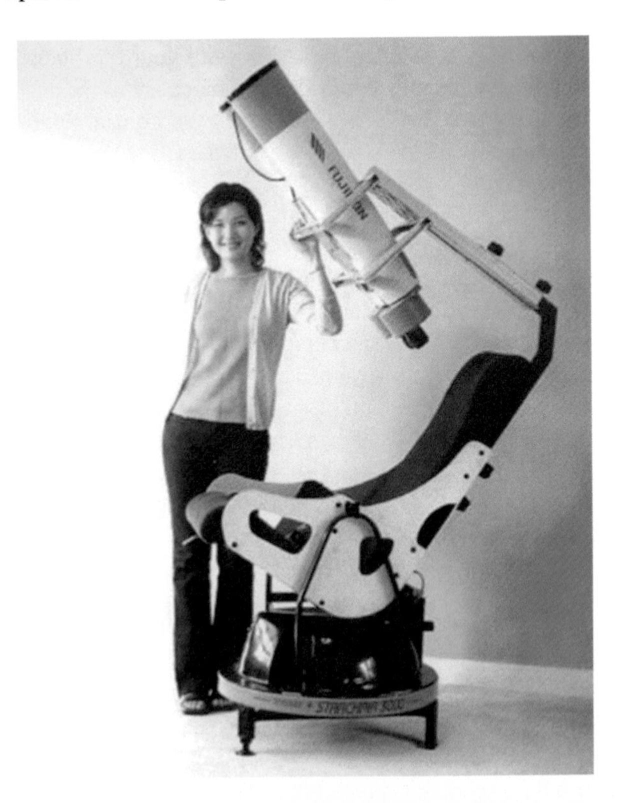

Fig. 3.37 The StarChair3000 is a commercial motorized binocular chair manufactured in Australia. Note: STARCHAIR NOW is temporarily in suspension. A new improved model underway. If all goes well (Image credit: www.starchair.com)

"Noted" comments on my part, an informative review about an exciting product for binocular astronomy.

The Sky Rover StarChair™ is a very well-designed robust portable motorized binocular chair that allows the observer to view the night sky through a variety of front-mounted binoculars of different sizes and magnifications.

Using a simple joystick, the observer can rotate the chair on its base through a full 360° and at the same time move the chair from the horizon to zenith in a smooth fashion. As a result, the binocular observer can view the entire heavens in seated comfort with the use of the hand-held joystick controller. The addition of optional computer Sky Vector hardware allows the StarChair™ binocular observer to quickly and easily position any one of several thousand celestial objects in the binocular field of view. The StarChair™ revolutionizes the use of binoculars for viewing the night sky.

The Sky Rover StarChair™ is powered by a standard 12-V motorbike battery or car battery. The binocular observer uses a joystick to control the motion of the chair. The motorized chair has a variable speed, low power motor that rotates a central worm gear and a unique altitude actuator that act together to move the chair in the desired direction. There is no slippage while traversing, nor directional backlash or flexing on halting. As a result, binocular images always remain steady.

The design ensures the electronic wiring does not suffer umbilical cord fouling. This allows base rotation beyond 360°. Binoculars are held rigidly in place as part of the chair, but remain fully adjustable to allow an observer to tailor the position of the eyepieces to their individual body height and size. The Sky Rover StarChair™ is constructed from high-quality materials and components. It is designed for outdoor use and will not deteriorate if left outdoors for extended periods of time. User safety has been a paramount goal from the initial design stage. The resulting StarChair™ design puts safety first and instills user piece of mind. The construction is very strong—will not break—and easily supports 1000 lb of weight! StarChair™ 3000 model is basically composed of three easily manageable components that fit together in around 3 min ready for use. The main base section is only 26 in. wide and can be conveniently carried through any standard doorway.

The StarChair™ weighs only 45 lb and can be lifted by most average users. The seat section is light weight and is quickly inserted into the main base. The material of the specially designed seat is made from laser cut 6-mm thick extremely strong aluminum plate and quite rigid, yet light weight. It can easily support 1000 lb! The inside width of the seat is 20 in.

The plush thick upholstery is ergonomically contoured and covered with appealing UV-treated navy blue velour material. The front posterior upholstery is so "obtusely rounded," there is now no need for a foot rest. "Under-the-knee section" is now just so comfortable. The third section is the new heavy-duty back hook bar and binocular head attachment. This new designed section is now capable of supporting all large binoculars from 70-mm up to the giant 150-mm Fujinons, totally adjustable to cater to all sized stature observers. A hand-held supplied joystick will control the speed and direction to all points of the celestial dome. Maximum speed is around 8° per second! Minimum speed is around 1 min of arc per second! Both axes can be activated simultaneously! That is, chasing and centralizing fast moving satellites is possible and exhilarating. So are fast moving aircraft. Sky divers and Glider aircraft are also suitable for close observation, using the variable speed joystick and StarChair™ combination in complete steady reclined luxury comfort.

When it comes to computer control, the StarChair™ has very accurate inbuilt fittings in its design, so that standard optical encoders will quickly snap on to StarChair™. The optical encoders will give a full read out of positions of the celestial dome as well as feedback to most PC programs that will show where the StarChair™ is pointing. Deep-sky objects can then be accessed by using the "Guide To" or "Identify" function mode. When it comes to its portability, the StarChair™ will fit into any small sedan and be driven to any dark sky site. The StarChair™ setup time is approximately 3 min! And packing it up after an observing session is also about the same amount of time. The StarChair™ has extensively field tested this factor many times and finds this as one of the most favorable features of the motorized chair, especially, in cold weather climates, which makes it a great plus for deep-sky observing! StarChair's power supply is curtailed by the need to have a "safe and secure" simple use of a 12-V dc car battery. This low power requirement using a 12-V dc power supply is a nice comfort to have when observing with temperatures hovering around 5° or less and at a "remote" dark sky site! A simple 9 A per hour dc battery will easily control the StarChair™ all night, and even the next night! And even the third night before recharging the battery, which only takes about 3 h. This is a very nice situation for the users. StarChair™ is advanced in this situation. In fact, at extreme locations, a small "solar power" cell can recharge during the day!

Note: Star Chair is temporarily in suspension. A new improved model underway. If all goes well (www.starchair.com).

A Motorized Binocular Chair

For those who enjoy observing the heavens in complete comfort while using a big pair of 100 mm or even larger big binoculars, consider the use of a motorized exam chair that can be converted and modified into a motorized binocular chair. A new, used or refurbished exam (or ophthalmic chair) could become your dream motorized binocular chair project (Fig. 3.38). There are also many resources on the internet where you can purchase a reasonably priced one.

One additional component that could be added to your chair is a "Joy Stick" that is wired into the original existing electronics of the motorized binocular chair. The "Joy Stick" can normally be found at any electronic store or website. Even the use of a foot pedal electrically attached to the foot rest would serve as a convenient way to move the motorized binocular chair in the desired direction. If you are thinking about progressing to a new level of binocular astronomy, consider a refurbished exam or ophthalmic chair for your next ATM astronomy project.

Let's see what kind of project our two favorite ATMs Larry and Gordon are working on these days (see Figs. 3.39, 3.40, 3.41, 3.42, 3.43, 3.44, 3.45, 3.46, 3.47, 3.48, and 3.49).

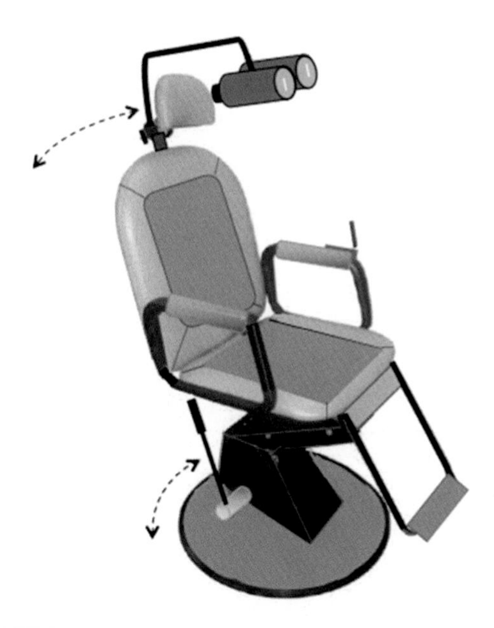

Fig. 3.38 An illustration of an exam chair converted into a motorized binocular chair (Image credit: the Author)

Fig. 3.39 (Cartoon credit: the Author)

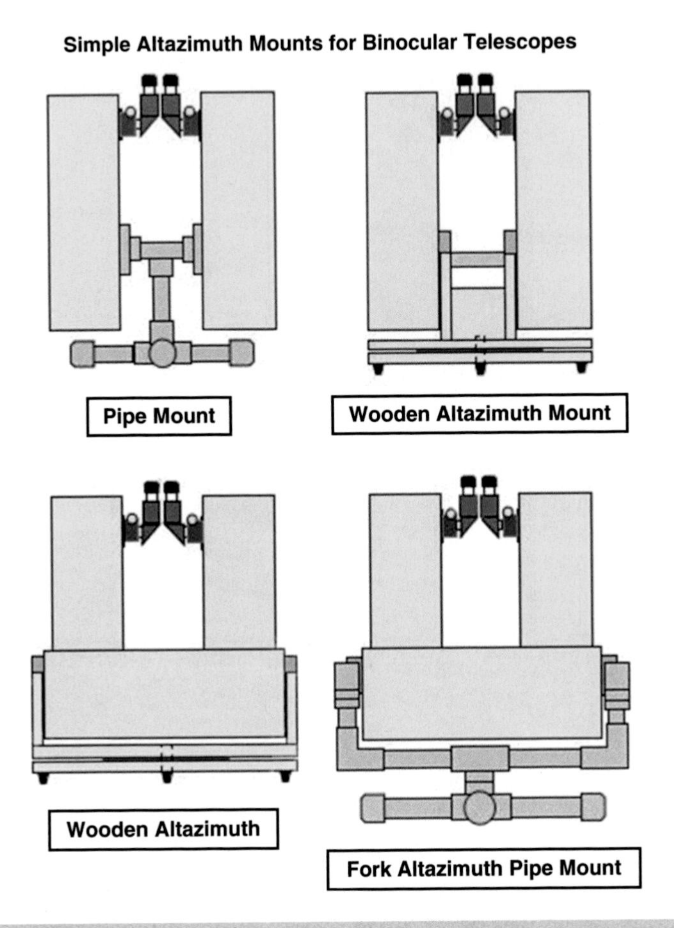

Fig. 3.40 Four examples of wooden and Altazimuth pipe mounts that can be simply made for a homemade binocular telescope. Note: Drawings are not to scale and are presented in this chapter for illustration only. (Image credit: the Author)

Fig. 3.41 Rollei 7×42 fixed-focus binoculars were based on the British AVIMO 7×42 mm made in a rubber-armored form during the 1980s for British troops, police and NATO forces. The Rollei version was intended for civilian use (hence the attractive silver finish) although they still have the military graticule. Today, the Russian BELMO factory in St. Petersburg makes copies of this fine binocular. (Image credit: Antony Kay—OptRep—www.opticalrepairs.com)

Fig. 3.42 A rare Carl Zeiss 5+10×24 mm "Marine-Glas m.Revolver" dual power turret binocular (circa 1896–1904). It was known as the "Admiral Togo Glass" because the commander of the Japanese in the Russo-Japanese war of 1904–1905 used this particular binocular. The construction of the body, objective lens cells and the method of securing the prisms are all similar to those used in the famous single power 'Feldstecher' binoculars. (Image credit: Antony Kay—OptRep—www.opticalrepairs.com)

Fig. 3.43 A Carl Zeiss 12–20–40 × 80 mm turret binoculars (Asembi) was used in WW1 and WW2 by the Kriegs marine. The design remained unchanged with one of the few changes being that the air-spaced objective lenses were eventually replaced with cemented ones in the later models. The prism housings can be swiveled to change the IPD and also be swung outwards to allow two observers to use it as two separate telescopes observing the same target. (Image credit: Antony Kay—OptRep—www.opticalrepairs.com)

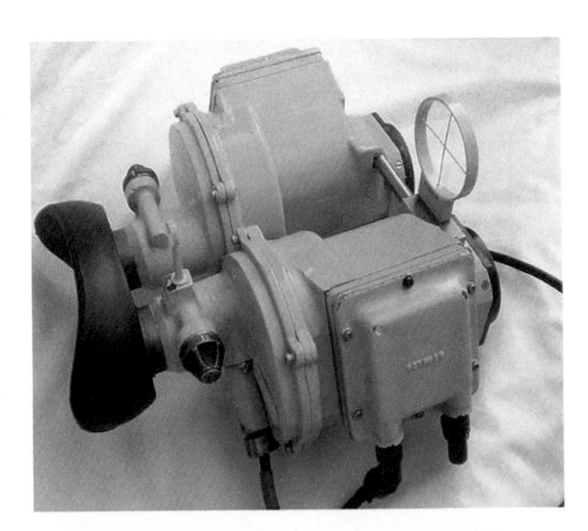

Fig. 3.44 A British AVIMO warship binocular, type G389/1 fitted with electrical heaters and internal filters. These binoculars, which have many unusual features along with the smaller G386/3, were the last types designed by AVIMO before the company was drawn in by Thales. This type of binocular is used by the Royal Navy and the navies of many other countries. (Image credit: Antony Kay—OptRep—www.opticalrepairs.com)

Fig. 3.45 A tiny Carl Zeiss 3.5×15 mm prismatic "pocket" binocular (Image credit: Antony Kay—OptRep—www.opticalrepairs.com)

Fig. 3.46 Folding opera glasses made out of brass and Mother of Pearl (Image credit: Public Domain—Wikimedia Commons)

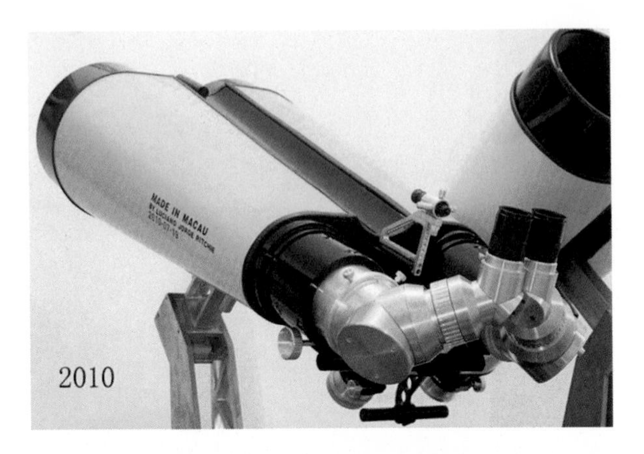

Fig. 3.47 A splendid looking giant pair of homemade 203 mm giant binoculars (Image credit: Luciano Jorge Ritchie of Macau)

Fig. 3.48 NGC 7635, which is often called the "Bubble nebula", in the constellation Cassiopeia (Image credit: Sara Wager—www.swagastro.com)

Fig. 3.49 (Cartoon credit: Jack Kramer)

Further Reading

Abrahams, P. History of the telescope and the binocular. http://home.europa.com

Butler, N. (2012, July/August) An ATM Go-To Astro-Bino Chair. www.astronomytechnologytoday.com

Kay, Antony, OptRep, http://www.opticalrepairs.com

Nabholz R. Downward looking binocular. http://www.homebuiltastronomy.com

Nikon Sport Optics. Exit pupil. http://www.nikon.com

Schmidt, W. The making of voyager. http://www.waynesthisandthat.com

Shaw Creek Bird Supply. http://www.shawcreekbirdsupply.com

Schott 2007 refractive index. SCHOTT Glass, P.O. Box 2480 Hattenberg 10 D-55122 Mainz, Germany. www.shott.com

Starizona.com, 757 Oracle Rd., Suite 103, Tucson, Arizona 85704. www.starizona.com

The Starchair3000. www.starchair.com

Vehmeyer, C. Binocular mirror mount. http://Vehmeyer.net

Wager, Sara, http://www.swagastro.com

Chapter 4

Riverside Telescope Makers Conference (RTMC)

It's always fun to attend a local star party or a yearly telescope maker's conference such as the popular Riverside Telescope Makers Conference (RTMC) founded by Clifford Holmes in 1969 and the Riverside Astronomical Society and is held each Memorial Day weekend at Camp Oakes in Big Bear City, California. The Texas Star Party is held in May/June and the Stellafane Telescope Makers Conference held yearly in August in Vermont. Amateur telescope makers and amateur astronomers from around the country always look forward to attending these yearly conferences to enjoy all the amateur-made telescopes, newest commercial trends, the popular swap meets, and the telescope merit award competition. There are scheduled talks held at these conferences, with subjects such as telescopes, optics, mirror grinding, astrophotography, CCD imaging, and, of course my favorite, topics about telescope making in general. There is also an invited guest speaker who is noted for their achievements in astronomical research in some interesting area of astronomy or even a discussion on planetary studies that will make for an informative talk for all those who attend the RTMC conference (Figs. 4.1 and 4.2).

One of the most popular things to do at these telescope makers' conferences is to walk around the telescope field during the day looking at all of the excellent homemade telescopes and visiting the vendor stands and swap meet tables that are scattered all around the field, and there are also various assortments of homemade telescopes hidden behind trees and campers that were crafted by some very creative individuals with some innovative ideas and some interesting materials they had in mind. Just walking around the telescope field and looking at all of these fine homemade hand-crafted telescopes is pretty inspirational, especially if you're a telescope maker.

Most if not all of those who attend an RTMC conference usually head for the swap-meet tables spread out near the conference room dining hall where all of the

N. Butler, *Building and Using Binoscopes*, The Patrick Moore Practical Astronomy Series, DOI 10.1007/978-3-319-46789-4_4

Fig. 4.1 The Los Angeles Astronomical Society members are ready for action at the 1989 RTMC (*left image*) and Steve Overholt (*right image*) has already setup his splendid looking dobsonian telescope and is getting ready for the 1991 RTMC's night observing fun to begin (Image credit: Pierre Desvaux—http://dobsonfactory.blogspot.com)

Fig. 4.2 Everyone including Mike Hill (*left image*) at the 1984 RTMC is anxious to show off his elegant looking dobsonian telescope including David Gilahia (*right image*) at the fun-filled 1987 RTMC (Image credit: Pierre Desvaux—http://dobsonfactory.blogspot.com)

good deals to be had are at. You can just about find anything from telescopes, accessories, telescope mounts, mirrors, eyepieces, T-shirts, meteorites, paintings, fossils books, magazines, and a host of other astronomy items and things that you could only find at an RTMC swap meet. And after a hard morning of looking for deals at the swap meet tables, the smell of hot dogs and hamburgers barbecuing on the open grill starts to fill the air and soon thereafter…you guessed it, lines start to form at the hamburger and hot dog stands. During the day sessions, lectures are given in the conference room dining hall by some very knowledgeable individuals on various topics in optics, astronomy, telescope making, etc.

The majority of the RTMC attendees like to camp out in their RVs and campers as well as those who enjoy sleeping in the dorms at Camp Oakes. All of the dorms have bunk beds and hot showers in the bathrooms. The dorms are all located in close proximity to the Camp Oakes dining hall and conference room. Another nice thing about attending the RTMC conference is that you can get free coffee in the dining hall to help keep you going for the all-night observing sessions. At 7200 ft elevation, a cup or two of hot coffee at Camp Oakes will keep you going. The merit awards are special to all of the RTMC amateur telescopes makers and to the conference attendees too. Everyone who entered their homemade telescope in the merit award competition is hoping they'll receive a merit award for their telescope making efforts or, at the very least, an honorable mention. For the most part, it's a great feeling when your name is called in front of your friends and fellow telescope makers to receive a merit award and also to mention the Clifford Holmes Award, along with the Clyde Tombaugh and Warren Estes Memorial Award that are given out each year too. It's an especially good feeling that all of your hard work in building your telescope is appreciated by your peers, and those at the RTMC conference who are also quite capable of building an excellent homemade telescope will also be very appreciable of your telescope making merit award.

After the merit award ceremony come the drawings for door prizes. Tickets have been passed out, and the crowd waits with eager anticipation inside the crowded conference dining hall and also a hundred people sitting outside wearing heavy winter clothing sitting in lawn chairs anxiously waiting for the drawing to begin. Whether it's cold or not, no one really cares about the weather at that point. It's winning a door prize that counts. Sometimes someone wins a new 8-in. Celestron or Meade GOTO telescope, a sky atlas, 2-in. TeleVue eyepieces, a new astronomy book, or something electronic. When a number on a ticket is called out, everyone has their fingers crossed that they have the winning number. Actually, everyone's chances are good that they might win something. Lots of door prizes are given out. Even the kids can win a big door prize too. After the drawings are over, the crowd slowly disperses back onto observing field to continue to enjoy a great night of observing at the RTMC conference. Those who won a merit award can expect some lines of people waiting eagerly to view through their award-winning homemade telescope. Chances are they won't be disappointed.

One of the best things about attending an RTMC conference is getting the opportunity to talk with others who have some of the same interests that you have in astronomy, such as telescope making, mirror grinding, eyepieces, comet hunting, binocular observing, or CCD astronomy. Of course there are many other exciting and interesting things to do at an RTMC Memorial Day weekend conference. Horseback riding, swimming, hiking, and exploring around Camp Oakes can be a lot of fun, not only for the family but for everyone too. Just walking around and looking at all stuff at the vendor stands are fun to do. If you're looking for that special astronomy T-shirt or hats, you'll find it there. Everything from telescopes, accessories, meteorite collections, to fossils, you won't have to look long before you'll find something you want or need to buy for yourself. One can look around

Fig. 4.3 Don Machholz at Camp Oakes with his trusty 10-in. f/3.82 Newtonian telescope on the morning of his comet discovery (Image credit: Don Machholz)

for a neat gift for that birthday coming up for an astronomy friend or family member. And that's just some of the many fun things to do when you attend an RTMC conference. If you enjoy birding with your binoculars, then you'll really enjoy the different types of birds that live in and around Camp Oakes.

How about those Wood Peckers and Blue Birds? Got to love those pesky squirrels that scamper about the camp grounds, especially in morning when they're hungry and foraging for nuts. If it's fishing you like, then there's the "pond" at Camp Oakes or Big Bear Lake just a few miles down the road.

For those of you who are ardent comet hunters, here is a little bit of inspiration for you. During the 1985 RTMC telescope maker's conference at Camp Oakes, Don Machholz, the noted comet discoverer (nine comet discoveries), discovered a new comet while searching the night sky at Camp Oakes. It was named Comet Machholz 1985e. The following is the story in Don's own words of his comet discovery at the 1985 RTMC (Fig. 4.3).

The Discovery of Comet Machholz 1985e

By Don Machholz, May 27, 1985

Now fully dressed, I opened up the back window of the camper shell and, leaving behind my sleeping wife Laura, I stepped out into the cold air. The moon was still up and casting light upon the landscape. Soon it would be setting, and I would begin searching for a new comet.

This weekend had been a tiring but refreshing one at the Riverside Telescope Maker's Conference at Camp Oakes, near Big Bear City in southern California. Over 1300 individuals had attended the Memorial Day weekend gathering, held every year since 1969. Talk had been about telescopes, photography, computers, and Halley's Comet, which most of us will be observing before the end of the year.

I walked down the dirt road, known as "telescope alley," and onto the telescope field. Many of the 100 telescopes were gone, their owners having already left or packed for an early getaway in a few hours. Those telescopes remaining were being put to good use, some searching for faint galaxies or nebulae, others being used to show heavenly wonders to friends. Many new friendships are made each year at the conference, often under these dark skies, where faces are not seen and we learn to recognize the voice. Here is a place where astronomy is both enjoyed and shared.

I turned and started walking back toward telescope alley; my telescope was set up near the end of the alley among the bushes on the right. The instrument is a reflecting type of telescope; it uses a mirror 10 in. across which focuses at a distance of 38.2 in., meaning a focal ratio of 3.82. This is a short ratio of focal length to minor diameter but allows for a wide field of view. In this instance, however, I had placed a cardboard cutout in the eyepiece, giving a field of view 1.6° square. Telescopes usually have round fields, but I believe a square field has some advantages for comet hunting. The eyepiece gives a magnification of 32. While the optics were commercially made, the rest of the telescope is homemade. Its first construction took place in 1975, and it was redesigned in July 1981. The complete optical system is

Fig. 4.4 Don's telescope in the discovery area (Image credit: Don Machholz)

Fig. 4.5 The comet discovery site—Southern California's Camp Oakes (Image credit: Don Machholz)

now mounted on an Altazimuth mount made of lead pipes. With this type of mount, I can scan (or sweep) parallel to the horizon. This allows for more efficient comet hunting since comets are often found near the horizon (Fig. 4.4).

Three years ago, at my first trip to Riverside, this telescope won the Warren Estes Award, given each year for a telescope made from simple materials. This year, I did not enter the telescope. I had brought it only so that I could continue my comet seeking while on my 4-day vacation.

At 1:25 on this morning, May 27, 1985, I began comet hunting session no. 1385. This started as any other, with anticipation and excitement, because I never know just what I will find while comet hunting. Two mornings ago, I had picked up a very faint nebula which I had never seen before (Fig. 4.5).

My first sweep was at an altitude of about 45°. My goal was to cover the eastern sky from the celestial equator to 40° north. This required peering through the telescope as I pushed it northward to about 40° north. During that time, I would be looking for anything faint and fuzzy gliding through the field of view. Such an object could be a new comet; more often these would be clusters, nebulae, galaxies, or small groups of faint stars.

Nearly 8 years before, on the morning of September 12, 1978, I uncovered an object which turned out to be a comet. Being newly discovered, it was named Comet Machholz, another designation being "1978L." It was the twelfth comet recovered or discovered in 1978. That find, for which I used many parts of my present 10-in. telescope, had taken 1700 h of searching over more than 3½ years. The comet was faint when found and did not get much brighter after discovery. I was able to observe it for 1 month before it disappeared below my southern horizon. It is in a type of orbit by which it will never return (Fig. 4.6).

Fig. 4.6 Don at the KRON Channel 4 studio (Image credit: Don Machholz)

My first comet was found from a mountain called Lorna Prieta, in the Santa Cruz Mountains, 22 min south of my home in San Jose. But this morning, I am over 400 miles south of my home, at an elevation of 7700 ft. The telescopic views were great. I swept up the Veil Nebula in Cygnus with all its delicate beauty. I also saw a globular cluster named NGC 6740 and a wonderful open star cluster known as M11. With each sweep, I moved closer and closer to the horizon, which, ideally, I would reach just as the sky was beginning to brighten.

Shortly after 3:00 a.m., I stepped back from the telescope, removed my eye patch, put on my glasses, and looked around the night sky. This was a perfect night for comet hunting. Being far from city lights, the stars and Milky Way stood out in high contrast to the dark background. Only under such conditions can astronomers carry out many of their programs. I commented to my friend, Darwin Poulos, on the darkness of the sky. He was now observing with his 8-in. reflector about 20 yards from me, examining objects in the Southern Milky Way. He was having a good night too.

As I continued to sweep, I observed a faint galaxy known as NGC 185, and moving over a field, I also examined NGC 147. Because of the lights of San Jose, I do not often see these two objects from my site at Lorna Prieta. But they were easily visible from here. On the next sweep, I saw the majestic Andromeda galaxy; it more than filled my field of view. But I could not gaze for long, I had to keep sweeping, and dawn was approaching.

About three sweeps later, at 4:13 a.m., I picked up a fuzzy object, not too faint, which suddenly aroused my suspicion. It was in a part of the sky where I knew there were no galaxies or clusters, and it appeared pearly white, like a comet. I stopped my sweeping and started my work of determining the nature of the object.

Foremost, I had to know exactly where my telescope was pointed. This is needed for two reasons. First, to check my charts to see if the Skalnate Pleso chart showed no known objects and I plotted a small pencil mark at its position on the map. Next I had to check for motion. A comet will move against the background stars, so I drew a map and hoped to detect movement. Meanwhile I made a quick measurement of its position and checked my more extensive catalog. It did not list any known object in that position.

With dawn nearly upon us, I made a few more checks. A Barlow lens, doubling the magnification, showed the object to be fuzzy and elongated, and not a small group of stars. Next I tried a "Comet Filter" (with it some comets will appear more visible as the background is darkened. Any other object will appear fainter). This object was more visible.

By now it was difficult to see the object, as the sky was rapidly brightening. Furthermore, I had not detected movement. The 27° temperature was nipping at my hands because I had removed my gloves when I spotted the object some 20 min earlier. I felt quite sure that I had discovered a comet, and yet the lack of motion prevented me from being absolutely certain. I had searched 1742 h since my first comet find (Fig. 4.7).

I went to the truck and woke up Laura. "Laura," I said, "I think I found a comet." She woke up immediately.

Reporting the comet was nearly as hard as finding it. I merely had to call the Smithsonian Astrophysical Observatory in Cambridge, Mass., the clearing house for comet discoveries. We knew of only one phone at the camp, a pay phone at Coombs Lodge. After writing up the telegram and walking a quarter mile to the phone, I had trouble getting through to the operator. The phone finally jammed and we had to move on.

Fig. 4.7 Don films a piece with a Channel 5 KPIX news team in his backyard (Image credit: Don Machholz)

Fig. 4.8 Don receives a plaque from Cliff Holmes (*left*) to mark the site at in Camp Oakes, CA, where Don discovered Machholz 1985e (Image credit: Don Machholz)

We tried looking for a few of the comet hunters there, but at 5 a.m., they could not be found. Perhaps they would know of any known periodic comets in that area of the sky. Or, perhaps they would know of a recent discovery in this region. I could hardly believe that a bright comet of magnitude 9.3 would still be undiscovered. Unable to find anyone, we had no choice but to pack up and go to our motel in Big Bear City, 10 miles away (Fig. 4.8).

Following an hour of telescope disassembly, Darwin's car occasionally quitting, and a hurried drive, we arrived at the Motel 6 lobby. Our luck was unchanged. Western Union, which would send the telegram to the Smithsonian, was not answering the phones. Meanwhile, this being a holiday, no one was at the Smithsonian Astrophysical Observatory to take the discovery message. I tried to call the Smithsonian director, Dr. Brian Marsden, at home, but his number was not listed. Suddenly, among my notes, Laura found his home phone number. I tried it.

Dr. Marsden said he had no reports of a comet being found, and we could not recall any known periodic comets in the region. He took my message and said he'll try to get someone to confirm that this was indeed a new comet. I said I'll try to do the same.

Then I was back to our rooms for a quick shower and packing and then out to breakfast. Our 9-h drive home brought us from the clear morning skies of Big Bear to the cloudy evening skies of San Jose.

Upon arriving home, I called Gerry Rattley of Phoenix, Arizona; Jack Marling of Livermore, California; and the observers at nearby Lick Observatory, asking them to observe the object and confirm it.

Fig. 4.9 In this Lowell Observatory photo, Comet Machholz 1985e makes it way past M45 (Image credit: Don Machholz)

The next morning saw Rich Page, Laura, and I on Loma Prieta, sitting in the clouds. Finally it began to clear, and near 3:30 a.m., we set up our telescopes and began to search. After a few minutes, I found it 1½° ENE of the previous day's position. It was a comet! Rich then found it too, and we showed it to Laura. We all were happy!

Meanwhile, Charles Morris and Alan Hale of the Los Angeles area had received word from the SAO via Stephen Edberg of my possible discovery. Following only a few hours of sleep, they went to a nearby observing site and confirmed the existence of the comet (Fig. 4.9).

A preliminary orbit shows the comet will be nearest the sun on June 8, 1985, at a distance of only 10 million miles. Until June 15, it should continue to brighten in the morning sky, and then we will lose it in the solar glare. About July 10, it will emerge into our evening sky and then fade rapidly.

According to my recent study, *A Decade of Comets* (24.5 MB Word document), now in booklet form, I found that none of the comets found by amateurs during the past decade came as close to the sun as this comet. And, while this might be one of the brightest comets in several years, for much of the time, it will be too close to the sun for easy observation from Earth.

Don Machholz

Adapted from "The Discovery of Comet Machholz 1985e" with permission from Don Machholz (http://www.donmachholz.com)

Each outdoor dorm has several bunk beds with mattresses. There are eight other heated indoor dorms with bathrooms, hot showers and various camping sites that are available, plus a nice big outdoor swimming pool for everyone to enjoy at Camp Oakes (Figs. 4.10, 4.11, 4.12, 4.13, 4.14 and 4.15).

Telescope makers all owe a debt of gratitude to Sir Isaac Newton, who in 1688 invented his first reflecting telescope. As a result of his remarkable invention, reflecting telescopes bear his name "Newtonian' reflector." Sir Isaac Newton's first reflecting telescope with its tiny 1.3-in. f/5 spherical speculum mirror, though small by today's standards, still sets the mark for all telescope makers to try and emulate. At all the major telescope makers conferences such as RTMC, Stellafane, and The Texas Star Party that are held each year around the country, the word "Newtonian" is heard frequently in conversations at these telescope making conferences. Sir Isaac Newton is almost considered a deity by telescope makers, amateurs and professionals alike, especially by those who want to build telescopes as ingenious as was his first Newtonian reflector. And that's what makes the Riverside Telescope Maker's Conference and the others so interesting and fun to attend.

Fig. 4.10 Pictured above is one of the several outdoor dormitories that are available for the RTMC conference attendees to sleep in during the long Memorial Day weekend (Image credit: Long Beach YMCA)

Fig. 4.11 The Camp Oakes dining hall and conference room is where most of the major RTMC talks, lectures, and merit award ceremonies take place (Image credit: Long Beach YMCA)

Fig. 4.12 If you like canoeing and fishing, the pond at Camp Oakes is the place you can really enjoy with the family (Image credit: Long Beach YMCA)

As amateur telescope makers (or more commonly called ATMs), we all seem to have a little bit of Sir Isaac inside of all us. And from time-to-time, we like to show it off with our own homemade telescope creations, and the annual Riverside Telescope Makers Conference is the perfect place for that (Figs. 4.14, 4.15, and 4.16).

The Riverside Telescope Makers Conference merit award has a special meaning for those who have made their own telescope and want to enter it in the merit award competition. There are several categories that one can enter their homemade telescope in. For example, at the 1981 Riverside Makers Conference, some of these categories included engineering, unusual design, best wood design, best wood

Fig. 4.13 A replica of the original Newtonian reflector (Image credit: Solipsist/Andrew Dunn-Wikimedia Commons)

Fig. 4.14 Looks like it didn't take James Stevens to long to get his ultra-portable dobsonian telescope into action at the 1991 RTMC (Image credit: Pierre Desvaux—http://dobsonfactory.blogspot.com)

Fig. 4.15 Here's one big Dob that could probably sleep three at the 1993 RTMC (Image credit: Pierre Desvaux—http://dobsonfactory.blogspot.com)

Fig. 4.16 Here's a big dob that's pointed in the right direction at the 1994 RTMC (Image credit: Pierre Desvaux—http://dobsonfactory.blogspot.com)

Fig. 4.17 RTMC merit award (Image credit: the Author)

construction, workmanship, and original design. If you won a merit award in any of these categories, you can feel proud of your accomplishment. When you're handed an RTMC merit award plaque, you can attach it to your telescope and display it with pride (Fig. 4.17).

At night when the observing starts, the crowds disperse and the telescopes they looked at with interest and perhaps with a little envy in the daytime now become the center of attention for the RTMC conference goers. If one of the big binocular telescopes is setup in "Telescope Alley," then expect it to have the longest lines of people waiting eagerly to get a chance to take a look through it. And when they do, they will have some of the nicest views that they ever had through these big homemade binoscopes. Even though not everyone who enters their telescope in the telescope making competition will receive a merit award or honorable mention, it doesn't mean that their effort was in vain, quite the contrary. In fact, one of the premises behind amateur telescope making is sharing your ideas with others who may wish to build a homemade telescope of their own someday. Letting others see what you have made at a telescope makers' conference will, in fact, allow them to build a good telescope using a bit of your telescope making knowledge and experience (Figs. 4.14, 4.15, 4.16, 4.17, 4.18, 4.19, 4.20, and 4.21).

Dann McCreary's entire Pedagogical Equatorial Platform system is so uniquely designed that it is capable of accurately tracking celestial objects no matter where the telescope or binoscope is positioned on the platform. If you are thinking about a new type of equatorial platform drive system for that big binocular telescope project you're planning or planning to build, then visit Dann's subarsec.com website for more detailed information about his unique equatorial platform (Figs. 4.21, and 4.22).

Now that this year's RTMC telescope maker's conference is just about ready to finish up, all of the conference goers have enjoyed their 4-day event-filled Memorial Day Astronomy weekend. Plus, just about everyone, there has had a chance to observe through a multitude of fantastic homemade telescopes while camped out under the stars and the beautiful night skies of Camp Oakes. There were the interesting and informative lectures on amateur telescope making, astronomy, and space science as well as other related astronomical topics, the homemade telescope

Fig. 4.18 Looks like Mike Clements and Dennis Young are getting ready to enjoy the night sky at the 2011 RTMC with a nice big pair of 100 mm WWII vintage binoculars (Image credit: Dean Ketelsen)

Fig. 4.19 That looks like a big pair of 5-in. WWII Japanese vintage binoculars that Dean Ketelsen is using for his night of observing the heavens with (Image credit: Dean Ketelsen)

Fig. 4.20 At the RTMC conference at Camp Oakes, there are a lot of good spots available to observe from with your telescopes (Image credit: the Author)

Fig. 4.21 Dann McCreary standing beside his homemade and quite unique 2012 RTMC merit award-winning "Pedagogical Equatorial Platform" (Image credit: Dann McCreary)

Fig. 4.22 Dann McCreary presenting his merit award-winning "Pedagogical Equatorial Platform" program to an appreciative RTMC audience (Image credit: Dann McCreary)

merit awards ceremony, and the ever-popular door prize drawings. There is horseback riding, swimming, birding, fishing, and canoeing at the pond. There are other out-door activities to do at Camp Oakes as well, such as wildlife and scenic photography, hiking along the trails, and nature study. All of it is outdoor fun to enjoy with the entire family.

But all good things must come to an end, so as the end of the conference nears, there are the last minute deals at the swap meet tables. The vendors have already started to take down their tents and tables and begin to load up their cars and trucks. By Monday noon, everyone has just about finished taking down their telescopes and camping equipment and started to pack up for the drive home. Many conference goers have driven from several states and even some have come from other countries to enjoy the annual RTMC conference weekend. This year's RTMC conference was again a success, and the Riverside Astronomical Society (RAS) is already looking forward to next year's annual Riverside Telescope Maker's Conference at the Long Beach YMCA's Camp Oakes in Big Bear. Don't forget to bring your telescope! (Figs. 4.23, 4.24, 4.25, 4.26, 4.27, 4.28, and 4.29)

Fig. 4.23 The RTMC attendees in the conference hall are anxiously awaiting the start of the merit award ceremony and the door prize drawings (Image credit: Tom Koonce—Astro-Tom)

Fig. 4.24 The RTMC attendees gathered outside the conference hall eagerly awaiting the start of the merit awards and door prize drawings (Image credit: Tom Koonce—Astro-Tom)

Fig. 4.25 People shop around for a good deal at the swap meet tables at the 2001 RTMC (Image credit: Tom Koonce—Astro-Tom)

Fig. 4.26 Attendees walk around the observing field and check out the vendor stands at the 2001 RTMC (Image credit: Tom Koonce—Astro-Tom)

Fig. 4.27 The annual RTMC conference is a great place to find some fantastic deals on new and used telescopes and accessories (Image credit: Tom Koonce—Astro-Tom)

Fig. 4.28 The yearly RTMC telescope maker's conference is definitely a "GOTO" event (Image credit: Tom Koonce—Astro-Tom)

Fig. 4.29 M-101, better known as the "Pin Wheel galaxy", is a relatively large face-on spiral galaxy in the constellation Ursa Major (Image credit: Sara Wager - www.swagastro.com)

Further Reading

Astro-Tom. http://www.astro-tom.com
Desvaux, Pierre. http://dobsonfactory.blogspot.com
Ketelsen, D. The Ketelsens!RTMC! http://theketelsens.blogspot.com
Long Beach YMCA, 3605 Long Beach Blvd. Suite 210, long Beach, CA 90807. http://ww.lbymca.org
Machholz, D. http://www.donmachholz.com
McCreary, D. Pedagogical equatorial platform. http://subarcsec.com
RTMC YMCA Camp Oakes, Big Bear city, CA 92314. www.astronomyexpo.org
Wikipedia Commons. http://wikipedia.org
Wager, Sara. http://www.seagastro.com

Chapter 5

Homemade Binoscopes

When it comes to designing and building your own binoscope or binocular telescope, you generally start with several ideas in mind before you actually settle on one. Sometimes it takes a while to really decide which idea or design you want to go with. It becomes almost confusing sometimes when you look at recent astronomy magazine or book on telescope making and see so many excellent examples of homemade telescopes, Dobsonian or otherwise. Trying to decide which design, type, or style of telescope or binoscope you want to build really comes down to the amount of resources you have available plus the amount of spare time you can afford to spend building it. Resources as I like to call it can have two or three meanings: one is how much money are you able to invest in building it and the tools you have available that will be needed to build it with, and last but not least, how much desire and enthusiasm do you have to start and actually finish your telescope building project. All of these are important resources and you have to weigh each with an equal amount of importance before you start your telescope building project. Once you start your telescope or binoscope building project, you don't want to stop its progress because of something you hadn't planned for earlier. Once you have decided on a particular binoscope or binocular telescope design, then comes the fun part, getting started on it.

There are some splendidly handcrafted Dobsonian binocular telescopes and binoscopes that have been built over the past many years. Each one could be a model for excellence in wood craftsmanship. Likewise, the same goes for refractor binoscopes too. If you're striving for perfection in building your binocular telescope, then I wouldn't worry too much about trying to make it the most beautifully handcrafted wooden binocular telescope ever built. Instead, you would want to just concentrate on building it correctly, so that it meets your original

N. Butler, *Building and Using Binoscopes*, The Patrick Moore Practical Astronomy Series, DOI 10.1007/978-3-319-46789-4_5

design specifications and performance expectations. One can spend a lot of time and money trying to find that perfect nut and bolt combination or an expensive piece of mahogany sheet to use for the wood construction. That's fine if you want to spend the money and time for it, but ultimately a telescope or binocular telescope will be judged by its optical performance and how user-friendly it is, no matter how nicely crafted it may appear. It's the same old "saying" that applies to homemade telescopes too: "You can never tell a book by its cover." Most of us would rather just enjoy a wonderful celestial view through a basic plain looking Dobsonian telescope than be disappointed by looking through the eyepiece of a wonderful piece of hand-crafted Dobsonian artwork (Figs. 5.1 and 5.2).

On the other hand, there are in fact some very beautiful homemade Dobsonian and Newtonian telescopes out there that are crowd pleasers just by their beautiful wooden handcrafted appearance alone. Many of them are seen at the national telescope maker's conferences and astronomy conventions and even some of them also appear as stories in the popular astronomy and telescope making magazines. These beautiful handcrafted wooden and aluminum telescopes are no doubt a joy

Fig. 5.1 (Cartoon credit: Les Lamb)

TELESCOPE MAKING 23

ISSN 0190-5570

The magazine *for*, *by*, and *about* telescope makers! Summer 1984

Lee Cain's 17.5" Binocular at the Texas Star Party

Fig. 5.2 Lee Cain is pictured above on the cover of the 1984 issue of the popular "Telescope Making" magazine standing beside his twin 17.5 binocular telescope at the Texas Star Party (Image credit: Kalmback Publishing Co.)

to use, and many of them produce views that are optically superb. So when it comes to building a telescope, binocular telescope or binoscope, it really comes down to the individual who is building it and what he/she really wants in their telescope and how they want to build it. When it comes to how nice you want your telescope to look, then it could be said, "when it comes to making your own telescope, build it for you and your high standards and make it as user-friendly as possible." When it comes to performance in a telescope, just make sure your optics are diffraction limited, and then you should be very happy with the images you view through your telescope's eyepieces. And at the same time, if you've done your homework and your homemade refractor binoscope or your binocular telescope turns out to be a beautiful work of art too, then that's a bonus and, no doubt, something you should be proud of (Figs. 5.3, 5.4, and 5.5).

Fig. 5.3 Joerg Peter's 28″ f/4.8 big binocular telescope certainly stands out in a crowd of telescopes at star parties (Image credit: Joerg Peters)

Fig. 5.4 Joerg's big 28″ binocular telescope is the current largest ATM binocular telescope around (Image credit: Joerg Peters)

Fig. 5.5 One can imagine the stunning views that Joerg's big 28″ f/4.8 binocular telescope can provide (Image credit: Joerg Peters)

There is no doubt the Joerg Peter's big 28″ f/4.8 binocular telescope is the biggest homemade binocular telescope around. Just like the big Dobsonians, ATMs just seem to be building bigger binocular telescopes these days. It's probably safe to say that a 30-in. plus binocular telescope is just around the corner. But until then, it looks like Joerg's big 28-in. binocular telescope is currently the "king" of amateur-made binoscopes.

The mechanics is made completely as an aluminum-welded construction. It is a computer-controlled Alt/Az telescope with a Dan Gray/Mel Bartels servo controls. To adjust the interpupillary distance, one cage is made rotatable. Congruence of the two light paths can be adjusted comfortably by activating motors at mirror cell while viewing through the eyepieces. The mechanical structure is very precise, so that one tube can be used as a photo machine by rotating the secondary. But for this purpose, I have to make a field derotator. This is one project for the future, and now it's time for visual observing first. The advantage of bino telescopes is getting the same amount of light into the brain as from an equivalently large mono telescope aperture at a smaller minimal magnification. A much larger celestial area is visible.

Two tubes use different seeing cells, which results in much better perception even at poor seeing.

Joerg Peters
Germany

Fig. 5.6 Bruce Sayre's 14.5″ elegant portable binoscope certainly demands attention at star parties (Image credit: Bruce Sayre)

Adapted from "The 28 Inch f/4.8 bino telescope of Joerg Peters" with permission from Joerg Peters. http://www.stathis-firstlight.de

Bruce's portable binocular telescope's 14.5-in. dual primary mirror cell support system demonstrates his advanced design planning that makes for a very stable Altazimuth mounting with a very low center of gravity (Figs. 5.6, 5.7, and 5.8).

Bruce Sayre's 22-in. giant binocular telescope was not only popular with the merit award judges; it was also a big crowd pleaser too at the 2004 RTMC with its spectacular big binocular views of the heavens (Figs. 5.9 and 5.10).

One of the important points to remember when you build a big binocular telescope or refractor binoscope is that both focusers become a critical part of the final actual optical alignment for both telescopes. If there is any mechanical "slop" or excess movement in the focusers, where the eyepiece and/or the eyepiece focusing tube "wobbles" for even a few thousands of an inch when you're trying to focus, that could potentially throw off your original merged image(s), even though the primary and secondary mirrors are still in collimation and alignment. Enough can't be said about using precision focusers for your binoscope or binocular

Fig. 5.7 Here's another closer look at Bruce Sayre's portable binocular telescope's 14.5-in. dual primary mirror cell support system (Image credit: Bruce Sayre)

Fig. 5.8 Bruce Sayre's very open truss 22-in. giant binocular telescope won a merit award at the 2004 RTMC (Image credit: Robert Stephens RTMC)

Fig. 5.9 (Cartoon credit: Jack Kramer)

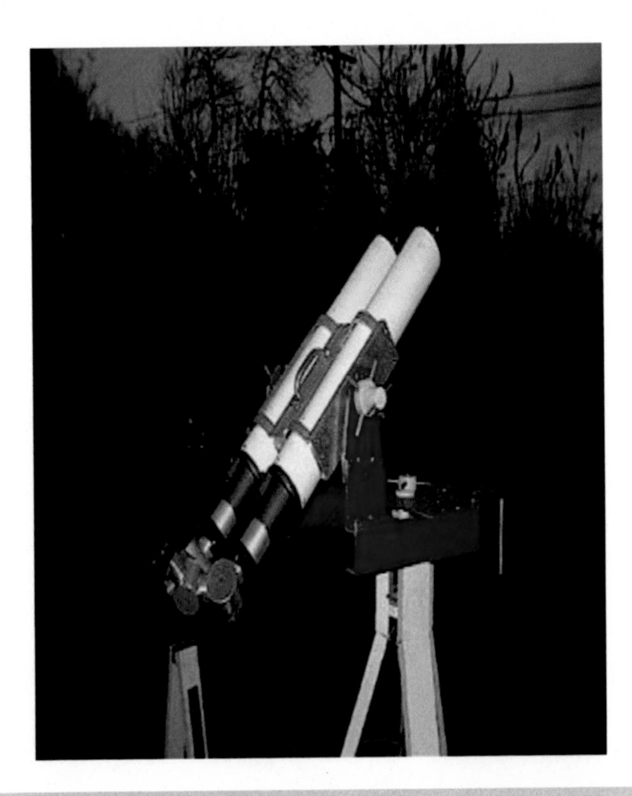

Fig. 5.10 Dave Trott made this nice looking 4-in. f/10 refractor binoscope combined with a pair of homemade binobacks on a simple Altazimuth mount (Image credit: Dave Trott)

telescope's focusing system. One needs to make sure that once they have either purchased or even made a pair of focusers for use in their new binocular telescope or refractor binoscope, there is "no" mechanical slop in the rack and pinion gear system or slide mechanism of the focusers. It's also very important to clean and maintain your precision focusers when their movement starts to feel rough or sloppy. One of the highly recommended lubricants on the market for lubricating your focusers is made by DuPont. It's called "Krytox EG2000." It's a bit pricey, but it works fine for rack and pinion focusing mechanisms or even sled/slide focusers that need a good thick grease to keep their precision movement. "Net Mercury, Inc." out of Dallas, Texas, sells it. Their website is Netmercury.net.

Judging from the photo in Fig. 5.11, it looks like that splendid looking observing chair sitting directly underneath his homemade 13.1-in. binocular telescope could also be some of Dave's handy work (Fig. 5.12).

Jim Burr of JMI offers reverse binocular telescopes from 6 to 16 in. It looks like Jim has just about thought of everything to make his reverse binoculars complete with Crayford focusers, very user-friendly, mobile, and transportable. Caster wheels on the back of each of the support legs make it easy to roll the big binocular telescope to its desired observing location with ease (Fig. 5.13).

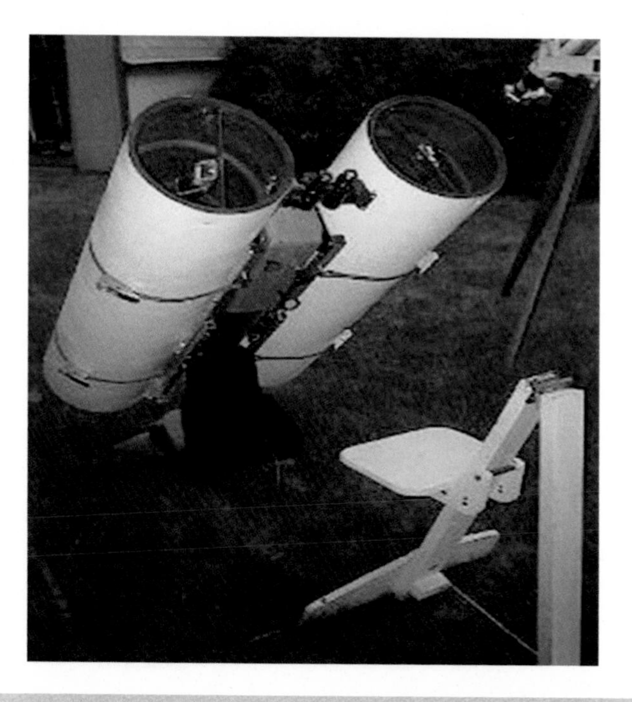

Fig. 5.11 Here's another example of Dave's handy work, a 13.1-in. Sonotube binocular telescope (Image credit: Dave Trott)

Fig. 5.12 Jim Burr of JMI offers a line of reverse binocular telescopes from 6 to 16 in. (Image credit: Jim Burr—Jim's Mobile, Inc.)

Fig. 5.13 Ian Gibbs has made an interesting dual 120 mm f/7.5 Equinox Sky Watcher ED refractor binoscope arrangement on a Altazimuth mounting. (Image credit: Ian Gibbs)

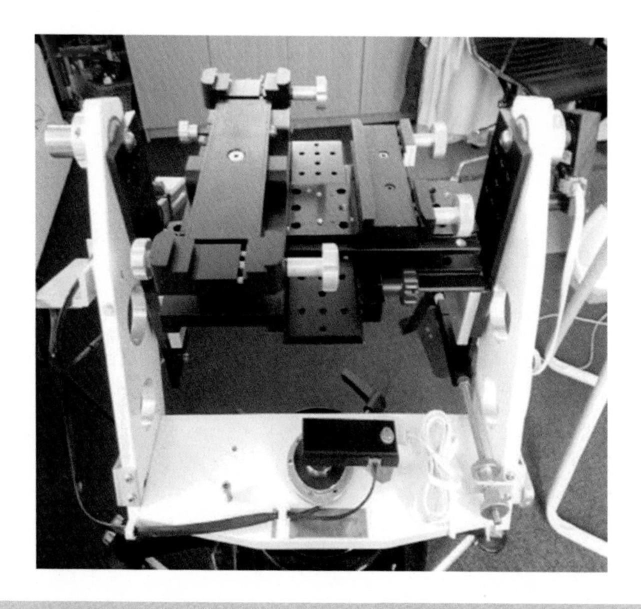

Fig. 5.14 Ian Gibbs' clever binoscope Altazimuth mounting arrangement for his dual 120-mm f/7.5 Sky Watcher ED refractors (Image credit: Ian Gibbs)

Ian Gibbs of New Zealand has made an interesting dual 120-mm f/7.5 ED binoscope setup powered by a pair 2-in. binobacks. It looks like one could do some serious astronomical observing with this system (Fig. 5.14).

Ian Gibbs' binoscope mounting arrangement for his dual 120-mm f/7.5 Sky Watcher ED refractors is obviously designed for maximum adjustment, support, and stability with his ADM Max-Guider (right-hand scope) with a Losmandy dovetail bar/clamp combination, plus a short Vixen-style dovetail mount and clamp for the left-hand scope for IPD adjustment (Fig. 5.13 and 5.14).

Pierre Schwarr was a popular Arizona amateur telescope maker who made some very unique merit award-winning telescopes in the 1970s, 1980s, and 1990s, some of which are still around today. One of his more unique projects was a dual 8-in. f/4 Newtonian DC-motorized binocular chair operated with a DC motor and a pair of car batteries that provided the power to drive the observer to different parts of the sky with just a touch of a button.

Pierre Schwarr was a clever innovative telescope maker and early pioneer in building unique motorized binocular chairs. Pierre's remarkable motorized binocular telescope chair in Fig. 5.15 and Fig. 5.16 was ahead of its time and his unique motorized binocular chair still sets a standard for amateurs to try and achieve even today.

Pierre Schwarr seated at the controls of his remarkable homemade dual 8-in. motorized binocular telescope really stood out in the crowd at the 1991 RTMC (see Fig. 5.17). One can imagine how nice it must have been to sit in

Fig. 5.15 Pierre Schwarr made this motorized with two homemade 8-in. f/4 binocular telescopes mounted on top (Image credit: Steve Dodder)

Pierre's motorized binocular chair and, with just a push of the button, move about the heavens effortlessly in complete comfort, observing each wonderful celestial object through a pair of his homemade 8-in. f/4 reflecting telescopes (Fig. 5.18).

Pierre Schwarr and Wayne Schmidt's original motorized binocular telescope chairs that were built in the early 1990s set the standard for future amateur telescope makers to try and emulate. Even today, amateur-made and commercially manufactured motorized binocular chairs are rare to see at star parties and at national telescope maker's conferences (Figs. 5.15, 5.16, 5.17, and 5.18). However, commercial motorized chairs like the StarChair3000™ are becoming real front runners for the first versions of commercial motorized binocular chairs. We also need to give some credit to Pierre Schwarr and Wayne Schmidt for their early and excellent motorized binocular telescope chairs and the groundwork they've set for ATMs to start building their own versions of a motorized or even GOTO binocular chairs (Figs. 3.28–3.37).

Fig. 5.16 Pierre Schwarr built this motorized 8-in. binocular telescope chair back in the 1990s and its innovative uniqueness is still talked about today (Image credit: Steve Dodder)

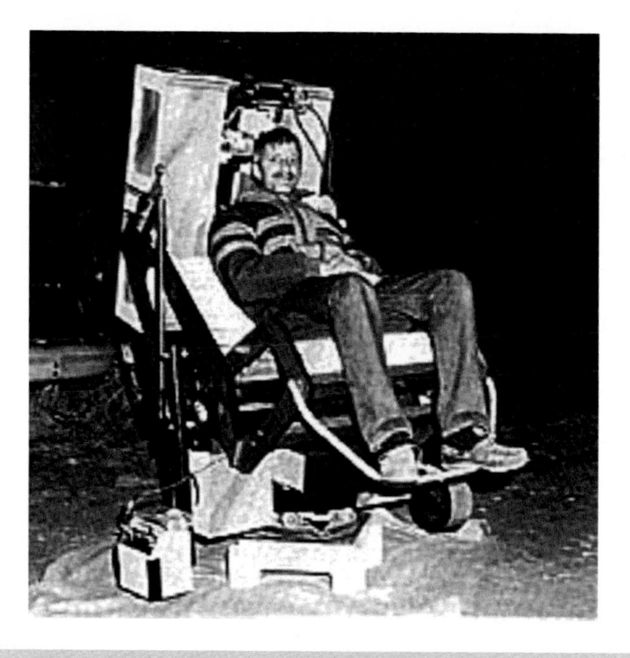

Fig. 5.17 Pierre Schwarr shown here at the 1991 RTMC was another early pioneer in homemade motorized binocular chairs (Image credit: Robert Stephens RTMC)

Fig. 5.18 Pierre Schwarr and his dual 8-in. f/4 Newtonian binocular telescope motorized chair (Image credit: Robert Stephens RTMC)

Swivel Bearings

If you intend to build your own binocular telescope, then one of the first important areas you will want to look before you start on your binocular telescope project is your Altazimuth mount bearing. The bearing must be able to handle a big load. It's always under a lot of stress and pressure. The bearing needs to be able to support the entire weight of your binocular telescope without binding while providing a smooth 360° rotational movement. It's important to choose the right swivel bearing with the right load capacity in order for your big binocular telescope to rotate smoothly. When you have determined the entire overall load weight of the Altazimuth mount combined with the two big Dobsonian telescopes, then carefully research what kind and/or type of bearing you want to use for the mount's rotational base. One of your choices to consider would be using a Lazy Susan bearing. For example, a 12″ flat round thick capacity Lazy Susan bearing shown in Fig. 5.32 can be rated for a load capacity up to a 1000 lb or more. It's always recommended to use a bearing that has a higher load capacity than the total weight of your entire binocular telescope (Figs. 5.19, 5.20, 5.21, 5.22, 5.23, 5.24, 5.25, 5.26, 5.27, 5.28, 5.29, 5.30, 5.31, and 5.32).

Fig. 5.19 A dual 200 mm (7.87 in.) nicely machined open tube binocular telescope (Image credit: David Ratledge—The Bolton Group—www.deep-sky.co.uk)

Another bearing configuration to consider would be a thick capacity Lazy Susan bearing (see Fig. 5.33) that is rated at 220 lb capacity. The author also recommends using three nylon furniture sliders of an appropriate diameter and thickness (nail type or adhesive) positioned on the wooden base at 120° intervals around the Lazy Susan bearing to help support the overall weight of the binocular telescope (especially if it's a big one). The furniture sliders provide just a little friction when the entire binocular telescope is rotated. The type of Lazy Susan bearing shown in the photo below ranges in diameters up to 40 in. with a maximum load capacity of 500 lb (Figs. 5.32 and 5.33).

Fig. 5.20 Cartoon credit: Jack Kramer

Fig. 5.21 Ulrich Vedder of Germany has built an all-wooden beautiful 10-in. f/5 open tube truss binocular telescope (Image credit: Ulrich Vedder)

Fig. 5.22 The rear section of the Ulrich's 10-in. f/5 binocular telescope showing the dual mirror cell support system and large wooden Altazimuth mount base (Image credit: Ulrich Vedder)

Fig. 5.23 Ulrich Vedder's beautifully machined dual eyepiece focuser system complete with two big Televue eyepieces (Image credit: Ulrich Vedder)

Fig. 5.24 (Cartoon credit: Jack Kramer)

Fig. 5.25 David Moorhouse of New Zealand proudly shows off his elegant homemade wooden 16-in. truss tube binocular telescope (Image credit: David Moorhouse)

Fig. 5.26 Here's another of Dave's binoscope projects: an 8-in. all-wooden "Baby Bino". This one looks like it would be a lot of fun to look through and easy to transport too (Image credit: David Moorhouse)

Fig. 5.27 Borg/Hutech offers this splendid looking 125SD (Haruka) model refractor binoscope on an Altazimuth mounting accompanied by a pair of impressive looking binobacks (Image credit: Hutech Astronomical Products)

Fig. 5.28 (Cartoon credit: Jack Kramer)

Fig. 5.29 Hutech offers a nice equatorial AOK platform that is mounted axially between the forks, providing the observer a way to adjust for good head and eye position as the binoscope moves equatorially across the night sky (Image credit: Borg/Hutech)

Fig. 5.30 (Cartoon credit: the Author)

Fig. 5.31 A front view of APM's big custom looking 12-in. f/7.5 triplet APO refractor binoscope on it's computer controlled mount (Image Credit: Marcus Ludes—www.apm-telescopes.de)

Fig. 5.32 A flat round thick capacity Lazy Susan bearing (Image credit: the Author)

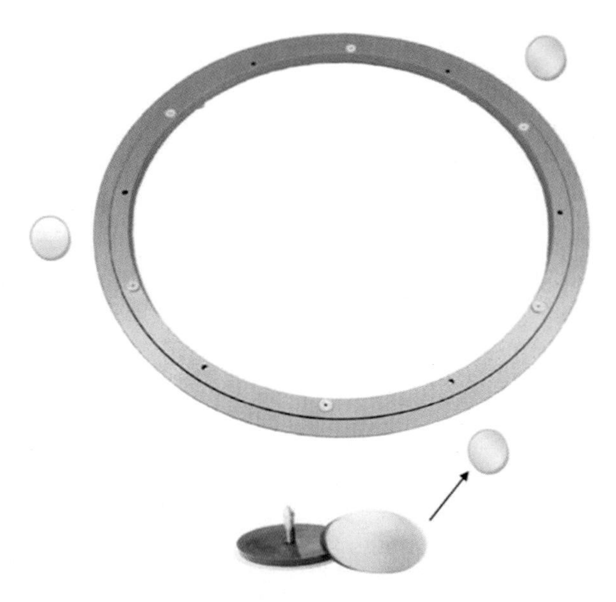

Fig. 5.33 Thick capacity swivel bearing in combination with three nylon furniture sliders (Image credit: the Author)

Optical Alignment of a 16-in. Binocular Telescope

Any telescope can be broken down into two main essential elements. Firstly, the optics, there are multiple different optics designs that are suitable for use in astronomical telescopes. The one that I have chosen to use is just a slight variation on the standard Newtonian design. The variation is that in a binocular Newtonian, you need the third mirror to fold the two separate light cones so they are once again parallel and able to be fed to two eyepieces to your eyes. The second major element in the telescope is the structure supporting the optics. This needs to fulfill a large number of often conflicting requirements. Its main job however is to allow all the optics to be held perfectly in alignment with each other while being able to be pointed at any part of the sky and often to also track the sky simultaneously. This is no mean task as often the structure flexes and bends under its own weight, putting the optics out of alignment. In the binocular telescope, the alignment of the optics is much more critical than with a standard monocular telescope. This more stringent requirement is mainly brought about by the need to have the optics aligned so that the eyes and the brain are able to merge the two separate images that are being presented to them. There have been two main schools of thought as to how to approach this problem. The first approach has been to simply get two separate optical assemblies that are both capable of having their individual optics collimated separately and then using a system of hinges and sliding platforms to maneuver the two independent tubes so that the merged image is presented at the eyepiece. The main difficulty in this type of construction is the complexity of the mechanisms needed to independently move while still holding rigid the two separate tube assemblies (Fig. 5.38).

The second approach, the one that was taken, is to have the top end of the telescope and the bottom end completely rigid structures. The top end is two rings joined together and the mirror cell is one solid unit. This makes the whole structure move as one and any sag in any part is exactly the same on the other side, keeping the optics aligned to each other. In order to make the optic paths parallel, the primary mirrors are able to move in X-Y direction. They are also held firmly in place by piano wire edge supports rather than that traditional sling system. I am now convinced of the merits of this style of design compared with the independent tube idea. The entire structure is a rigid, the movement is smooth, its rock solid with no vibration and easy for a number of users to view one after another with little or no adjustment. You do need a cool methodical approach to collimating this instrument, but once it is achieved, it is a pleasure to use. So how do you get the two images to merge? Radio control provided the answer. Once both halves of the binocular telescope are aligned with lasers, the next step is to look through the instrument for a bright star in the field of view. Chances are what you will see is two bright dots fairly well separated. So what do you do next? I use fine tweaks of one of the primary mirrors. When we first started, we used to do it with two people, one looking while the other adjusting; this is a real pain. A radio control plane servo system was added to the collimation bolts of the left-hand mirror; this made the adjustment easy to do yourself while viewing. The adjustment needs to happen with different viewing altitudes as the focuser boards, and the secondary holders I think must flex a minute amount. This idea works great and it's easy to merge the images. I usually start by purposely vertically misaligning the images and then getting the horizontal images correct and then rejoining the stars vertically (Fig. 5.39).

David Moorhouse
New Zealand

Adapted from "Dave's 16-Inch binocular Telescope Page" with permission from David Moorhouse (http://www.binoscope.com.nz)

When choosing the size and type of Lazy Susan bearing to use on your homemade Dobsonian mount, choose one that will provide the best results based on the following factors:

1. The overall diameter of the bearing (6–40 in.)
2. Load capacity – can it support 180 pounds or more?
3. The thickness (3/8–3/4 in.)
4. Type of bearing to use (Ball or radial bearing, roller skate, wheels, etc.)
5. How smooth is rotational motion? (Smooth, binds a little or sloppy)
6. Is it easy to attach? (Screws, bolts, etc.)

Once you have chosen an Azimuth bearing type to use on your Dobsonian mount, then you have more or less "bought the boat". After mounting it on the Dobsonian base, if it rotates smoothly without binding and there are no areas with unwanted friction during its rotational movement, then you choose wisely. If not, then you will probably have to re-evaluate and perhaps invest in another bearing to avoid future problem.

There are other bearing options besides the Lazy Susan bearings. The author has actually used skateboard wheels (Fig. 5.34) on the base of his homemade Dobsonian mount. A series of three dual skateboard wheels were mounted 120° apart on a Dobsonian mount. They provide a very smooth rotational movement for the entire Dobsonian telescope, and they have a very large load capacity too (Fig. 5.33). A ball bearing skateboard wheel has a long life span and good durability. So skateboard wheels are a good choice for the creative ATM to consider when looking for a bearing that can "handle the load" and provide a very smooth rotation (Fig. 5.35, 5.36, 5.37, 5.38, 5.39, 5.40, 5.41, 5.42, 5.43, 5.44, 5.45, 5.46, 5.47, 5.48, 5.49, 5.50, 5.51, 5.52, 5.53, 5.54, 5.55, 5.56, 5.57, 5.58, 5.59, and 5.60).

Fig. 5.34 Skate board and roller skate wheels are available commercially and can be used on your Dobsonian mount if you plan to build your own binocular telescope (Image credit: Public Domain—Wikimedia Commons)

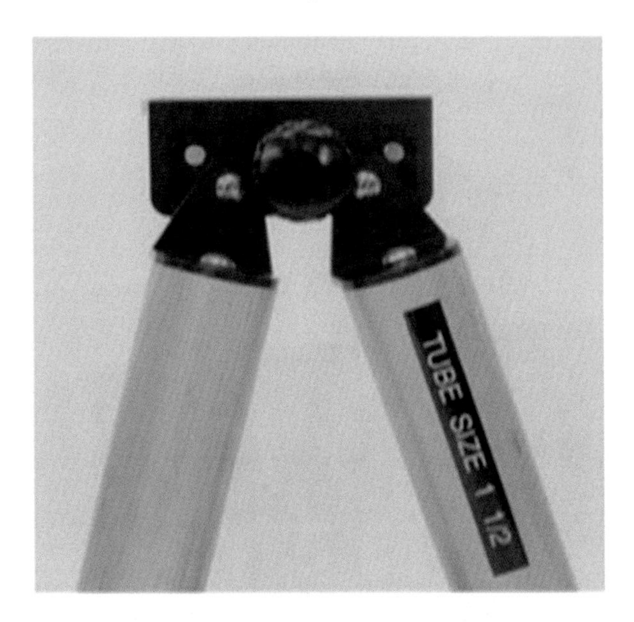

Fig. 5.35 Tube clamps, tube connectors, truss tubes, brackets, mirror cells, and various assortment of parts are commercially available if you plan to build your own truss tube binocular telescope (image credit: Aurora Precision)

Fig. 5.36 Tube clamps, tube connectors, truss tubes, brackets, mirror cells, and various assortment of parts are commercially available if you plan to build your own truss tube binocular telescope (Image credit: Aurora Precision)

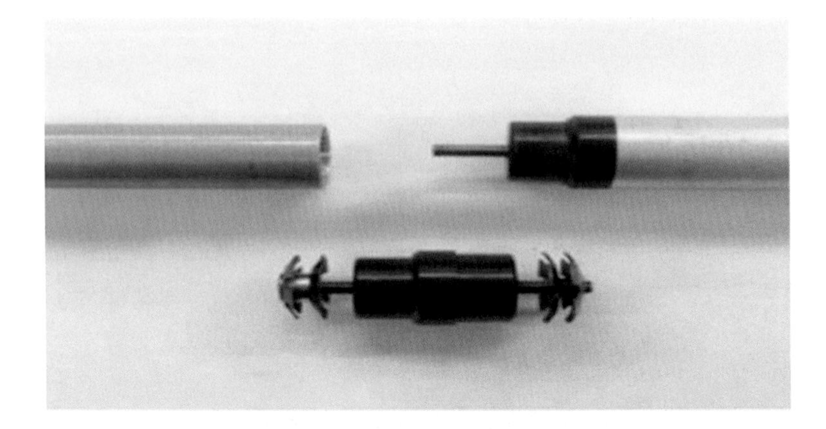

Fig. 5.37 Tube clamps, tube connectors, truss tubes, brackets, mirror cells, and various assortment of parts are commercially available if you plan to build your own truss tube binocular telescope (Image credit: Aurora Precision)

Fig. 5.38 The dual 16-in. mirror cells (Image credit: David Moorhouse)

Fig. 5.39 Close-up look at Dave's radio control servo system located on the right side of his 16-in. dual mirror cell system (Image credit: David Moorhouse)

Fig. 5.40 (Cartoon credit: the Author)

Fig. 5.41 An all pipe mount binocular stand. Sold as a kit called the "EZ Binoc Supermount" (Image credit: www.petersonengineering.com)

Eric Royer comments on the "First Light" views through his 14-in. F/4.6 binocular telescope (Figs. 5.61, 5.62, and 5.63):

> I finally had first light under a black sky, near Lake Pavin, Auvergne, France. The images were breathtaking. The improvement compared to the 6″ binoscope was enormous. I could see the spiral arms in M33 like never before, along with some HII regions. The globular cluster M15 was full of tiny stars. I don't have enough words to describe the veil nebula. Since then, I've used it to observe and sketch deep-sky objects, comets, and planets, with magnifications ranging from 70× to 500×. At star parties, people are always impressed by the clarity of the views through a large binocular telescope.

Fig. 5.42 A dual 300 mm (11.8 in.) binocular telescope with a splendidly machined skeleton tube system (Image credit: David Ratledge—The Bolton Group—www.deep-sky.co.uk/telescopemaking.htm)

Here is Eric Royer's initial impression of using his first homemade 6″ binocular telescope.

In 2003, I built a small 6 in. binocular telescope. At the time, I had never used a binoscope or even a binoviewer. I was not sure that I would be able to successfully align the two optical assemblies and get a single image. I only hoped to be able to use this instrument at low power for wide field observing. For First Light, there was a conjunction between Jupiter and M44. I had both in the same field of view, and there was definitely a sense of depth with Jupiter and M44. I had both in the same field of view, and there was definitely a sense of depth with Jupiter appearing closer than the star cluster. In fact, when I started to use it, I realized that merging images

Fig. 5.43 Jim Carlisle made this very unique patented homemade all-wood 10-in. Herschelian binocular telescope on an Altazimuth mounting that won a merit award at the 1992 RTMC. This is a splendidly crafted example of a homemade Herschelian binocular telescope (Image credit: Jim Carlisle)

in a binocular telescope was not difficult at all, even at 200× magnification. The only trick is to be able to adjust the merging while watching through the eyepieces. From this time on, I didn't use my 8 in. reflector for visual observing anymore.

Reproduced and adapted with the kind permission from Eric Royer (http://www.astrosurf.com/eroyer/)

The Amsterdam instrument maker Jan van Deijl, together with his son Harmanus, were quite busy in the production of achromatic microscope objectives towards the end of the eighteenth century. They spent most of their time trying to fulfill the increasing demand for achromatic telescope lenses. The splendid looking binocular telescope pictured in Fig. 5.64 is an excellent example of their innovative and masterful skills. By the end of the eighteenth century, Harmanus van Deijl

Fig. 5.44 Danny Strong of the UK has made this very impressive 16.1-in. f/4.7 Goliath of a binocular telescope (Image credit: Danny Strong)

successfully applied the concept of John Dollond's achromatic telescope lens to the small lenses of compound microscopes. This tackled the annoying colored edges seen by observers, but spherical aberration remained a problem with glass microscope lenses until Joseph Jackson Lister's discoveries in 1830.

The workshop where Jan van Deijl (1715–1801) and his son Harmanus (1738–1809) made their optical instruments was quite well known in Amsterdam during the eighteenth century. The van Deijls were known to make telescopes of very good optical quality. Harmanus Deijl claimed they had made some real improvements, such as the achromatic lens, which led to important developments in the optical quality of microscopes.

Fig. 5.45 Call it "Hexy". Here's a rear view of the big 16.1-in. f/4.7 binocular telescope (Image credit: Danny Strong)

After measurements were made on some of the existing instruments in the collection of the Utrecht University Museum and on one telescope in particular that is in a private collection. The measurements have shown that these claims were in fact, valid and that the Van Deijls had the skill and ability to work within somewhat narrow mechanical restrictions and in doing so, made appropriate changes and corrections for variations in the optical qualities of typical eighteenth-century glass that was commonly used at that time. The van Deijls' major contribution was the introduction of an achromatic lens in microscopes and the placement of an achromatic objective lens in telescopes.

Fig. 5.46 Arie Otte of the Netherlands has made a very nice, stable looking tube truss 13-in. f/5 binocular telescope on an Altazimuth mounting (Image credit: Arie Otte)

Fig. 5.47 Looks like Arie has built another fine binocular telescope with an even larger primary mirror. Looks like it could even be a GOTO model (Image credit: Arie Otte)

Fig. 5.48 Vixen optics has produced a very fine looking new 80-mm binoscope on an Altazimuth mounting. The 80-mm binoscope comes with two 25-mm eyepieces and, from the looks of this setup, will make any amateur astronomer take notice (Image credit: Vixen Optics)

Fig. 5.49 (Cartoon credit: Jack Kramer)

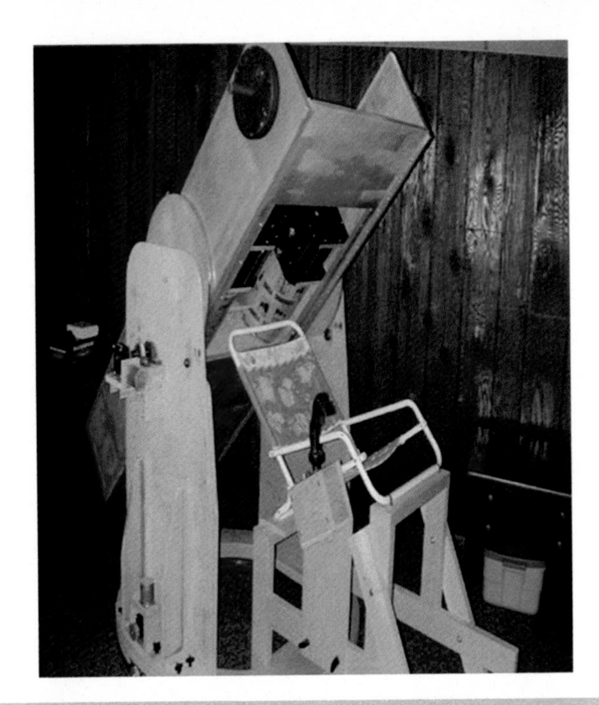

Fig. 5.50 Ray Jurevicius designed and built this splendid looking 8-in. f/6 all-wood binocular telescope that has a lawn chair mounted on top of the bottom section of the Altazimuth mount (Image credit: Ray Jurevicius)

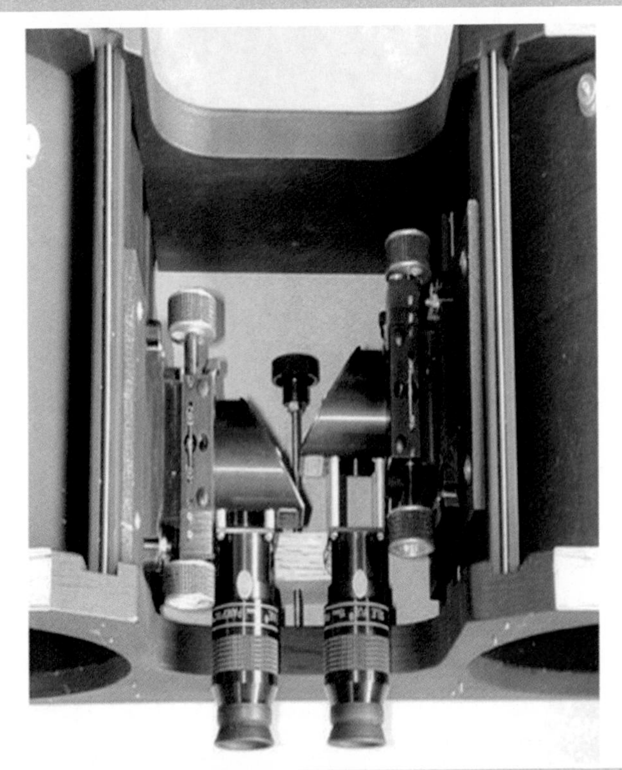

Fig. 5.51 Check out the dual offset eyepiece reverse Crawford focusing arrangement that Ray has made for his binocular telescope, which indicates the primary mirror on the right telescope is also offset by approximately the same distance (Image credit: Ray Jurevicius)

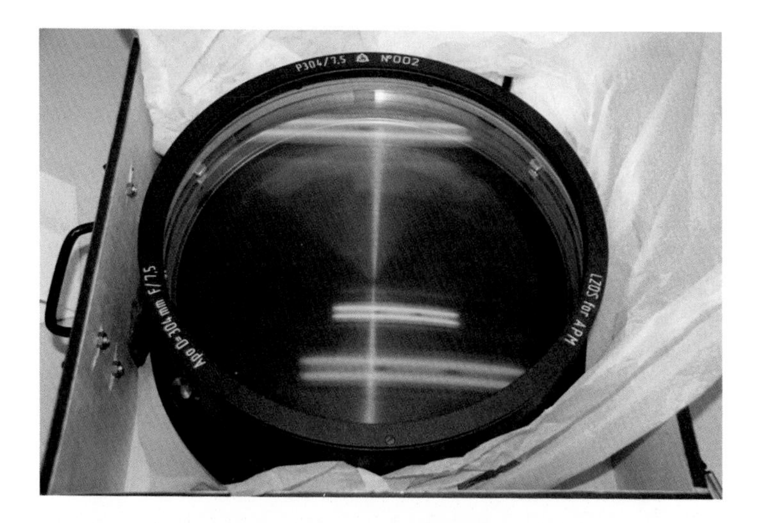

Fig. 5.52 Here's one big beautiful 12-in. F/7.5 APO objective lens from APM in Germany that anyone would love to have mounted in their big refractor telescope and their binoscope too (Image credit: www.apm-telescopes.de)

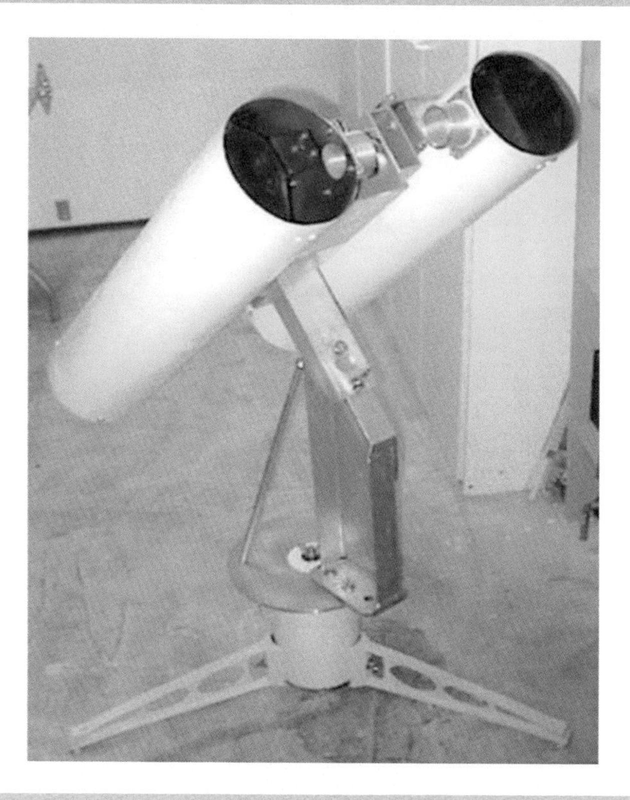

Fig. 5.53 Dave Trout built this splendid looking all-aluminum 6-in. binocular telescope on a very stable looking Altazimuth mounting (Image credit: Dave Trott)

Fig. 5.54 Here is the lower section view of Dave Trott's all-aluminum 6-in. binocular telescope showing the binocular telescope tube support system along with the Altazimuth mounting with the pier and tripod legs (Image credit: Dave Trott)

Fig. 5.55 Here's a photo showing Dave seated at the controls of his all-aluminum 6-in. binocular telescope (Image credit: Dave Trott)

Fig. 5.56 Here is Doug Chapman's homemade version of an open tube truss splendidly crafted 12.5-in. f/6 binocular telescope that utilizes a cutaway portion of a 55-gal oil drum to house his two primary mirrors and move it about in the vertical axis on a Altazimuth mount. Now that is a very clever and innovative use of an oil drum (Image credit: Doug Chapman)

Fig. 5.57 This photo shows the cutaway section of the 55-gal oil drum on the rotating Altazimuth base that makes up the primary mirror cell support for Doug's 12.5-in. binocular telescope (Image credit: Doug Chapman)

Fig. 5.58 An added treat about Doug Chapman's astronomy outreach program with his unique 12.5-in. binocular telescope allows two people (Bill and Olivia here) to view through each eyepiece at the same time (Image credit: Doug Chapman)

Fig. 5.59 A big pair of Australian navy ship binoculars from the HMAS Melbourne called the "Big Eyes" (Image credit: Public Domain—bg.wikipedia.org)

Fig. 5.60 (Cartoon credit: the Author)

Fig. 5.61 Here's a photo of Eric Royer's big homemade 14-in. f/4.6 binocular telescope. Notice the rectangular box-shaped primary mirror section (Image credit: Eric Royer)

Fig. 5.62 Another view of Eric Royer's splendidly crafted open tube truss binocular telescope. Note the secondaries and their support system along with his dual reverse focuser arrangement mounted in a secure fashion in the solid dual wooden rings (Image credit: Eric Royer)

Fig. 5.63 The rear section of Eric's big binocular telescope showing the 14-in. f/4.6 aluminum primary mirror cell support system, a very interesting mirror cell design indeed (Image credit: Eric Royer)

Fig 5.64 A Hans and Harmanus Jan van Deijl binocular telescope—Circa: 1789 (Image credit: Commons, Wikimedia.org)

Shingletown Shootout

On October 24–25, 2003, near Mt. Lassen in a test conducted at the Shingletown air strip, I was able to compare the 22″ binocular (Fig. 5.65) to Dan Gray's innovative new 28″ string telescope. We were joined by Shneor Sherman and Mel Bartels, both experienced observers. Our benchmark objects included the NGC 281 edge-on galaxy, the Andromeda galaxy, the Perseus double cluster, and the Great Orion Nebula. Although the seeing was poor and Dan's temporarily silvered mirror was slightly undercorrected and astigmatic, the consensus among the group was:

1. In general, views were essentially the same
2. The 28″ showed slightly brighter star images.
3. Contrast and brightness of galaxies and their dust lanes were similar.
4. Nebulosity in M42 was equally impressive, and the pink color in M42 was easy to detect in both telescopes.

Fig. 5.65 Bruce Sayre's 22-in. splendid looking very open trust binocular telescope. See his personal comments on its performance from website (http://brucesayre.net) (Image credit: Bruce Sayre)

5. The brain much prefers the comfort of binocular viewing—the sense of presence, the silkiness of faint, distended objects, the ability to concentrate, and the removal of "background noise" by the brain were all noted as improvements due to the binocular view.
6. Low power 95×, 80° field of views were unique to the 22″ binocular.

Adapted from "An amateur binocular maker's Web site" with the kind permission

From Bruce Sayre (http://brucesayre.net)

Author's Note:

If all things are considered and we compare the two telescopes in terms of light gathering power, then Mr. Sayre's 22-in. binocular telescope wins hands down with a "whopping" 968 square inches compared to Mr. Dan Gray's 28-in. "String" telescope, which produced nearly the same views that tipped the scales at 784 square inches. If the 28-in. "String" telescope were to be equal in light gathering power, it would have had to have a primary mirror diameter of approximately 31.113 in. But equal or not, a big binocular telescope is hard to beat when it comes to comfortable celestial observing using two eyes as pointed out in item number six in the comparison.

A 12.5 in. Binocular Telescope

When you start grinding mirrors, you end up with more mirrors than you know what to do with. After I made the mirror for my 12.5″ trackball, I picked up a second mirror of nearly the same focal length at a swap meet. I joked about making a big binocular scope, then put away the mirror and worked on other projects for most of a year. From what I'd heard, a binocular scope was way too difficult for someone of my skill level. Conventional wisdom held that they were fussy, temperamental things that would make your hair fall out and increase the national debt.

But Frank Szczepanski, a telescope maker in my hometown of Eugene, Oregon, built an impressive 8-in. binocular, and he said it was relatively easy to make and to use. He later made another one out of a pair of 4″ mirrors he got from Surplus Shed. He built it in a week.

Frank has the talent to put together a scope in one afternoon, and his success encouraged me to give it a try. Firstly, I needed to make both mirrors the same focal length, so I shortened the second one's focal length until it was within 3/4″ of the first. I could have gotten it closer, but I naively thought it would shorten by in the processes of polishing and parabolizing. This did not happen. They wound up 57.9″ for the first one (f/4.63) and 58.6″ for the second one (f/4.69).

I then started thinking about the design of the scope. It was really tough to keep the scope rigid enough that both optical trains would point at the same spot in the sky from zenith to horizon. The designs I had seen were generally two types: two telescopes strapped together side-by-side or two optical trains in a framework that looked a little like a truncated Eiffel Tower. After reconsidering my plan to come up with something completely new, I went with a more conventional design. I used two large boxes stacked on top of each other and connected them with eight trusses, two to a side, resulting in *eight* triangles to provide the necessary stability.

In order to align both optical paths, I would need to be able to adjust the aim of at least one, if not both, optical trains. The most reasonable advice that I could find on the internet was that I should make the primary mirror mounts horizontally adjustable and tiltable. Horizontal motion combined with a little tilt to both the primary and the secondary would move the optical train's aim while keeping the mirrors collimated. In order to do that I had to make floating mirror cells. I came up with all sorts of Rube-Goldberg ideas, but finally settled on a simple design: make the collimation screws run all the way through the mirror platform to rest on the bottom of the mirror box, and use springs to pull the mirror platforms down so they wouldn't fall out when the scope was tilted toward the horizon. Adding lateral motion then became as easy as putting in four screws, two on each mirror that could push the mirrors sideways. They double as the mirrors' side supports, which should be 90° apart from one another anyway. I can move each mirror at least an inch or so in any direction and even more if necessary.

The mirrors are full-thickness Pyrex, so I only need three points of support. That means the "cells" are just plywood triangles with cork pads for the mirrors to rest on. I have clamps that reach up over the top to keep them from falling out. I put metal plates underneath the collimation screws so they wouldn't dig into the bottom of the box. The metal was shiny, so I covered it with construction paper disks. They don't look very black in photos, but they're darker at night.

The collimation adjustment rods are just all-thread. They stick up a ways so I can reach them easily when I'm merging the images, which I will discuss later. I made the right ones taller than the left because I only use one mirror to merge the images. I leave the left one alone so that optical train stays collimated and becomes the reference train for the other one. If I didn't set it up this way, I could wind up moving both optical paths out of collimation and chasing a phantom sweet spot on a bad merging day.

I still had one major decision: How was I going to change the interpupillary distance? People's eyes aren't all the same distance apart, so you need to be able to move the eyepieces closer together or farther apart to account for that. With hand-held binoculars, you just twist them together or apart, but I figured that might be a little difficult to do with two 12.5″ scopes. That left me with two choices: move the eyepieces sideways with focusers or rotate the secondary cages.

Using focusers had its appeal. For one, I could bolt down the secondary cages and keep my rigid framework all the way to the top. But the more I thought about it, the more complicated it seemed. It would require four focusers, two to move the eyepieces sideways and two more aimed vertically to actually focus the images. Worse; every time you changed the interpupillary distance, you would have to refocus. And you would have to refocus a long ways. Plus, all four focusers would have to be rigid as rocks. Any flexibility there would be just as bad as flex in a moveable secondary cage.

Rotating the secondary cages began to look like the better option. I had already come up with a design that I liked for my 12.5″ trackball, and one of its features was a nice round bottom ring. I had made that one out of lightweight foam to save weight, but I wasn't worried about weight for this scope so I decided to make its cages out of plywood. I used 1/2″ Baltic Birch for the rings and 1/8″ doorskin for the panels. I cut the rings with a router on an arm pinned to the center, which made them nearly perfect circles (Figs. 5.66, 5.67, 5.68, 5.69, 5.70, and 5.71).

The secondary cages are held onto the top platform with only four bolts each. The bolts have nylon spacers between the wide washers on top and the platform, so the bolts don't pinch the bottom rings of the cages. The cages rest on Delrin pads so they can slide smoothly. There's a thousandth of an inch or so of play to the cages, but that barely affects collimation or image registration. Your eye compensates way more than they move.

The wire spiders give me another neat advantage: I can move the secondary mirrors around in the cages just like I can move the primaries around in their box. I can move them up and down, sideways, and change their tilt just by tightening and loosening various wires. I also have tilt adjustments at the center hub so I don't have to mess up the other adjustments for simple changes.

Note the flattened faces on the top rings just above the focusers (Fig. 5.72). This is forehead space. I wanted to keep the light path between secondary mirrors and

Fig. 5.66 (Cartoon credit: the Author)

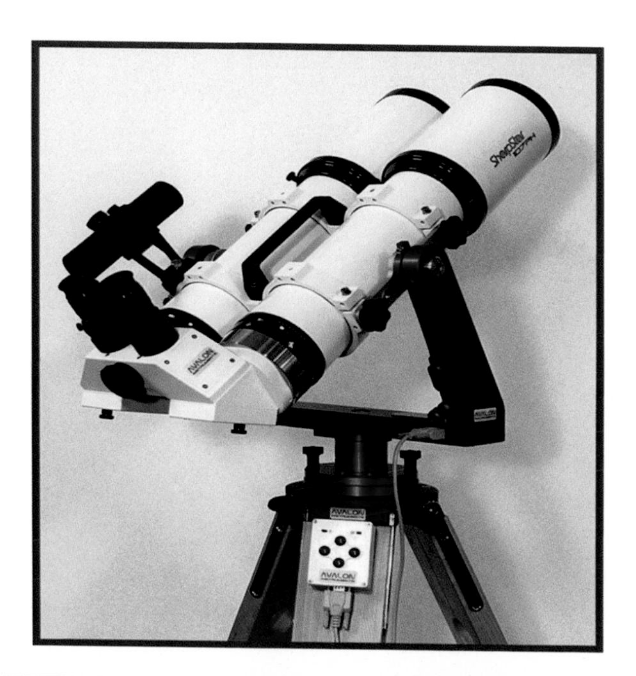

Fig. 5.67 A fine example of a commercial 107 mm APO refractor Binoscope made by Avalon Instruments in Italy (Image credit: www.avalon-instruments.com)

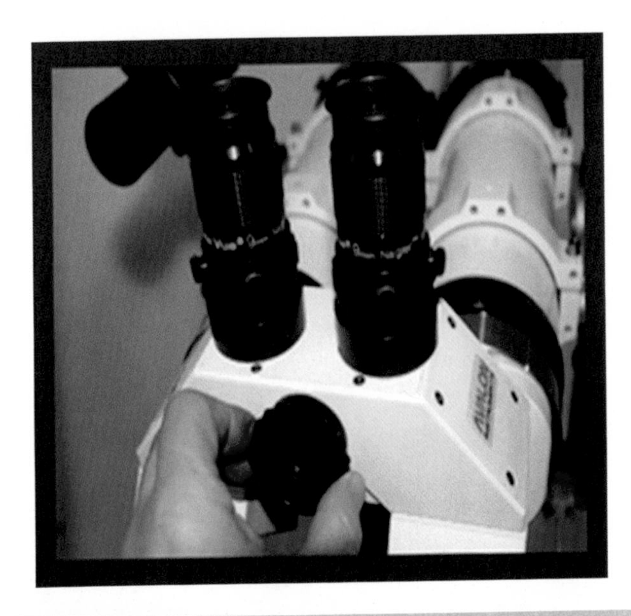

Fig. 5.68 The IPD adjustment is easily made with just a twist of the knob. Eyepieces are independently focused (Image credit: www.avalon-instruments.com)

Fig. 5.69 The "King" of the Binoscopes! The LBT (Large Binocular Telescope) with its twin 8.4 M mirrors sits on Mt. Graham in Arizona. That is one big binocular telescope! (Image credit: NASA—Wikimedia Commons)

Fig. 5.70 Jerry Oltion with his homemade dual 12.5 in. f/4.6 Dobsonian style binocular telescope (Image credit: Jerry Oltion)

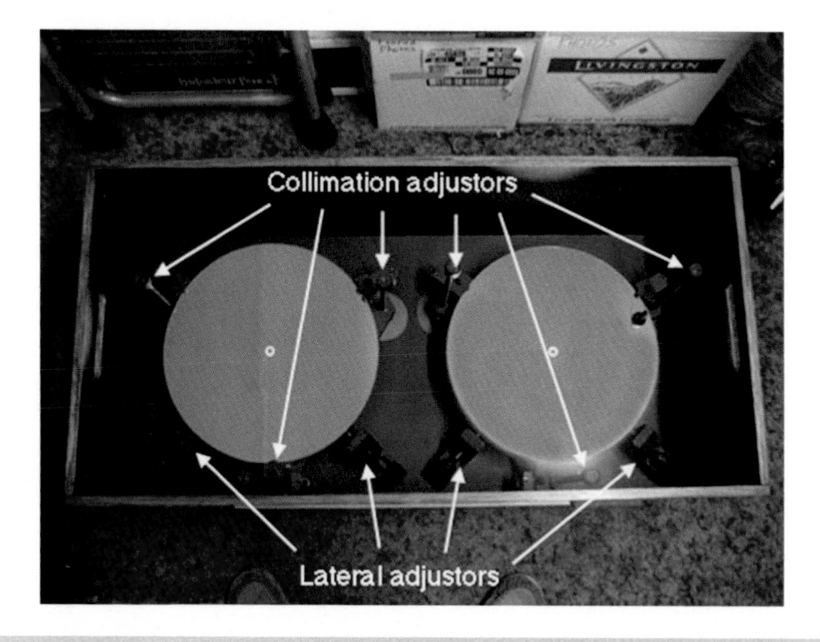

Fig. 5.71 The dual 12.5 in. mirror collimation and alignment system (Image credit: Jerry Oltion)

Fig. 5.72 The dual wire strung spiders secondary support and adjustment system (Image credit: Jerry Oltion)

eyepieces as short as possible (to keep the secondaries a reasonable size), and that meant getting the observer's head in as close as possible. I ran into one snag early on: the tertiary mirrors — actually just diagonals like you'd use in a refractor — wouldn't come close enough together for most people's eyes. The corners banged into one another, so I ground off the corners.

I made the tertiary mirrors adjustable by putting foam tape between the mirror plate and the diagonal housing. I can tighten or loosen the mounting screws and get a little tilt that way; enough to help collimate the entire system. I collimate secondaries and primaries first, and then adjust the tertiaries to match.

At first, I tried just rotating the secondary cages by hand, but it was way too easy to over or undershoot, so I started thinking about how I might be able to make a screw driver for it. I mentioned my problem to the Scopewerks guys, and a couple of days later Frank Szczepanski said, "I've found your adjustor." It was a cargo strap tightener. The far end of it used to have a big coil of cargo strap on it, and you had to tighten the strap by turning the handle and pulling on the strap anchors. I cut the strap holder off and left the tightening arms, and connected the arms to the secondary cages with a couple of aluminum bars and some vertical rods made out of a knitting needle. The links allow the upright rods to get closer together or farther apart as the cages rotate. It's intuitive enough that I can tell people at star parties, "This is where you adjust it for the width of your eyes," and that's all the instruction they need.

The focusers are odd-looking components. I didn't really want to buy two commercial Crayfords because in addition to being expensive, they are also heavy. I wasn't worried about weight at first, but as the design progressed, I began to

Fig. 5.73 Photo showing the dual eyepiece holders, tertiary mirrors focusing and clever focusing system (Image credit: Jerry Oltion)

Fig. 5.74 Photo showing Jerry's very clever mechanical lead screw mechanism used for the IPD adjustment (Image credit: Jerry Oltion)

Fig. 5.75 The beautiful stained wood Altazimuth mount. This image still shows the Teflon pads (Image credit: Jerry Oltion)

realize that I was creating a monster and I decided to use light-weight focusers (Figs. 5.73 and 5.74).

I had made an extremely light focuser for my 12.5″ trackball, but that design was too flexible to hold up under the weight of a diagonal and an eyepiece. Next, I considered a square set on its side. The bearings could go in two of the faces, and I could use a stiff wire for an axle. This design was actually discussed in an article printed in Sky & Telescope.

The flex rocker is pretty conventional. I used three roller bearings, one Teflon pad on the bottom and four Teflon pads on the top. The latter turned out to be a little too stiff, so I replaced the front two pads with roller bearings (Fig. 5.75).

The altitude bearings are half-inch Baltic birch with aluminum strips for runners. They're removable so the mirror box can sit in the back seat of a car. I made them a little smaller in diameter than they should have been for perfect balance, figuring that would reduce the amount of "fin" sticking out in front. The operator has to stand in between them to view but the finder is accessible from the back, and I didn't want to be tripping over the altitude bearings all the time in the dark. I compensate for the lack of balance with springs on either side, and that works really well.

The ground ring is from the 20″ scope that Kathy and I bought from Mel Bartels. I was glad to see that we could reuse that, saving me the trouble of building (and storing) a second one.

I stained the wood "Early American" and varnished it with Varathane. The prime finish is one of the things that pleasantly surprised me most about this scope.

In a fit of energy not long after, I built an observing chair and stool to match. I added glow-in-the-dark tape around the edge of the stool (see Fig. 5.75).

How It Works

It takes me about 15–20 min to set up the binoscope. The collimation stays pretty tight even after tear-down and re-assembly, so I normally just have to tweak that a little at the beginning of the night, stick in a couple of eyepieces, and merge the images by tilting the right-hand primary mirror. That throws off the collimation of that side a bit, but hardly enough to matter. If it ever gets too bad, I shift the primary to the side a little and recollimate. It took me a while to figure out which direction to move things in order to improve things rather than make them worse, but it later became intuitive. The challenge with binocular scopes is merging the images while keeping the scopes collimated, but it's actually not that big of a deal.

I don't even have the adjustment rods extended up to where I can reach them while looking in the eyepiece. I know from experience how much motion and in which direction a turn of each screw will give me, so I look in the eyepiece, figure out where the right-hand image needs to go, and reach down and move it that much. I usually have to fine tune it, but this is a fast process. I will probably raise the rods someday just to stop having to bend down (Fig. 5.76).

Once the two images are merged, the sky takes on a depth and richness like nothing you've seen before in even the largest single-mirror scope. Star clusters look three-dimensional, rather than like salt spilled on velvet. Nebulae look like they're floating right out there in front of you. And the Moon looks unbelievable. If

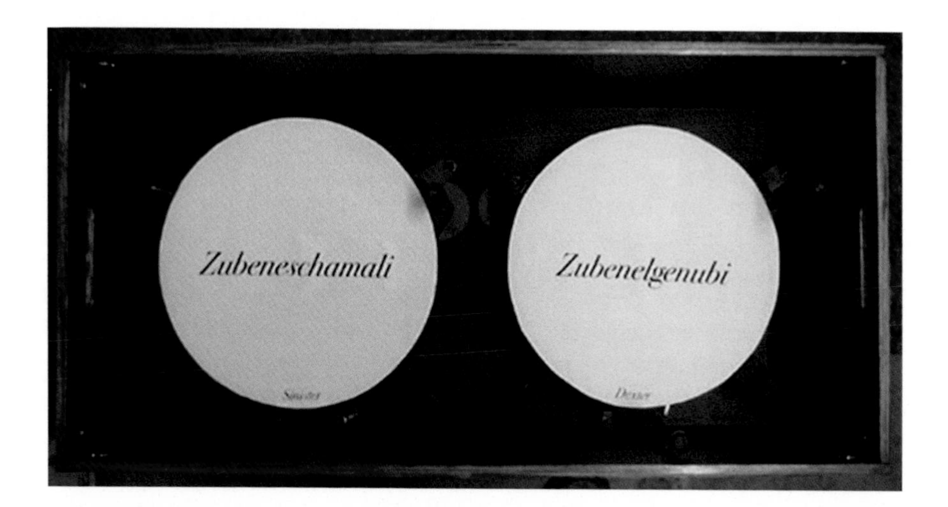

Fig. 5.76 The two mirror covers with the names of Oltion's two favorite stars in the constellation Libra on the covers (Image credit: Jerry Oltion)

that was the only thing I ever looked at with this scope, it would be worth it. It looks like a big bumpy rock floating about five feet away. The horizon curves away from you, and the shadows along the terminator have *depth*. I used to feel disappointed when the Moon would get into first quarter, but now I'm eager to get out and observe on moonlit nights just so I can look at it in stereo.

The distance even to the Moon is too great for real parallax, of course, but the image processing parts of your brain don't know that. You've got two separate images going into two separate eyes, so the wetware does its magic and you see everything in stereo.

The rigid design does its job: the images stay merged from zenith to horizon. They stay merged when I swap out eyepieces, too, except for my highest-power ones (4.7 mm Explore Scientifics, which give me about 315×). For some reason, those eyepieces don't like the same settings as all my other ones, so I have to tweak the merging when I use them. It takes a few seconds to do that, so it's no big deal, but I keep trying to figure out why they're different. I think it may be their tapered barrels interacting with the set screws differently than the straight barrels of the other eyepieces.

I occasionally have to adjust the merging even when I don't use the high-power eyepieces. Stuff gets bumped, or temperature changes shift the tolerances, or cosmic rays addle my visual cortex. I've learned that different people's eyes have varying abilities to merge. Some people are slightly cross-eyed, while others are slightly wall-eyed. Most of us can meet somewhere in the middle, but it doesn't take much deviation before you start dropping the outliers off the bell curve. I've learned to ask them "Which way does the right-eye image need to go?" and I run the merging knobs while they call out "left, right, up, down" and we quickly dial it in for them. After this adjustment, it's usually wrong for everyone else. Fortunately this only happens for maybe 1 out of 20 people.

The light-gathering ability seems to be about the equivalent of a 16 or 18-in. mono scope. I've been able to see 16th-magnitude galaxies through it, which I can't do with just a single 12-in. mirror. Planetary nebulae like the Ring and the Dumbbell are amazingly rich in detail through this scope.

All in all, it's an easy scope to use and the view it provides is way more exciting than that of a mono scope. It's rapidly becoming my favorite. People kept telling me I should name the scope "Gemini" and call one mirror "Castor" and the other one "Pollux." I liked the idea, but my favorite star names are over in Libra instead. I put the names on the mirror cover.

I'd love to hear from people who are interested in this scope design. Please feel free to email me at the address on the right. I'll do my best to answer everyone who writes with a genuine question or comment about the design.

(Story and images are reproduced here with the permission of: Jerry Oltion http://www.sff.net/people/J.Oltion/binoscopes.htm)

Further Reading

APM Telescopes Goebenstrausse 35 66117 Saarbrueceken Germany. http://www.apm-telescopes.de
Aurora Precision, 20420 Boomer Ferry Rd. N.E. Aurora, OR 97002. www.aurora.com
Avalon-Inatruments. http://www.avalon-instruments.com
Burr, J. (JMI) Jim's Mobile Incorporated. http://www.jmitelescopes.com
Kalmbach Publishing Co., 21027 Crossroads Circle, P.O. Box 1612, Waukesha, WI 53187. http://www.kalmback.com
Oltion, Jerry. http://www.sff.net/people/j.oltion/binoscopes.htm
Sayre, B. Large amateur binoculars. http://brucesayre.net
Schmidt, W. The making of voyager. http://www.waynesthisandthat.com
Trott, D. Six inch binoculars. http://davetrott.com
Vixen Optics. http://www.vixenoptics.com

Chapter 6

One of a Kind

Building a GOTO Refractor Binoscope

Adapted from: "An ATM GO-TO Binoscope" Astronomy Technology Today—July/Aug. 2011 Warren Estes Memorial Award—2011 RTMC.

By Norm Butler

Looking at all of these splendid looking homemade binoscopes and binocular telescopes in the book provokes one with enough incentive to want to start on a binoscope project of their own. It's almost hard to decide sometimes what type of binoscope you want to make. Whether it is a big binocular telescope or a refractor binoscope, it really depends on what kind of visual observing you really want to do with your binoscope. A big binocular telescope or refractor binoscope with its dual light gathering power can provide wonderful views of nebula and deep-sky objects that will surpass a single telescope of the same aperture. But building one will require a lot of work and some expense. Building a binoscope requires that both optical systems be identical, not only in just the optical aspects, such as the objective diameters, focal ratios, and the exit pupil diameters, but also the mechanical aspects. Plus, one must consider its portability and how easy will it be to transport, set up, and use (Fig. 6.1).

Something that was always on my ATM list of things to build someday was a GOTO refractor binoscope. I knew the technology was available and already out there. All I had to do was to decide what kind of hardware and software to use to build it with. In undertaking a one-of-a-kind project like this, certain critical design elements, such as adjustment of the interpupillary distance and optical alignment, focusing, the electronics and software, and the gearing to drive the GOTO mount,

N. Butler, *Building and Using Binoscopes*, The Patrick Moore Practical Astronomy Series, DOI 10.1007/978-3-319-46789-4_6

Fig. 6.1 2011 RTMC Warren Estes Memorial Award winning Dual 102 mm f/6 Celestron GOTO binoscope (Image credit: the Author)

must be thoughtfully planned for. Each of these important elements presented its own engineering challenges and difficulties. In an engineering sense, it was certain it was going to be challenging, and it wasn't going to be as easy as originally thought and of course it never is.

So after asking myself the same question a few times, "Are you ready to build a GOTO refractor binoscope or another binocular telescope?," and after thinking about it for a while and pondering the engineering challenges of such an ambitious project, it was decided, "Sure, why not? Let's get started." Having already built a 10-mirror 6-in. f/15 Cassegrain binocular telescope on a clock-driven equatorial mount over 35 years ago, I now was ready for something different and perhaps an even greater engineering challenge. And this new binoscope project was certainly going to be challenging enough, especially with a GOTO system. As an ambitious ATM who always had a desire to build a refractor binoscope, using a pair of 102 mm Celestron f/6 optical tube assemblies and putting them on an Altazimuth mount using a Lumicon Universal Pier seemed simple enough in it's design to mount the dual 102 mm f/6 Celestron OTAs on and that's part of the reason why I chose to build it because all of the parts needed to construct it are available, either commercially or easily fabricated with many of them being purchased off the shelf at a local hardware store. The second reason is because the 102 mm Celestron refractor with a focal ratio of f/6 can perform very good optically for visual astronomy. It's relatively fast and has a wide enough field of view for observing all kinds of celestial objects. The third reason is it is mechanically simple in design, plus they can be easily transported and assembled and both OTAs can be aligned in quick fashion on top of the Lumicon platform using the same Celestron mounting hardware.

To get the GOTO refractor binoscope project started, after the decision was made to use a pair of Celestron 102 mm f/6 achromatic refractors for the refractor binoscope optical tube assemblies (OTAs), then mounting them on a Lumicon Universal Pier mount seemed like a natural choice and then that became the engineering priority (see Fig. 6.2).

Before we get started on our Celestron binoscope project, let's define our major important design elements first:

- The optical alignment of the dual optical tube assemblies
- Adjustment of the interpupillary distance (IPD) and focusing system
- A Celestron NexStar 4SE GOTO electronic drive system and software to benefit our visual observing program

All of these are important considerations and should be part of the advanced planning and each one needs to be completed in order for our Celestron binoscope project to be successful. Obviously, each of these important engineering elements will be challenging, and with our advanced planning, and our various assortment of proper tools and materials, we should be able to complete our dual 102 mm f/6 GOTO Celestron refractor binoscope project in a timely and successful fashion without encountering any major problems or issues that would slow us down during the construction and fabrication of our binoscope project.

Fig. 6.2 Dual 102 mm f/6 Celestron GOTO binoscope on the Lumicon Universal Pier (Image credit: The Author)

The optical and image quality of the Celestron 102 mm f/6 refractor with its excellent achromatic objective had always impressed me each time it was used for observing. What really impressed me the most was the wide field of view that it rendered using a standard 20 mm eyepiece. So it was a good choice to choose this combination of objective size 102 mm and focal ratio f/6 to use for the refractor binoscope project. For those who have used the Celestron 102 mm f/6 optical system before, using it in the format of a binocular telescope would yield very pleasing celestial views. Plus, the added benefit of using an all lens system instead of mirrors would make it much simpler and quicker to align optically and setting it up on its Altazimuth mount for observing as well. The major thing that all telescope makers want to achieve with their telescope making projects is to make their optical instrument as user-friendly as possible. A very user-friendly telescope makes for a more enjoyable observing session. The less complicated you can make your telescope in terms of design and functionality, the easier it is to set up and observe with. And thus, it will give you more time to enjoy the celestial views through your new homemade telescope or binoscope along with the appreciation of others who get a chance to view through it too.

So let's look at the Lumicon Universal Pier first. Among its biggest attributes is it's simple design, adjustable height, rigidity, and stability. After checking it out mechanically at a local telescope store, the decision was made to use it instead of a big tripod for several reasons. First, the Lumicon Universal Pier is adjustable up to a height of 72 in. and lowers down to 56 in. The observer would be able to raise or lower the entire dual GOTO 102 mm f/6 Celestron binoscope to a comfortable observing height for viewing without needing help. Located within the Lumicon Universal pier is a heavy-duty spring that helps make the height adjustment easier to accommodate when it comes to raising and lowering the dual 102 mm f/6 Celestron binoscope load mounted on top of the Lumicon Universal Pier's cradle. Second, the Lumicon Universal Pier is easy to lift and carry, transport, and set up in a very short period of time. The three legs on the Lumicon Universal Pier are detachable and are made out of heavy-duty black plastic with each leg having its own leveling pad. The Lumicon Universal Pier was designed to support and lift and lower a heavy load with ease. The best part about the Lumicon Universal Pier is that it is reasonably priced and well worth the investment to use for our Celestron binoscope project. There are two large round locking knobs located midway on both sides of the Universal pier to secure the movable inner pier in place once the desired observing height is reached (see Fig. 6.2).

Getting Started

In considering possible solutions for building a GOTO binoscope, using mirrors was not an option. It was decided instead to use an identical pair of the very good 102 mm f/6 Celestron refracting telescopes (just the optical tube assemblies) and go ahead and purchase them at a local telescope store. This particular binoscope will not use the traditional diagonals and set about finding a suitable alternative while keeping the IPD adjustment in mind. One weekend, I happened to visit a local swap meet and found an older Nikon microscope that was in somewhat good condition, at least optically speaking. I took one look at the two microscope eyepieces and prism assemblies and instantly knew it was possible to use them on the GOTO binoscope. The fact that the microscope prism assemblies would provide the correct orientation no matter how they were rotated was critical to success of the project. Plus, each microscope prism assembly could be rotated either individually or together to adjust for the observer's individual interpupillary distance (IPD). So after bartering with the seller a bit, the microscope was purchased for $10USD and that took care of my first major engineering requirement in building the GOTO binoscope project. It was relatively easy to adapt the microscope prism assemblies to a 1.25-in. eyepiece adapter for each Celestron telescope. A pair of 1¼-in. eyepiece adapters was ordered from Orion Telescopes and both microscope prism assemblies fit perfectly within the 1.25-in. Orion eyepiece adapter barrels without having to make any change in their barrel dimensions. After the eyepieces were installed, they needed to be tested by looking at a distant object. As an initial static test, a 25-mm eyepiece was placed in each microscope holder in both Celestron 102-mm telescopes and focused on distant trees. Each image appeared to be in a correct orientation no matter what position the microscope eyepiece holder prism assemblies were orientated. This simple solution using a pair of Nikon microscope eyepiece holders for the IPD adjustment made my project even more satisfying with such a positive result.

The Altazimuth Mounting

The dual Celestron binoscope mounting was the next challenge. An equatorial mounting was not considered for the dual Celestron binoscopes. As equatorial mountings go, it would have presented too many problems, especially when it comes to stability and rigidity. Mounting the two Celestron 102-mm f/6 refractors in the same way as traditional binoculars are mounted was also not an option. Instead, it was important to mount both telescopes on a stable platform where one could install certain mechanical collimation and alignment features to make them easily accessible when alignment of one or both telescopes was needed. After taking a final look at what kind of tripods and piers were available on the Internet and finding nothing really suitable for my GOTO binoscope project, then the only obvious choice would be the Lumicon Universal Pier Altazimuth adjustable height tripod. It not only was designed for use as an Altazimuth binocular tripod, but its height could be easily adjusted via a big, heavy-duty spring inside the inner tripod tube. With a clamp-style locking knob on each side of the tripod tube, the tripod height could be easily locked into place. What was important for my GOTO refractor binoscope project purposes and was one of the main advantages of the Lumicon Universal Pier was that it has a 3/4-in. thick black-plastic platform for mounting binoculars up to 80 mm. However, after acquiring the Lumicon adjustable pier and really having a chance to check it out, it was necessary to reverse the Lumicon mounting platform 180°, which enabled me to easily mount both Celestron 102 mm f/6 optical tube assemblies (OTA) on the top of the reversed Lumicon platform.

The first initial modification made to the Lumicon mount's platform was to drill a couple of 1/2-in. holes at the appropriate separation distance to provide pivot bolts for the OTA optical tube connecting base mounting hardware. It was necessary to drill two 1/2-in. holes approximately 1.75 in. from the center of the first two holes for a 1/4-in. diameter by 3/4-in. length bolt that would protrude through each hole. After the two 1/4-in. diameter holes were drilled, another 1/4-in. diameter hole was drilled at the same distance from the center into the bottom of each of the two aluminum Celestron telescope mounting clamps. A single 1/4-in. diameter bolt was epoxied into the bottom of each round mounting clamp to act as the pinion rod for another 1/4-20 by 2.5-in. bolt to push against that would rotate each mounting clamp a few degrees in either direction for individual binocular telescope alignment in the horizontal axis. Another hole of the same size was drilled at the same distance and centered on the exact opposite side of the black-plastic platform. Then two 1/2-in. holes were drilled and tapped at the same distance on the rear side of the mounting platform for the two horizontal axis 1/4-20 by 2.5-in. long alignment bolts. A pair of the mounting clamps was attached that Celestron typically uses on its NexStar 4 SE telescopes. These were ideal for securing the two Celestron 102 mm f/6 refractors in place on top of the mounting platform. A pair of 1.75-in. wide by 1/2-in. thick by 3.0-in. long aluminum stops was added on the front of the black-plastic platform to position both 102 mm f/6 refractors always at the same place and relative position each time during setup to make the optical and mechanical alignment of both optical tube assemblies a somewhat simple step each time the dual Celestron binoscope was set up for viewing.

The Twin Refractor OTAs

Choosing the Celestron 102 mm f/6 refractors was probably the easiest part of the GOTO binoscope project. The optical quality of the Celestron 102 mm made it my first choice and combining them with the two swap-meet Nikon microscope eyepiece holders and prism assemblies made it a simple task to provide for interpupillary adjustment; simply rotating the tubes about their individual optical axes took care of that issue. But at the same time, I also needed to decide on what electronics, software, and drive system hardware to use in order to make this project into a real working GOTO binocular telescope. My initial plan was to keep my GOTO binoscope dual-axis drive system as simple as possible. After giving some serious thought on which commercial GOTO electronics package would work best for my project, it was decided that the Celestron NexStar 4SE had all of the electronic, software and mechanical design, and hardware elements needed to make my dual GOTO Celestron binoscope work according to my original specifications that were planned for it. So it was back to the swap meet where, on a lucky weekend, found a complete Celestron NexStar 4SE with its hardware and electronics in good working order and at a very reasonable price. As a result, I got a good deal and purchased it and added it to my "must need" items list. After returning home from the swap meet with my newly acquired NexStar 4SE, it was totally stripped down and any and everything was removed that was needed to use for my GOTO Celestron dual- axis drive system, including the electronics.

The Azimuth Gear

Next important item was to install the NexStar 4SE gears first. Starting with the azimuth gear and needing only to drill a 1/4-in. hole into the exact center of the tapered pivot bolt that is located on the top of the Lumicon tripod head. This particular bolt is what secures the Lumicon rotating Altazimuth head via an approximately 2.0-in. diameter flat roller bearing to the top of the tripod. After drilling the 1/4-in. hole, a 1/4-20 TPI bolt with some liquid epoxy (J-B Weld) spread evenly on its threads was screwed into the hole, and after a few hours, the epoxy was set up and then it was time to proceed with the next stage of the project, the Azimuth gear (Fig. 6.3).

It was important to have a center shaft for the Azimuth gear to rotate around for it to operate properly as a "Grab and Go" azimuth slip clutch system that had been planned for it. So after thinking about what would work the best for this particular

Fig. 6.3 The azimuth gear and knob and clutch system (Image credit: the Author)

part, my metric socket set just happened to be sitting nearby. Upon seeing the socket set, an idea came to mind, which made me want to try out several metric sockets to see if any would fit with a close enough tolerance in the azimuth 180 tooth gear central hole. After trying out a few of the metric sockets for the correct O.D. size in the central hole of the 180 tooth azimuth gear…."Bingo." A 14-mm socket fits perfectly within the azimuth gear central hole. The next thing to do was to saw it off to the correct socket height dimension using my trusty saber saw with the 14-mm socket locked securely in a small machinist vise. It was sawed off to approximately 3/8 of an inch and then sanded it perfectly flat. After checking its height dimension for the final time and then sanding the entire metric socket's O.D. to a semi-rough surface finish, it was then placed on the electric stove's heating element until it got very hot. After grabbing it with a pair of pliers and placing it on the top dead center of the large Altazimuth black plastic knob. The red hot metric socket slowly melted its way down to the proper depth, stopped it, and then let it cool down after it was in place. It took a couple of tries and a couple of black plastic knobs to get it right. To ensure that the azimuth gear locked solidly in place, a 2-in. round piece of 180-grit sand paper was epoxied in place on top of the round 2.0-in. diameter tripod post head. Likewise, to give the 180-grit sand paper something to "grab," a round 2-in. diameter 1/16-in. thick piece of cardboard with a 5/8-in. central hole was epoxied on the mating side of the azimuth gear. Once the round cardboard pad was firmly epoxied in place, an even amount of cyanoacrylate glue was spread over the entire surface and let it soak in. This was done in order to make the entire round cardboard pad very hard and durable and ensuring that it was strong enough to last a long time without wearing out quickly. After the cyanoacrylate glue had set up, the next test was to lock the azimuth gear in place. Once the black plastic knob on the azimuth gear shaft was tightened, it worked perfectly. I was able to achieve the solid "Grab and Go" clutch action that was needed in order to sweep the skies with the gear still engaged to the NexStar 4SE servomotor and still tracking, just by unlocking the DIY azimuth knob. After that small victory, attaching the NexStar 4SE azimuth servomotor to the gear was next. After drilling a couple of holes for the NexStar 4SE's original servomotor mounting plate, a couple of 1/2-in. diameter by 3/4-in. height aluminum spacers were used to position the servomotor with its drive gear firmly in place with the azimuth gear. The alignment and height for both the azimuth and servomotor gear was nearly perfect and without any slop between the gears. That we were getting much closer now to our goal of a working GOTO Celestron refractor binoscope.

The Altitude Gears

After completing the assembly of the azimuth gear drive system, tackling the altitude "Grab and Go" drive gear system was the next order of business. Retrofitting the NexStar 4SE altitude drive gear and servomotor assembly would prove to be more challenging. First of all, in order to make the NexStar 4SE altitude drive gear system work efficiently and effortlessly, the next thing to do was to balance out the entire dual 102-mm binoscope assembly, which ultimately took a total of approximately 16 lb of counterweights that were mounted on a 12-in. long 1/2-diameter shaft on the opposite side of the Lumicon binocular platform. So by using another piece of semi-round 3/4-in. thick black plastic, a 1/2-diameter hole was drilled in the center with a threaded 1/2-in. nut embedded on the opposite side for the counterweight shaft to attach to. Also, the counterweight shaft had to be just long enough to clear the inside of the Lumicon Altazimuth head components without hitting anything, including the newly installed azimuth servomotor assembly. After the black-plastic counterweight holder had been secured to the bottom of the binocular platform with three 1/4-20 tapered head shoulder bolts, the counterweight shaft was attached and immediately started to add some counterweights to balance out the dual Celestron OTAs. Starting first with 5 lb and ultimately ending up using 16 lb of counterweights in order to get a nearly perfect balance of the entire Celestron binoscope. After a few swings for testing, the balance simplified everything. The same basic mounting gear "Grab and Go" clutch assembly was used as with the azimuth drive gear system. A 1/4-20 TPI bolt was made as the center shaft with a 14-mm center pivot just as with the azimuth drive gear assembly. The same kind of black-plastic knob was used as before on the azimuth drive gear "Grab and Go" clutch system, but this time using a 1/4-in. thick stepped steel flange insert that was laying around in my coffee can of miscellaneous spare nuts and bolts. It fit perfectly inside the plastic knob, allowing me to apply more clamping torque to the gripping surface of the altitude drive gear if it were needed. The same 180-grit sandpaper was used as before and then epoxying it to the clamping surface of a 3-in. diameter by 1/8-in. thick flat steel washer that was epoxied to the another plastic knob that had been sanded down and attached to the binocular platform altitude rotation axis plastic shaft. This arrangement seemed to work out just fine. After epoxying a round 3-in. diameter red composite pad with a 3/4-in. diameter central hole to the NexStar 4SE altitude 180 tooth drive gear, everything was now in place for the first clamping test. This assembly allowed me to clamp the altitude gear with enough force to make the entire gear and the binocular platform rotate about it's axis by using my hand to grip the altitude gear and rotating it. There was no slippage; it worked better than one had expected. The only thing left to complete this part of the project was to install the NexStar 4SE servomotor and mount it in a correct and secure position next to the altitude drive gear. This was accomplished using a couple of 1/2-in. diameter by 1/2-in. long brass spacers. After drilling and tapping a couple of 10–32 holes at the appropriate position into the upper support arms of the Altazimuth mount, the altitude drive gear was mounted in position. At this point, it was time for the final project assembly (Figs. 6.4 and 6.5).

The Electronics Package

My plan was to use the same electronics (hand controller, printed circuit boards [PCBs], switches, cable harness, and software package) as used on Celestron NexStar 4SE, but also wanted to attach them to the upper part of the Lumicon mount

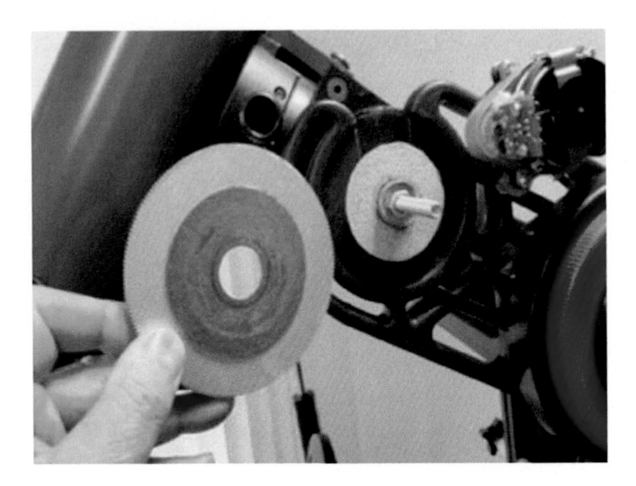

Fig. 6.4 The altitude gear, knob "Grab and Go" clutch system (Image credit: the Author)

Fig. 6.5 The side view of the dual 102-mm GOTO Celestron binoscope showing the altitude drive gear, electronics package, hand controller, and counterweight system (Image credit: Dann McCreary)

Fig. 6.6 Two RTMC amateur telescope makers discussing the author's homemade dual 102 mm f/6 GOTO Celestron binoscope at the 2011 RTMC (Image credit: Dann McCreary)

and as close to the servomotors as one could get them. It was important to avoid modifying any of the NexStar 4SE electronics or wiring to make them fit. Fortunately, without any modification, all of the PCBs were secured together with small screws within a circle diameter of 6 in. on the right Altazimuth arm of the Lumicon mount. All of the wiring cables and cable attachments from each servo-motor fit perfectly with no cables stretching for length. Once all of the cables were plugged into their own individual PCB connectors, it was ready for the first electrical test (Fig. 6.6).

After plugging in the power from my Celestron 7 Ah Power Tank, the appropriate messages flashed across the NexStar hand controller's screen asking for the location, time, date, etc., and so the SkyAlign screen was chosen to begin the first actual test for the servomotors. With eager anticipation, each of the four directional arrows on the NexStar hand controller was pressed. Upon hearing each servomotor answer, the directional command with their familiar buzzing sound became music to my ears. Success at last! Next thing was to test my "Grab and Go" clutch system. After tightening the clamping knob on the azimuth gear first and then pressing the direction arrows to see the Lumicon mount begin to rotate in one direction and then the other with splendid ease, the same operation was performed with the altitude servomotor, watching it rotate the NexStar 4SE altitude drive gear and binocular platform move about its vertical axis. After several tests, the GOTO Celestron binoscope design convinced me that it will live up to my expectations. For the ultimate test, I mounted both Celestron 102-mm f/6 optical tube assemblies and clamped them firmly in place on top of the Lumicon binocular platform, and with the entire dual Celestron optical tube assemblies in nearly perfect balance, the electronics was fired up again and repeated the same SkyAlign commands. With just a touch of the NexStar hand controller's directional arrow buttons, the entire Celestron binoscope system moved about in perfect harmony with its unique mechanical and optical design. There is a lot of satisfaction watching your homemade GOTO binoscope come to life and perform splendidly on command. That night after it had gotten dark enough, the dual 102-mm f/6 Celestron binoscope was "fired up" and it didn't disappoint me. Throughout the night, the Celestron SkyAlign program and other programs in the NexStar 4SE software package work perfectly, guiding my binocular telescope from object to object. It tracked splendidly and performed each GOTO command flawlessly. One couldn't have been happier with its performance. What was originally anticipated as being a fairly daunting task turned out to be an almost bolt-on project requiring surprisingly little custom fabrication. To provide for carrying the battery and to better eliminate any unnecessary cables that might get wrapped around the tripod pier during operation, the Power Tank was placed in a dual hook basket on the front of the Altazimuth mount along with another 12 lb of weights secured on a 1/2-in. shaft attached directly below the basket. The additional weight didn't stress the robust servomotor drive system in the least. After using the GOTO Celestron binoscope and watching it hop around the sky searching for various celestial objects in its 40 K object database and finding them made this entire Celestron GOTO binoscope project very worthwhile (Fig. 6.7).

Fig. 6.7 The side view of the dual 102-mm GOTO Celestron binoscope showing the altitude drive gear, electronics package, hand controller, and counterweight system (Image credit: Dann McCreary)

To sum it all up, the entire Celestron GOTO binoscope project only took about 3 months to design and build. All of the necessary hardware such as the nuts, bolts, and screws were conveniently found at a local Home Depot. All of the electronics, gears, stepper motors, software, and the hand controller that make up the GOTO system were cannibalized from a Celestron NexStar 4SE purchased at a local swap meet for a very low price. Also something to make note of was using a pair of 1¼-in. prismatic microscope eyepiece holders cannibalized from an older Nikon microscope purchased at a local swap meet made the entire project even more satisfying. Once the dual microscope assemblies are mounted on the dual Celestron binoscopes using a pair of 1¼ adapters (see Fig. 2.20), just a slight rotation of either microscope holder will not change the original image orientation and will allow for just about any IPD adjustment needed. And with just a slight twist of a lock screw will secure either one in place for a night of comfortable binoscope viewing.

This was a fun binoscope project to build. The end result was a one-of-a-kind dual GOTO 102-mm f/6 Celestron binoscope with a 40 K object database that almost everyone who gets a chance to view through it and uses it for the first time wants to build one too.

Six-in. f/15 Cassegrain Binoculars

By Norm Butler

Throughout my early telescope building years, the urge to build a big pair of 6 in. of achromatic refractor binoculars or a pair of 6-in. Cassegrain binoculars was always in the back of my mind. Until the late 1970s, the largest telescope the author made was a 6-in. f/4 square tube wooden Newtonian reflector on an equatorial fork mount for a friend of mine. The urge to build a 6-in. f/15 refractor binoscope was always there, but the cost of the objectives and the long f/15 optical tube length made me decide against it. So as a compromise, I decided to make a pair of 6-in. f/15 Dall-Kirkham Cassegrain binoculars instead. The f/15 optical design accommodated for a very compact tube length and that helped me make the decision to go with the Dall-Kirkham optics from Coulter Optical Co. instead of the more expensive 6-in. f/15 achromatic objectives (see Fig. 6.13). The primary mirrors in a Dall-Kirkham optical system (Created by Horace Dall in 1928) are concave elliptical in shape and use a convex spherical secondary. At the time, only lunar and planetary observing really interested me. So it was pretty easy to imagine that building a dual pair of 6-in. f/15 Cassegrain binoculars would provide some very spectacular views of the moon, planets, and planetary nebula. An amateur project like this would indeed be breaking new ground if a pair of 6-in. f/15 Dall-Kirkham binoculars were mounted on an equatorial clock-driven mount and be able to observe through them in a comfortable fashion while they're tracking in the equatorial mode across the night sky (Fig. 6.8).

This was going to be one of my more difficult engineering challenges. As if my project didn't have enough engineering challenges already trying to build the entire

Fig. 6.8 The author with his homemade dual 6-in. f/15 Dall-Kirkham Cassegrain binoculars won the Engineering Merit Award at the 1981 (RTMC) Riverside Telescope Makers Conference (Image credit: Rick Schmidt)

big Cassegrain binocular telescope and still try to simplify the engineering design as much as possible. After looking at the possibilities, the use of a dual steel ring rotation system was chosen instead of mounting them on a platform that would be mounted axially between the forks (see Fig. 6.9). It was easy to see that as the big Cassegrain binoculars moved across the equatorial sky, the head and eye position for observing with two eyes would obviously change, and without somehow adjusting the dual zoom eyepieces for a more comfortable observing position, it would soon would become an uncomfortable issue. So after some serious thought on the subject and how to solve it, the decision was made to go with a dual steel ring OTA rotation system. With a rotation system, one could rotate the entire dual Cassegrain binocular system together as a unit a full 360° within the dual steel rings for a comfortable head and eye position no matter where the big binoculars were pointed as there was tracking in the equatorial mode.

After I made the decision to use a dual steel ring rotation system, making them was another story. I drew up an 18-in. diameter by 1/2-in. thick steel ring design and had them made at a local machine shop, likewise four aluminum support brackets too. However, to cut costs, I drilled and tapped all of the holes in the dual steel rings

Fig. 6.9 The dual steel ring rotation system that allows the dual 6-in. f/15 optical tube assemblies to rotate a full 360° of free rotation in either direction for good head and eye position (Image credit: Rick Schmidt)

and aluminum support brackets myself. I also had the machine shop make a couple of 1½-in. diameter stainless steel shafts of the proper length. They are used to support the entire dual ring system and Cassegrain binocular telescope when it was mounted in the fork assembly on the big equatorial mounting via a pair of large pillow block bearings attached to the top of each fork assembly (see Fig. 6.9). After the dual steel rings were completed, a pair of 7-in. diameter 4-ft. long aluminum tubes was ordered from the Jaegers and cut to length to house the 6-in. f/15 Dall-Kirkham primary and 2.36-in. diameter secondary mirror system. After cutting the two aluminum tubes to an appropriate tube length of approximately 28 in., they were attached together permanently. But before this, the angle of convergence had to be calculated so the two aluminum OTA tube assemblies could be welded exactly at that particular angle so their mechanical axis would converge at infinity or very close to it. To do this, a pair of aluminum shims was used, one in the front (0.031″) and one in the rear (0.062″), and they were heliarced in place in the front and rear middle of the aluminum tube assemblies. After the two aluminum OTAs tube assemblies were welded in place, it was time for the next step. Cut out two rectangular pieces of 1/8-in. thick sheet aluminum to be heliarced to the middle top and bottom section of the aluminum tubes for the Y-shaped bracket support plate.

After both top and bottom support plates were welded in place, the aluminum OTAs were basically finished, at least enough to plan for the installation of the primary and secondary mirrors.

A pair of 6-in. primary mirror cells was ordered from Jaegers, and after they arrived, the Coulter Optical 6-in. f/4 Cassegrain primary mirrors were mounted in the mirror cells and placed aside for a while anyway until the two 2.36-in. diameter Coulter Optical spherical convex secondary mirror mounts were completed. Their design was actually quite simple. Two round aluminum plates were machined from

2½ bar stock and machined down to 2.36 in. in diameter. A 29/64 in. diameter hole was drilled in the center and tapped to 20 TPI. Then the 3/8-in. thick round plate was parted off followed by 1/2-in. thick round diameter plate. Two of each were made and then three holes drilled 120° apart in each of the 3/8 aluminum plate for a ¼-28 TPI secondary mirror alignment bolts. The 2.36-in. diameter secondary mirrors were epoxied to the front of each 1/2-in. plate and that completed the secondary mirror support holder at this stage. The next part of the secondary holder system was to make a 5-in. long by 1-in. diameter round aluminum holder with a 5/8 diameter central hole for a 6-in. long nickel-plated 1/2–20 bolt (threaded approx. 1 in. at one end) and screw it into the taped hole of the 1/2-in. thick diameter secondary holder. The 1-in. diameter round aluminum holder has eight 1/4-28 TPI holes drilled 90° apart on each end of the round aluminum holder for adjustment and alignment of the secondary mirror.

The final thing left to complete the secondary holder assembly is to drill and tap a total of eight 10/24 tap holes approximately 1 in. apart in the middle of 5-in. long round aluminum holder for attaching the four spider vanes. It is possible that this secondary holder design may be unique in terms of a secondary holder system. The advantage the design has over other secondary holder designs is that it provides the secondary with a longer path to move either forward or backward for correct positioning. Plus, it allows one to remove the entire secondary holder and replace it again without disturbing its original horizontal or vertical alignment (see Fig. 6.10).

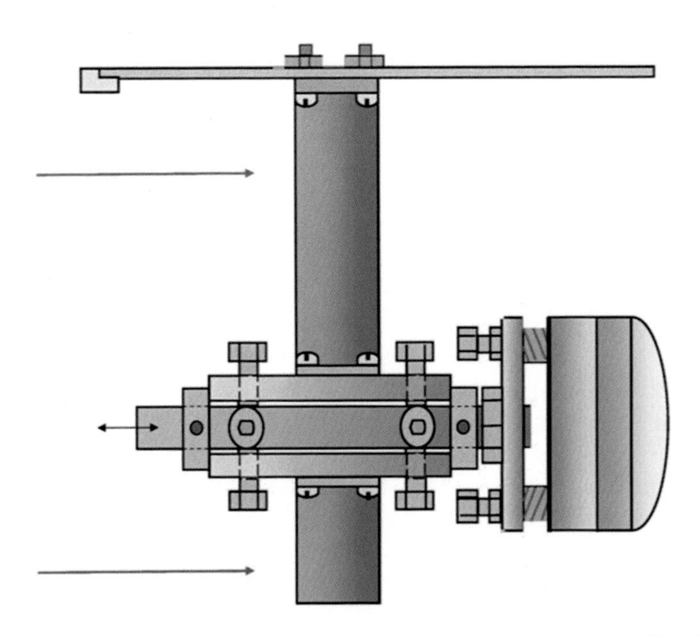

Fig. 6.10 Drawing represents the design of the 2.36-in. secondary holder for the dual 6-in. f/15 Dall-Kirkham Cassegrain binocular telescope (Image credit: the Author)

With the secondary holder complete, it's time to assemble the rest of the dual optical tube assemblies. Basically, to complete the entire dual optical tube assembly, it would take making a Y-shaped support bracket for the top and bottom section of the OTAs. The main reason for the Y-shaped bracket is for the installation of eight V-groove steel rollers on adjustable brackets with support braces that are attached to both top and bottom Y-shaped brackets. This is so that the dual Cassegrain OTA system could rotate within the steel ring system for good head and eye position while the big Cassegrain binoculars were tracking in the sidereal mode. I had decided to use 1/8-in. thick aluminum sheet bent at 45° angles on both sides of the 20-in. long Y-shaped bracket. Fortunately for me, a friend who worked at a local machine shop cut out the bracket and bent it at the desired 45° angle for me for a very modest price, which helped out a lot in terms of saving time and some expense too.

After the Y-shaped brackets were completed, twelve 3/8-in. diameter holes were drilled in each of the two brackets approximately 1/2 in. apart from the beginning of the 45° angle bend and were spaced 6 in. apart (see Fig. 6.11). Six 3/8-in. diameter holes were drilled on the top and bottom Y-shaped bracket support plates for mounting the Y-shaped brackets. After the brackets were mounted, it was time to move on to the eyepiece section at the rear of the dual Cassegrain OTAs. This too was to be made out of sheet aluminum bent at right angles to form a rectangular housing for the remaining first surface mirrors and diagonals to be mounted in (see Fig. 6.12). Fortunately, the aluminum housing was fabricated at the same local machine shop that made the Y-shaped brackets, so the cost was held to a minimum and so was the material. The basic dimensions of the aluminum eyepiece housing

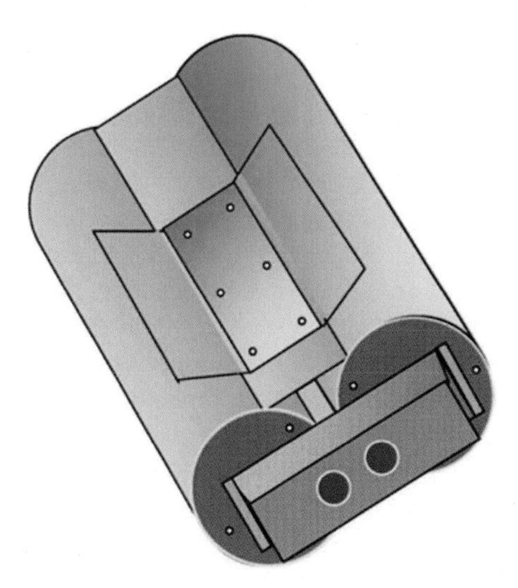

Fig. 6.11 Dual 6 in. f/15 Cassegrain optical tube assembly drawing (Image credit: the Author)

Fig. 6.12 A pair of Bushnell "Zoom" binocular eyepieces provides 105× to 315× magnification range (Image credit: Rick Schmidt)

are 12 in. × 4 in. × 4 in. with two separate aluminum side plates that acted as access covers and are secured on each side of the housing by set screws.

After the aluminum eyepiece housing was assembled on the rear surface of the Cassegrain primary holder plates, the two first surface mirrors and their holders were secured in place with one on each side within the housing just behind the housing covers. Each first surface mirror reflects the light at nearly a 45° angle coming from the spherical Dall-Kirkham secondaries. There are also two 1¼-in. mirror diagonals that were used to divert the light from the first surface mirrors to the two 1¼-in. secondaries that are secured in position in the front of the aluminum housing. They are spaced at approximately 65 mm and house the "zoom" eyepiece system. Focus is achieved by a 3/8-24TPI threaded bolt that is coupled at the top of the Bushnell "zoom" eyepiece hinge section. A 1½-in. aluminum knurled knob is used for focusing the "zoom" eyepieces (see Fig. 6.12).

The somewhat massive clock-driven equatorial mount that is pictured holding the 6-in. f/15 Dall-Kirkham Cassegrain binoculars was obviously built with stability and rigidity in mind. When you're observing at over 250×, you need that kind of stability when observing celestial objects at high magnification. Without that kind of stability and rigidity, the images would be bouncing around all over the place and making a very uncomfortable observing session. But even so, the big Cassegrain binoculars was built in the late 1970s when there was no commercial GOTO electronics and drives in place on the Altazimuth mountings of that period. Had there been, then an Altazimuth mounting would have been the obvious choice to mount these big Cassegrain binoculars on. Just putting this

6-in. f/15 Cassegrain binocular system on an Altazimuth mounting is not good enough by itself for serious observing. It needs to have an accurate tracking system, and at that time, a clock-driven equatorial with a full 360° dual OTA rotation system was the answer.

The 6-in. Dall-Kirkham Cassegrain optical system with its f/15 focal ratio makes it ideal for lunar and planetary observing. But how to utilize the 6-in. f/15 focal ratio to the greatest extent possible was an easy choice and a good idea to want to try and use a pair of "zoom" eyepieces instead of individually changing eyepieces in the dark. The big question was how to do it and make it work on a dual 6-in. Cassegrain binocular telescope? The answer is to use a pair of 7 to 21× "zoom" eyepieces from a pair of 40-mm Bushnell binoculars. There were a couple of trade-offs involved once it was decided to use a "zoom" eyepiece system with an f/15 focal ratio. Forget about low power observing and a wide field of view. The lowest power with this kind of "zoom" eyepiece system is 105× with an exit pupil diameter of 1.45 mm, boasting a 38.7° apparent field of view. Yet when observing over 300 in. magnification, the combined stereo effect and the fantastic extended resolution using two eyes viewing the lunar surface and the planets becomes your observing playground. And that pretty much describes what you can expect to see if one decides to build a Dall-Kirkham Cassegrain binocular telescope. For the original story on their design and construction, see Sky & Telescope—Nov. 1982.

It's interesting to note that in the past 33 years or so since the construction of the dual 6-in. f/15 Cassegrain binoculars, we haven't seen more similar Cassegrain binocular telescopes or even Schmidt Cassegrain binocular telescopes being built in the ranks of amateur telescope makers and commercial telescope manufacturers. With the increased popularity of refractor binoscopes and binocular Dobsonian telescopes, the dual Cassegrain binocular telescope has just not taken off yet and one wonders the reason why? For example, it is now possible to mount (with a little modification) a pair of 8-in. f/10 Celestron Schmidt Cassegrains on a C-14 Altazimuth GOTO mount. It's a little surprising why we don't see some enterprising amateur astronomer build one and give it a try. Fantastic new commercial telescopes with their advance coatings combined with a multitude of superb high-quality eyepieces and accessories are here today. The astronomers demand for more light gathering power and increased resolution using two eyes has never been greater than it is currently. It's probably safe to say that it won't be too long before we see a new Schmidt Cassegrain f/10 binocular telescope on a GOTO Altazimuth mount (commercial or otherwise) being introduced at a national star party someday soon, then we can say that Binoscope technology as indeed risen to a higher level (Fig. 6.13).

"One-Of-A-Kind 6 inch f/15 Dall-Kirkham Cassegrain Binoculars" (Circa: 1980)

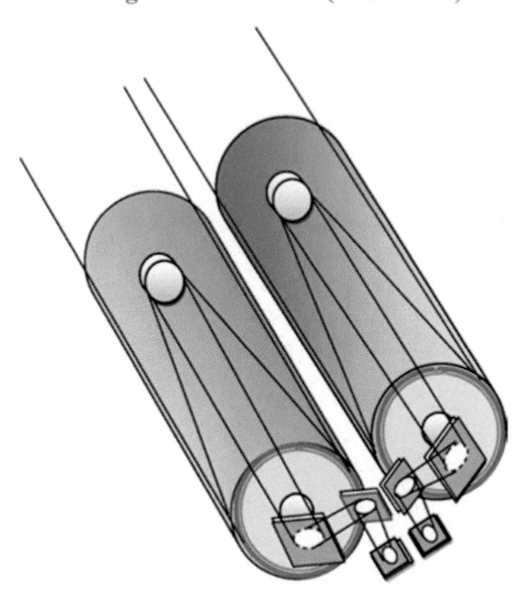

Fig. 6.13 Dual 6-in. f/15 Cassegrain binocular telescope optical diagram (Image credit: the Author)

The 10-in. f/4 "Dob Buster"

By Norm Butler

The 10-in. f/4 "Dob Buster" with the primary mirror located above the Dec. axis makes it a little unusual as Dobsonians go. It also uses a 16-lb plus bowling ball to counter balance it (see Fig. 6.17). It has four brass collimation knobs on each corner of the primary mirror box so that the observer can collimate the "Dob Buster" primary mirror while looking through the horizontal focuser's eyepiece holder. The horizontal focuser uses a lead screw design that has several certain advantages over a typical rack and pinion focuser that is normally not found on an average Newtonian telescope. It has a very smooth horizontal focusing movement. It has no lock screw to secure the eyepiece focuser and holder (which is made out of a 2¼-in. PVC coupling) in place. No matter how large and massive the 2-in. eyepiece, the horizontal focuser will keep its initial focus position in a rigid fashion without moving. And there are some pretty big 2-in. eyepieces out there. Some of them weigh more than kilo (more than 2 lb). The very smooth action of a lead screw when turning the focus knob makes for a nice smooth focus. Others who attempt making a similar horizontal single stalk focuser of this type might want to consider using an 18–20 TPI lead screw as a faster compromise for the focusing speed.

Fig. 6.14 The 10-in. f/4 "Dob Buster" with its 16-lb plus bowling ball counterweight (Image credit: the Author)

The 28 TPI is preferred because it is very much like micro-focusing. The single stalk secondary holder is made out of a steel rib from a paint roller. It was bent in two places in a right-angle configuration and attached to a 1/4-24 TPI threaded bolt by drilling a hole in the head of the bolt and epoxying the rod inside the drilled hole. After the epoxy hardened, it made for a strong and rigid single stalk. It is attached to the bottom of the wooden slide to hold a 3.14-in. Parks secondary mirror and its secondary holder, which itself is made out of a 2-in. plastic pipe normally used for yard sprinkler systems (Fig. 6.14).

The box that contains the 10-in. f/4 primary mirror is made out of 1/4-in. plywood with four sides (12.5 in. long × 12.5 in. wide × 12.5 in. high) with the bottom section having a 10-in. diameter access hole for attaching the bowling ball counter weight copper shaft to the pipe connector. Two round pieces of 5-in. O.D. aluminum pipe were used for the altitude axis trunnion bearings. They were cut off and sanded down to 1.5 in. in height. A round wooden plywood block insert was cut and fashioned and both press-fitted into each of the aluminum pipe altitude axis trunnion bearings. Likewise, I cut the same diameter/thickness plywood block, glued, and attached each (glued and wood screws)) on the inside of the box, directly opposite and centered behind each of the two aluminum trunnion bearings that are positioned on the outside of the box. Then two opposing pipe flanges were attached to a 1-in. pipe (threaded on each end) and the center pipe T-connector that would hold and center the position of the copper shaft of the 16-lb bowling ball counter weight. It all worked out quite well for this type of counterweight system. It made it very easy to install and remove

Fig. 6.15 The Dob Buster "sled" focuser (Image credit: the Author)

the bowling ball counterweight shaft by simply screwing it into the brass-threaded pipe T-connector. Plus, it has easy access to the interior of the mirror box to adjust, if needed, the pulley and O-ring collimation system (Fig. 6.15).

The horizontal single stalk lead screw focuser provides a very smooth platform to focus just about any 2-in. eyepiece, no matter what its size, mass, or configuration. It works as well as the traditional rack and pinion focusers that are typically seen on commercial telescopes. Thanks to the two spring-loaded steel ball bearings, there is no slide side play and no focusing slope or binding. It's a very smooth focusing system. The sled focuser is fun to use and, most important, turned out to be a very user-friendly system, and it pleasantly exceeded my expectations. The 16-lb plus bowling ball counterweight was great for this kind of Newtonian telescope design. I even added some more weight to it, adding two or three extra pounds of copper-coated BBs (the kind used in BB and pellet guns). I simply poured them down into the 1-in. diameter copper water pipe and through the threaded brass connector now embedded into what was the original thumb hole that was drilled into the bowling ball. Next was to pour some super glue into the top of the copper water pipe via the threaded connector to secure the BBs in place. The three extra pounds helped to balance out the entire wooden optical tube system in perfect fashion. Almost the entire telescope project in terms of materials including plywood, PVC, aluminum tubing, and plastic water pipe can be found in your local hardware store. Then it's the saber saw, electric drill, and wood glue. Toss in some wood screws for good measure and hand sand and finally stain it. And last but not least, add a big brass door handle and attach it at the balance point on the mirror box end, making it very simple to handle, transport, and set up. A little brass trim was added to give it some old fashion appeal. When you see it for the first time, it doesn't look like your basic Dobsonian. The "Dob Buster" really stands out in a crowd of telescopes at a local star party, and the low power wide field of view that the "Dob Buster" provides is very pleasing. For being a homemade all wooden Newtonian telescope, it is every bit the user-friendly telescope one could ever want or imagine. The "Dob Buster" is definitely a "one-of-a-kind" telescope and is a lot of fun to use (Figs. 6.16, 6.17, and 6.18).

Fig. 6.16 The Dob Buster single stalk secondary mirror holder (Image credit: the Author)

Fig. 6.17 The "Dob Buster" O-ring combination wooden pulley collimation system with the bowling ball counterweight shaft mounting flange in the center (Image credit: the Author)

Fig. 6.18 The Andromeda galaxy (M-31) as viewed through a small telescope or binoculars (Image credit: Public Domain—Wikimedia Commons)

Further Reading

Books and Monthly Periodicals

Amateur Astronomy (monthly periodical) 511 Derby Downs Lebanon, TN 37087. *Astronomy* (monthly periodical) P.O. Box 1612, Waukesha, WI 53187.

Astronomy Technology Today (monthly periodical) 1374 North West Dr. Stafford, MO 65757. English, N. (2011). *Choosing and using a Dobsonian telescope*. New York/London: Springer. Harrington, W. (1990). *Touring the universe through binoculars*. New York: Wiley.

Sky and Telescope (monthly periodical) 90 Sherman St. Cambridge MA 02140. Texereau, J. (1984). *How to make a telescope* (2nd ed.). Richmond: Willmann-Bell, Inc.

The Sky at Night (monthly periodical) Vineyard House, 44 Brook Green, Hammersmith, London W6 7BT.

Tonkin, S. (2014). *Binocular astronomy* (2nd ed.). London: Springer.

Chapter 7

1920s Binocular Telescope

Homemade binoscopes or binocular telescopes (as we know them today) have been around since the late 1920s, when Mr. Hilmer Hanson of Holdrege, Nebraska (one of the earliest pioneers of binocular telescopes), had his original homemade Newtonian binocular telescope introduced in Amateur Telescope Making Vol. 1 in 1931. Most if not all of the big binocular telescopes that we see today that are being built by some pretty clever amateur telescope makers can, in fact, trace their initial binocular telescope design or sometimes called Binocular Telescope's DNA all the way back to Mr. Hanson's original homemade binocular telescope that he made back in the 1920s. Mr. Hanson even ground and polished his own 6-in. mirrors, and considering the amount of materials he had available to him at the time, it was no small feat. When compared to his humble 6-in. f/8 binocular telescope (see Fig. 7.1), one can see little difference in the design of binocular telescopes that the amateurs are making today, at least when it comes to the smaller Newtonian-sized Newtonian binocular telescopes.

With today's amateur telescope makers, the trend is building very large Newtonian Dobsonian telescopes on Altazimuth mounts. There are some big Dobs out there that even exceed 42 in., and one big amateur size Dobsonian in Texas hits the scales at 48.9 in. There are even some commercial companies (Orion) that offer Dobsonian telescopes up to 50 in. The same trend seems to be catching on with big amateur-made binocular telescopes. At one time, homemade 17.5-in. binocular telescopes were considered the largest. They now are exceeding 22 in. If the same

N. Butler, *Building and Using Binoscopes*, The Patrick Moore Practical Astronomy Series, DOI 10.1007/978-3-319-46789-4_7

Fig. 7.1 Hilmer Hanson and his 1920s homemade 6-in. f/8 binocular telescope (Image credit: Reproduced with permission. Copyright (C) 1931 Scientific American, a division of North America, Inc. All rights reserved)

Fig. 7.2 Giant 49-in. Dobsonian verses a 6-ft. man's height (Image credit: Astronomy Technology Today)

trend continues for these big binocular telescopes, it's very possible that we will see some that are 36 in. or larger being built in the near future (Fig. 7.2).

Observing with a "giant" Dobsonian requires a long sturdy stable ladder that will allow the observer to stand confidently near the top and observe through the eyepiece without the fear of falling 15 ft to the ground. The larger the telescope, the more safety factors have to be observed. Another concern to be aware of would be birds flying overhead. As humorous as it may sound, a big 36-in. mirror, for example, would make a large exposed target for a wayward seagull or crow. If something like that happens, the big mirror would have to be recoated at great expense.

A good astronomy friend once told me "As a precaution, do not uncover your telescope's primary mirror until the birds have settled in the trees in the evening." That makes some good sense. The real cost of these huge Dobsonians is wrapped up in the cost of the large mirrors. In reality, few amateurs can afford a big 50-in. Dobsonian telescope, let alone a 36 in. and haul it around to various observing sights and set it up and then dissemble it and finally haul it back home again. One could only imagine if someone will ever make a big 50-in. binocular telescope someday, and if they ever do, it will certainly be a historic telescope to say the least. The mirrors are out there, but they're not cheap. So far, two or three 22 in. have been made and now an even bigger dual 28-in. f/4.8 binocular telescope has been built by Joerg Peters in Germany. One can almost foresee a big 30-in. plus binocular telescope being built someday and you can imagine the fantastic views that a binocular telescope of that size will provide. They will be nothing short of spectacular! Most people who are lucky enough to have made or bought a big binocular telescope have said that when you compare the images visually in a single big Dobsonian telescope to a big binocular telescope, it's like comparing "the Soap Box Derby to the Grand Prix." And that more or less sums it up in terms of what a binoscope is and what they can do. In short, a binoscope proves that two telescopes are better than one when it comes to visual observing! Using two eyes to view through binocular telescopes and refractor binoscopes will enhance the following areas of visual perception. The following are some of the reasons why:

1. Objects present more detail (acuity).
2. Fainter objects appear brighter (threshold detection).
3. Structure in many objects is more visible (contrast sensitivity).

When it comes to observing nebula, as in this case of viewing the Orion Nebula (M-42), individuals who have used a large binocular telescope like Bruce's Sayre's 22 in. have reported:

1. More color is seen.
2. More extensive nebulosity.
3. More apparent visual detail.
4. Red stars are in front of blue stars (as in Jewel Box cluster).
5. Nebula behind stars is visible.
6. More contrast is seen (e.g., Omega Nebula—M-17) (Fig. 7.3).

Fig. 7.3 Andromeda Galaxy (M-31) (Image credit: Adam Evans—Creative Common)

Binoviewers

Binoviewers are for the most part great for using two eyes to view celestial objects comfortably and with their added stereo effect and enhanced contrast, lunar, planetary, and galactic observing becomes a pleasant dual eye viewing experience.

Generally speaking, a binoviewer splits the incoming light beam in half, sending 50% to each eye. As a result, some feel a binoviewer is cutting the potential light grasp in half (see Fig. 7.4). Beyond that, binoviewers provide the observer with pleasing views that can't be had using a single telescope.

The one thing that just about all observers can agree is that when it comes to observing with a binoviewer, using them for general astronomical viewing, they do render some pleasing celestial views using both eyes. Binoviewers can produce a perceived stereo effect, especially in lunar and planetary viewing, but because they

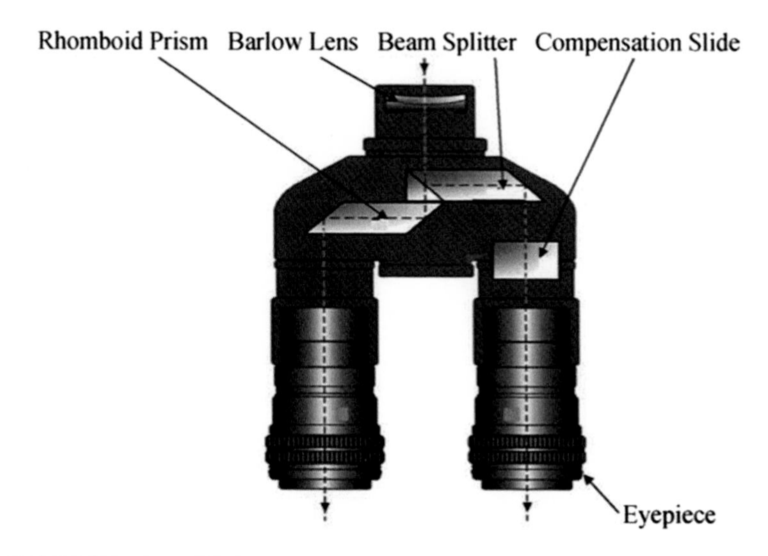

Fig. 7.4 Binoviewer optical diagram (Image credit: Tamasflex-Wikimedia Commons)

are prisms, they do have some light loss, up to 50% within their prismatic optical system. Some of it is made for in the gain in signal-tonoise ratio of approximately 41%. But is it enough to go out and spend $200 or more, plus purchasing a dual set of eyepieces too? Most of the commercial binoviewers offered using 1 ¼-in. eyepieces, and one would have to consider the appropriate eyepiece focal lengths and exit pupils in order to get the "optimum" performance out of a binoviewer, especially when it comes to observing deep-sky objects and various other celestial objects, like planetary nebula and close double stars.

One of the potential downsides about a binoviewer is if it were dropped or damaged, resulting in it being out of collimation or optical alignment. Obviously without instructions, you wouldn't be able to re-collimate it yourself and would have to send it back to the manufacturer to have it repaired at some expense and some time involved while waiting for your binoviewer to be returned in order to use it again. Also, if your telescope's drawtube does not have enough travel for focusing, you may need to use a Barlow lens to focus your new binoviewer. Binoviewers are fun to use and they have their advantages. I wouldn't want to discourage anyone from using them, especially if you do not have a binoscope or binocular telescope and you want to enjoy viewing the heavens using both eyes. Here are some companies that sell Binoviewers: William Optics, Celestron, Orion, Denkmeier, Siebert Optics, EarthWin, and Maxbright. Binoviewers are the next best thing to a binoscope for casual or even serious astronomical viewing using both eyes (Fig. 7.5).

Fig. 7.5 Celestron binoviewer (Image credit: Celestron.com)

Eyepieces

Today there are so many excellent choices for eyepieces; it's almost difficult to choose which one would work the best in your binocular telescope or refractor binoscope. Your best choices of high quality eyepieces depends mostly on what type of objects you want to observe and that will determine the magnification range you need to select your eyepieces for. Planets require relatively high magnification if you want to see any detail at all through your binocular telescope or refractor binoscope. Deep-sky objects require low and medium powers to view them with. You also want to consider the field of view, eye relief, and if you should wear glasses or contact lenses and the optical quality of the eyepieces. For a Dobsonian binocular telescope, it's better to have a range of eyepieces that produce magnifications that are in the low, medium, and high magnification range, and it's recommended "not" to use magnifications over 200× as a maximum for any Dobsonian binocular telescope. Again, use eyepieces that will produce high magnifications for the moon and planets, but keep the magnifications on the low side for nebulae, star clusters, and galaxies.

Some of the best brands of eyepieces to choose from are indeed the most expensive (especially the wide angle eyepieces) and usually have the highest quality. However, when it comes to eyepieces, many individuals who own Dobsonian and Newtonian telescopes spare no cost to have the best eyepieces for their telescopes. Some of the most popular brands of 1 1/4-in. and 2-in. eyepieces are TeleVue,

Nagler, Orion, Pentax, Leica, and Zeiss and are considered the "premium" brands of eyepieces (and there are others too) and would easy exceed the cost of your telescope. Some good quality name brand eyepieces today are listed in the low range of $115. Mid-range higher-quality eyepieces are listed for $300 and the highest price top-quality eyepieces (Nagler 31 mm type 5 Panoptic) list for $695. For example, here is what you can expect to pay for some of the more popular name brand two-in. eyepieces:

$300 Nagler 11 mm Type 6
$370 TeleVue Panoptic 27 mm
$115 Orion 21 mm Stratus Wide Field
$150 32 mm TeleVue Plossl

If you can afford some of these brand name eyepieces, then you'll have purchased some of the higher quality eyepieces that are available in today's telescope accessories market. It really pays to shop around for a good deal, especially when you're in the market for new eyepieces. But how does one choose the right magnification range for their eyepieces? First, establish a magnification range that you want to observe your favorite objects with. Then you need to calculate the magnification of the eyepieces you intend to buy. To do this, simply divide the focal length of your Dobsonian telescope by the focal length of the eyepiece you're interested in. For example, a 10-in. f/6 Dobsonian has a focal length of 60 in. × 25.4 mm = 1524 mm. The eyepiece you intend to purchase is a 32-mm TeleVue Plossl. Simply divide 1524 by 32 = 47.63 power. Once you have determined the range of magnification(s) you want to use with your binoscope or binocular telescope, selecting the eyepiece and their appropriate focal lengths in millimeters will not be so difficult.

Before you go out and spend several hundred dollars on eyepieces, it's best to check them out first. If you plan to attend a local star party, chances are there are club members who have accumulated a pretty good selection of eyepieces and they would be happy to have you try them out in your own telescope. For example, if they're using a group of brand name eyepieces in their Dobsonian telescopes, check out the eyepiece's eye relief first. Is it comfortable for you? Can your eye accommodate the image rapidly? One cannot underestimate the importance of "eye relief" in terms of viewing. Eye relief is a very important aspect of many eyepieces. Eye relief is the distance from the eyepiece to the observer's eye. The shorter the eye relief distance, the more difficult it is to observe comfortably. Also, if the observer wears glasses, short eye relief eyepieces can be very difficult or almost impossible to use. Long focal length eyepieces (usually considered low magnification) generally have long eye relief, so they do not need to be specifically designed to make them easier to use in terms of eye relief (Fig. 7.6).

The eye needs to be positioned at a comfortable distance behind the eyepiece lens to see the images through it properly. This distance is called the eye relief and usually ranges from 2 to 20 mm depending on the design and construction of the eyepiece. Eyepieces with a longer focal length usually gave somewhat ample eye relief, but eyepieces with a shorter focal length are a little more uncomfortable to use. Until just recently, it was still common to see eyepieces of short focal length

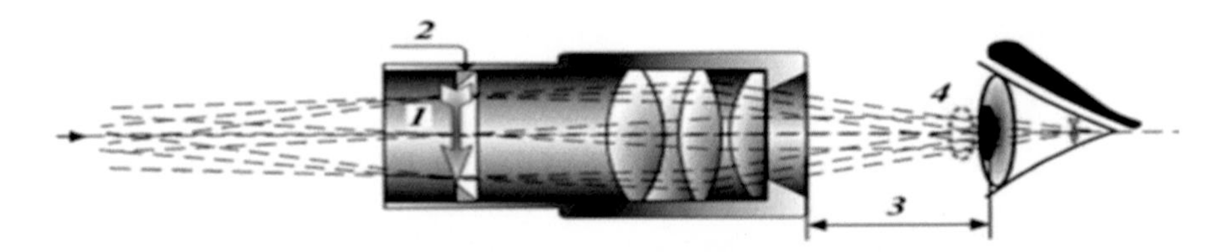

Fig. 7.6 An optical ray relief diagram showing the eye relief through an Orthoscope "Abbe" eyepiece (Image credit: Tamasflex-Wikimedia Commons)

having a short eye relief. Good design guidelines suggest an eye relief with a minimum of 5–6 mm to accommodate for the eyelashes of the observer for comfortable observing. Modern eyepiece designs with many elements, however, can make observing more comfortable for individuals who wear spectacles and who may need up to 20 mm of eye relief for their glasses. The eye of a spectacle wearer is typically further from the eyepiece and, as a result, needs a longer eye relief in order to still see the entire field of view.

Note:
An exit pupil larger than the pupil will allow more movement without vignetting. An exit pupil smaller than the pupil will cause vignetting of the image, which will make your viewing experience through a telescope or binocular's eyepiece somewhat uncomfortable. It's important that you chose eyepieces for your astronomical viewing that have good eye reliefs.

To quote a quite familiar old phrase, "A telescope is only as good as its eyepieces". This statement is in fact quite true, and if you have designed and built a truly wonderful optical telescope, you would naturally want to use some very good eyepieces with it in order to enjoy the wonderful celestial views it can provide. So selecting the right kinds of eyepieces with the right magnifications and fields of view is just as important as selecting the perfect high-quality primary mirror with an appropriate focal ratio and secondary for your telescope. In order for us to make the best choices for the kinds of eyepieces we need for our observing pleasure, let's include some historical discussion about the earlier types of oculars that were used at the dawn of the eyepiece era. Some of these early eyepieces are still found in rare use today, even though they are very much outdated compared to the new modern eyepieces that are commercially available. But even so, they are fun to use and to talk about and still quite capable of providing some pleasing astronomical views.

One of the first kinds of telescope oculars that appeared was in fact used on an early microscope. It was a simple convex lens used by Zacharias Janssen with his first compound microscope invented in 1590 (see Fig. 7.7).

Galileo used a single negative lens as an ocular in his first simple refracting telescope in 1609. We still use them and they are more commonly called "Galilean" eyepieces and they can still be found today in opera glasses, binoculars, and cheap telescopes (see Fig. 7.8).

A simple negative lens used as an early ocular was placed before the prime focus of the objective in a refracting telescope and produced an erect image but with little

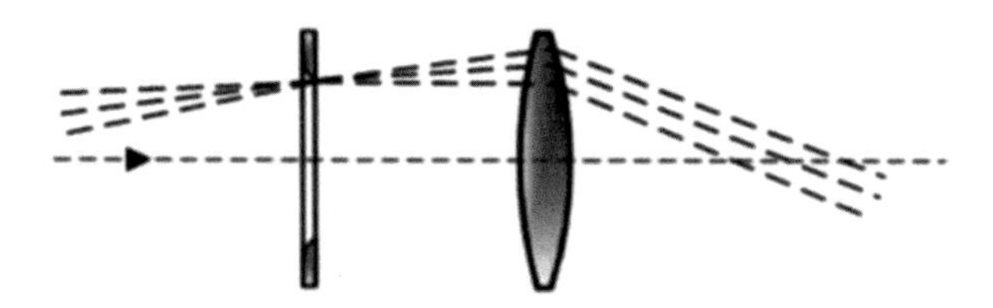

Fig. 7.7 A simple convex lens was used as an eyepiece in the first compound microscopes in 1590 (Image credit: Tamasflex-Wikimedia Commons)

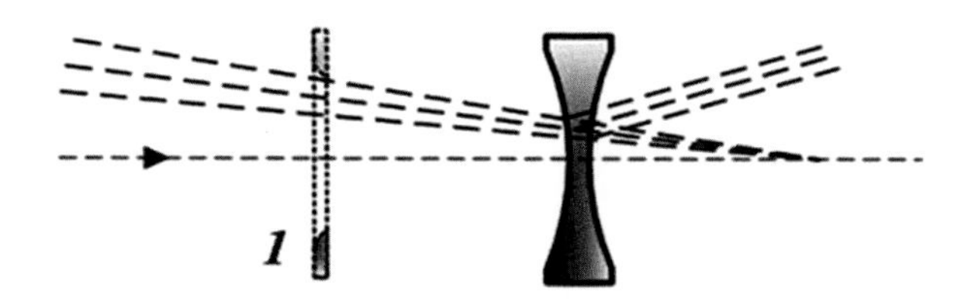

Fig. 7.8 A simple negative was the eyepiece used in Galileo's first refracting telescope in 1609 (Image credit: Tamasflex-Wikimedia Commons)

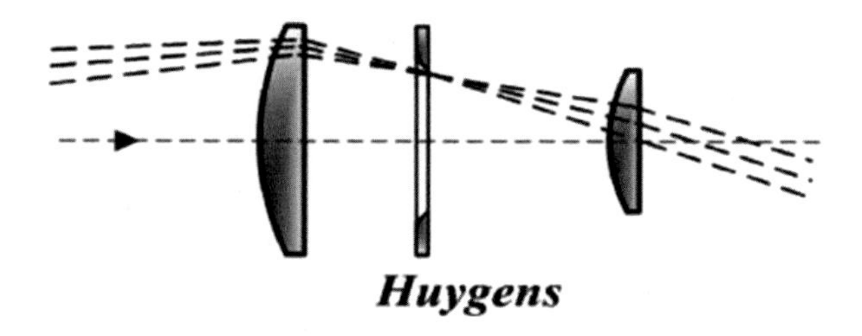

Fig. 7.9 The Huygens optical design consists of two plano-convex lenses with the plane side facing the eye (Image credit: Tamasflex-Wikimedia Commons)

apparent magnification. This type of ocular appeared in the first refracting telescopes in the Netherlands in 1608.

Christiaan Huygens invented the Huygens eyepiece in the late 1660s, which was the first of the compound (multiple lenses) eyepiece. The Huygens eyepiece consists of two plano-convex lenses with the plane (flat) side that pointed toward the eye separated by a short distance. The lenses are called the "eye lens" and the "field lens." The focal plane of the Huygens eyepiece is located between the two lenses. The Huygens ocular design is considered almost obsolete today and rarely used for astronomical observing, but because the lenses are not cemented, the Huygens ocular can still be used for solar projection on a screen without damaging the eyepiece and is suitable for use on long focal length telescopes (Fig. 7.9).

In 1782, Jesse Ramsden invented the Ramsden eyepiece which consists of two plano-convex lenses of the same glass type and with similar focal lengths. The two curved surfaces are facing inward toward each other (Fig. 7.10).

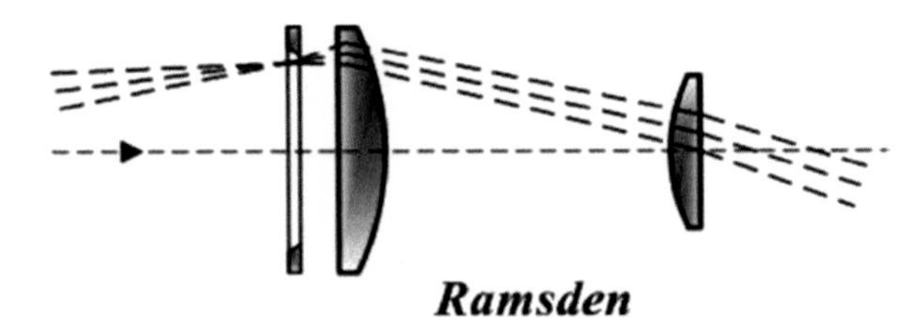

Fig. 7.10 The Ramsden two element eyepiece (Image credit: Tamasflex-Wikimedia Commons)

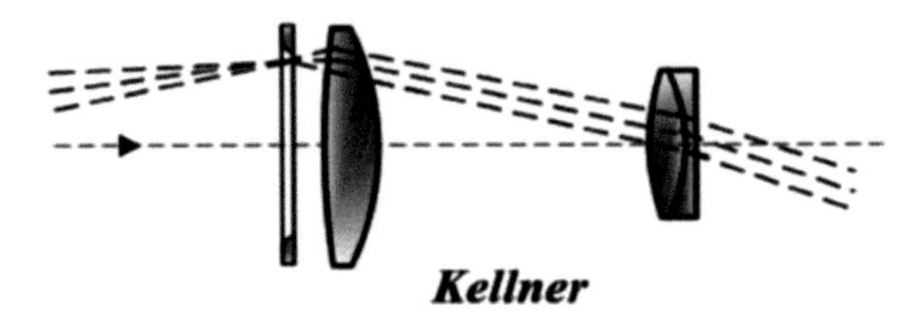

Fig. 7.11 The Kellner eyepiece is an achromatic doublet and is used to correct for residual transverse chromatic aberration (Image credit: Tamasflex-Wikimedia Commons)

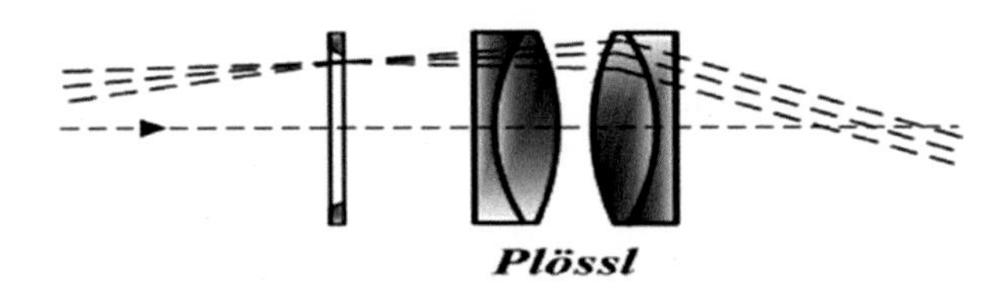

Fig. 7.12 The Plossl eyepiece is sometimes called a "Symmetrical" eyepiece and was invented by Georg Simon Plossl in 1860 (Image credit: Tamasflex-Wikimedia Commons)

The Ramsden eyepiece optical design is considered just a little better than the Huygens ocular but is still not quite up to today's modern eyepiece standards. It finds some limited use in various instruments that use, for example, near monochromatic lights sources, e.g., polarimeters.

Carl Kellner in 1849 designed this first modern achromatic eyepiece. It is sometimes called an "achromatized Ramsden" eyepiece and has three elements and is an achromatic ocular designed to correct for the residual transverse chromatic aberration. The Kellner eyepiece produces a good image from low to medium magnification range and is considered far superior compared to the Huygens and Ramsden eyepieces. With today's anti-reflection coatings, a Kellner eyepiece makes a smart economical choice for small to medium telescopes with focal ratios of f/6 or longer. The typical field of view in a Kellner eyepiece is approximately 40–50° (see Figs. 7.11 and 7.12).

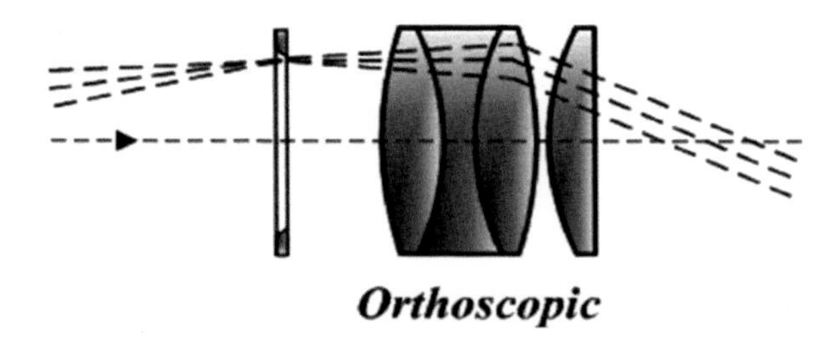

Fig. 7.13 The Orthoscopic eyepiece or "Abbe" is a cemented convex-convex triplet and plano-convex singlet lens eyepiece was invented by Ernst Abbe in 1880 (Image credit: Tamasflex-Wikimedia Commons)

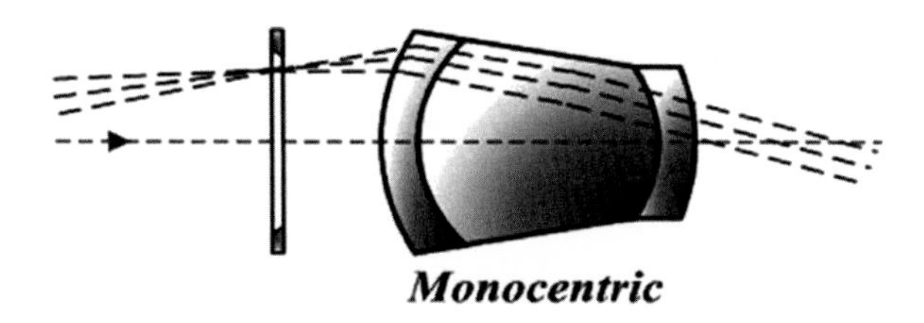

Fig. 7.14 The Monocentric eyepiece is an achromatic triplet lens that was invented by Adolf Steinheil around 1883. A favorite eyepiece among planetary observers (Image credit: Tamasflex-Wikimedia Commons)

The Plossl eyepiece is sometimes called a "Symmetrical" eyepiece and consists of two sets of doublets that are basically identical. The Plossl eyepiece provides a large apparent field of view 50° or more along with a large field of view (FOV) that makes it ideal for viewing a large variety of deep-sky celestial and planetary objects. It does have one notable disadvantage, which is having a short eye relief. But even in today's market, it is still very common and is considered a popular eyepiece for amateur astronomers to have in their inventory of telescope eyepieces (Fig. 7.13).

The four-element Orthoscopic or Orthographic or "Abbe" eyepiece as it is sometimes called consists of a convex singlet eye lens and cemented convex-convex triplet field lens achromatic field lens. This gives the image a nearly perfect image quality plus good eye relief too but a narrow field of view around 40–45°. The Orthoscopic "Abbe" eyepiece is considered a very good eyepiece for lunar and planetary observing due to their low degree of distortion. They are considered less suitable for applications which require a large amount of panning of the telescope. Until the advent of multi-coatings and the popularity of the Plossl eyepiece, Orthoscopics were the most popular for telescope eyepieces. Orthoscope eyepieces remain ever popular with even today's amateur astronomers and probably will be for a long time.

A Monocentric eyepiece (Fig. 7.14) invented by Adolf Steinheil somewhere around 1883 and is sometimes called the "planetary eyepiece" which is an achromatic triplet lens with two elements of crown glass cemented on both sides of

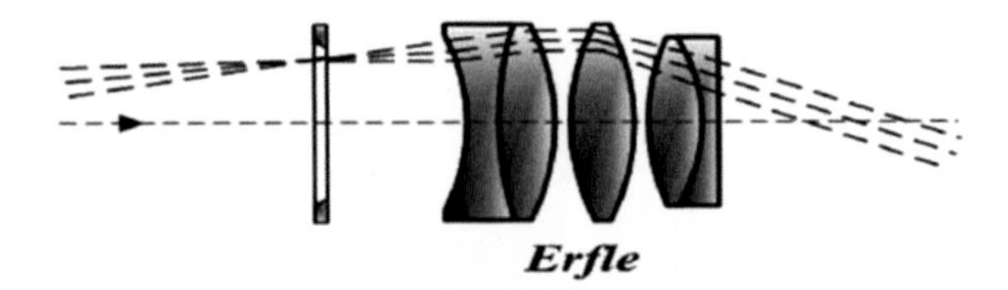

Fig. 7.15 The Erfle eyepiece was invented for military purposes by Heinrich Erfle during the World War I (Image credit: Tamasflex-Wikimedia Commons)

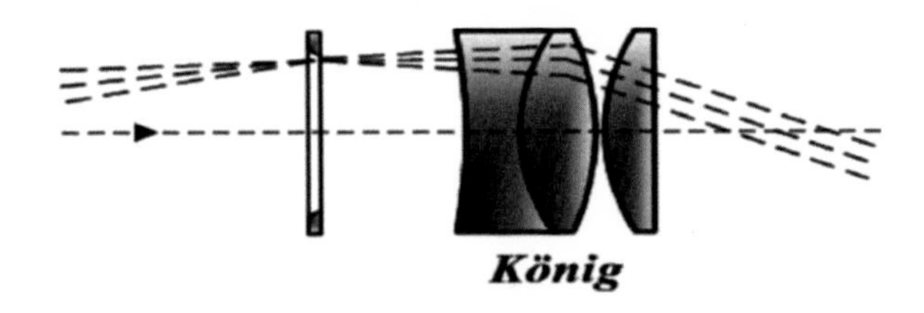

Fig. 7.16 The Konig eyepiece was designed in 1915 as a simplified Abbe by German optician Albert Konig (Image credit: Tamasflex-Wikimedia Commons)

a piece of thick flint glass. The glass elements are strongly curved and their curved surfaces have a common center giving it the name "Monocentric." The eyepiece is free from ghost reflections and produces a bright image with good contrast. The Monocentric eyepiece is considered a favorite among most planetary observers.

The Erfle eyepiece (Fig. 7.15) is a five-element eyepiece that consists of two achromatic lenses in between. It was invented by Heinrich Erfle during the World War I for military purposes. Erfle eyepieces are designed to have wide fields of view (about 60°) but unfortunately are not suitable at higher powers because they suffer from astigmatism as well as ghost images. However, with today's lens coatings, at low powers of 20 mm and up, they are suitable, and at 40 mm, they become exceptionally good eyepieces with their large eye lenses and good eye relief and are very comfortable to use when observing.

The Konig eyepiece (Fig. 7.16) is a three-element concave-convex doublet and a convex-flat positive singlet with strongly curved surfaces of the doublet and singlet lens that face and almost touch each other. The doublet has its concave surface facing the light source and the singlet has its almost flat and slightly convex surface facing the eye. The Konig ocular design allows for high magnification with excellent eye relief. Typical field of view in a modern-day Konig eyepiece using high quality glass and with various improvements in design is approximately 60–70°.

The RKE eyepiece was designed by Dr. David Rank for the Edmund Scientific Corporation and has an achromatic field lens and a double convex eye lens, a reversed adaptation of the Kellner eyepiece. The Edmund Scientific Corporation marketed the RKE eyepiece throughout the 1960s and the early 1970s. The RKE eyepiece provides a slightly wider field of view than the classic Kellner eyepiece and makes it similar in design to a widely spaced version of the Konig eyepiece.

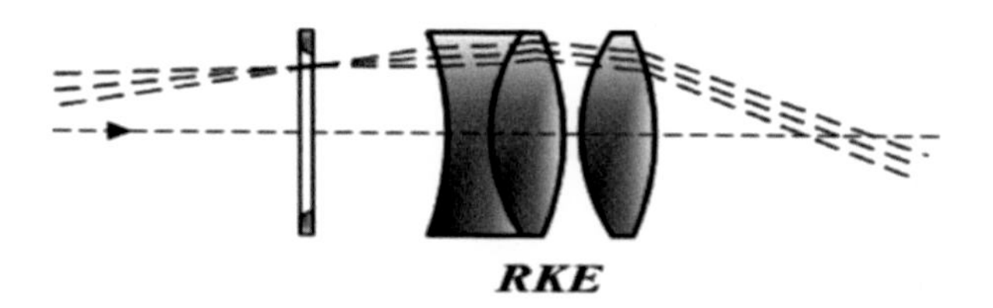

Fig. 7.17 The RKE eyepiece was marketed by Edmund Scientific Corp. in the 1960s and early 1970s (Image credit: Tamasflex-Wikimedia Commons)

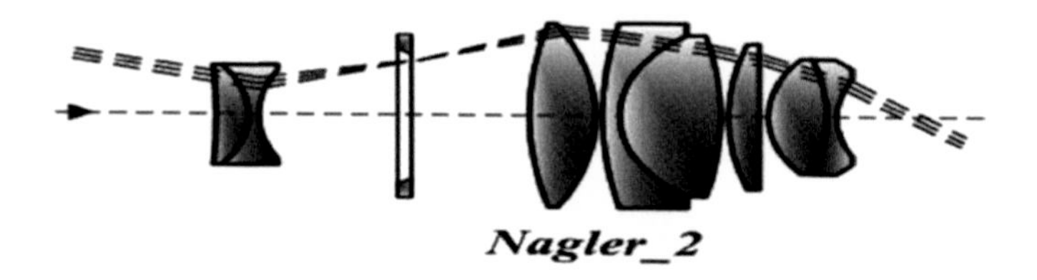

Fig. 7.18 The popular Nagler type 2 eyepiece diagram (Image credit: Tamasflex-Wikimeda Commons)

Edmund Scientific in accordance with a 1979 trademark amendment indicated that RKE stands for Rank, Kaspereit, and Erfle, the three designs that it was derived from (Fig. 7.17).

The Nagler eyepiece was invented by Albert Nagler and patented in 1979. The Nagler eyepiece design was optimized for astronomical telescopes to give an ultra wide field of view of 82°, and it has been corrected for astigmatism and other aberrations that were common to other eyepiece designs. The popular Ethos eyepiece, especially in the amateur astronomy market, is an ultra-wide field design optimized version with a 100–110 apparent field of view. The Ethos design is achieved by using exotic high-index glass and up to eight optical elements in four or five groups (Fig. 7.18).

There are five similar designs called the Nagler, Nagler type 2, Nagler type 4, Nagler type 5, and the Nagler type 6. The newer Delos design is a modified Ethos ocular version with a field of view of only 72° but comes with a long 20-mm eye relief. A Nagler eyepiece in reality is an improved version of a Barlow lens combined with a long focal length eyepiece. The Nagler initial weight of 0.5 kg (1.1 lb) and its long focal length is sometimes considered a disadvantage (Fig. 7.19)—(Adapted from: Eyepiece—Wikipedia).

A Nagler eyepiece comes with a high purchase price that is comparable to the price of a small telescope. But with today's modern telescopes, a Nagler eyepiece is highly sought after as a luxury eyepiece for anyone's eyepiece collection of quality telescope accessories.

When it comes to selecting good high-quality eyepieces, a good way to check out a potential eyepiece is to inspect the star images from the center to the extreme edge of the field of view. If the star images appear as pin points without any

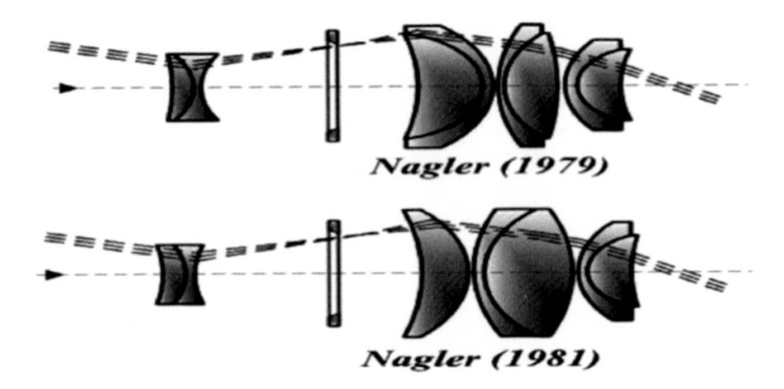

Fig. 7.19 The 1979 and 1981 Nagler eyepiece designs (Image credit: Tamasflex-Wikimedia Commons)

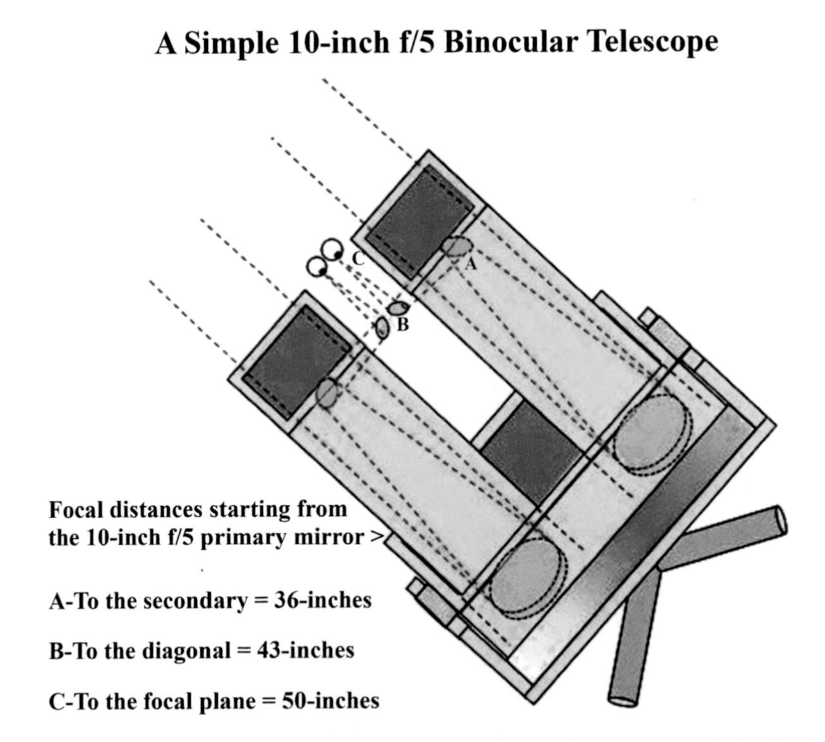

Fig. 7.20 An optical diagram of a simple 10-inch f/5 (50-inch F.L.) binocular telescope on a wooden Altazimuth mount. Note: Drawing not to scale and is used here for illustration only. (Image credit: the Author

noticeable aberration, then consider that a very good eyepiece, and depending on its focal length and apparent field of view, you might want to consider using an eyepiece with your own particular telescope someday (Fig. 7.20).

How Does a Barlow Lens Work?

The Barlow lens invented in 1833 is named after its inventor Peter Barlow. A Barlow lens is a diverging lens used in conjunction with other lenses in an optical system. A Barlow lens increases the effective focal length in combination with all the optical components that are directly after it in the optical system. A true Barlow lens is not a single lens, because that would create chromatic aberration and spherical aberration if the lens is not aspheric. More common configurations of the Barlow lens can use up to three or more optical elements for achromatic correction or apochromatic correction and higher image quality. In terms of its astronomical use, a Barlow lens may be placed immediately before an eyepiece to effectively decrease the eyepiece's focal length by the amount of the Barlow's divergence.

Since the magnification provided by the telescope and eyepiece is equal to the telescopes focal length divided by the eyepiece's focal length, this has the effect of magnification of the image by a factor of 2× or 3×.

An astronomical Barlow lens is rated for the amount of magnification they induce. Adjustable Barlow lens is also available. The power of an adjustable Barlow lens is changed by adding an extension tube between the Barlow and the eyepiece to increase the magnification. Generally speaking, the amount of magnification is one more than the distance between the Barlow lens and the eyepiece lens (Fig. 7.21).

What Does the Term "Field of View" Really Mean?

What's always been a little confusing to understand is the difference between the terms "true field of view" and "apparent field of view." So what's the difference? "True field of view" (also known as angular field of view) refers to the measurement in degrees of how much sky you can actually see when you insert an eyepiece in a telescope and observe with it. If you want to determine your actual "true field of view" in "real time," position a star (Fig. 7.23) near the intersection of the celestial equator and the meridian at the edge of your eyepieces field of view. Then turn off

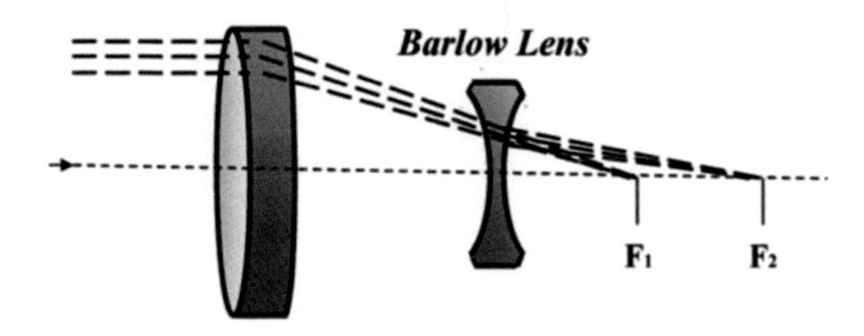

Fig. 7.21 Simple diverging lens placed behind the objective increases the focal length and magnification (Image credit: the Author)

your telescope's drive and let the star drift across the field of view (via the earth's rotation) to the opposite edge of the eyepiece's field of view, and then record the time in seconds. An example would be 1 min and 20 s. Then convert it to total seconds = 80 s. Then convert it to degrees = 1/2° "true field of view." By coincidence, is the same as the moon's angular diameter too.

Figure 7.22 represents three simulated images of the Andromeda galaxy (M-31) using different eyepieces. The image in the center uses an eyepiece of the same focal length as the one on the left, but has a wider "apparent field of view" resulting in a larger viewed image that shows more area. The image on the right also has a shorter focal length, resulting in the same "true field of view" as the left image, but at a higher magnification. The "apparent field of view" is the measure of the angular size of the image if you were looking through the telescope's eyepiece. This is actually a very useful measurement in terms of a telescope eyepiece, since the same eyepiece may be used in a variety of different telescopes and other optical systems. Apparent field of view allows you to compare the field of view (FOV) with different eyepieces. For example, if you are comparing a 15-mm Pentax eyepiece with a 52° apparent field of view (FOV) to an eyepiece with a 70° apparent field, the 70° eyepiece will produce the wider field of view (FOV), though both are 15 mm and both eyepieces produce the same magnification. Even so, this is not always true with eyepieces of with different focal lengths, because different focal lengths produce different magnifications and magnification also affects the field of view (Fig. 7.23).

True (actual) field Of view (FOV) can be measured in star drift time or sometimes called the drift method in seconds as the star drifts across the eyepiece's field of view (FOV) in a star diagonal from West to East.

True (actual) field of view (degrees) = time in seconds/240 s. Example: 120 s or 0.5 or 1/2°.

$$\text{Actual field of view} = \frac{\text{Apparent field of view}}{\text{Scope focal length}(\text{mm}) / \text{Eyepiece focal length}(\text{mm})\backslash}$$

Fig. 7.22 Simulated fields of view of the Andromeda galaxy through a telescope's eyepiece (Image credit: Stellarium—Wikipedia)

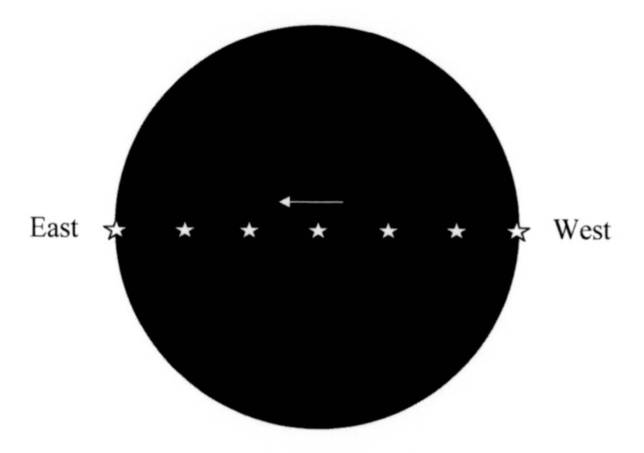

Fig. 7.23 Looking into the eyepiece (Image credit: the Author)

$$\text{Or} = \frac{\text{Apparent field of view}}{\text{Magnification}\backslash}$$

Below are some helpful equations and formulas.

One can calculate the magnification of a lens or a telescope optical system using the equations expressed below:

$$\frac{1}{f} = \frac{1}{di} + \frac{1}{do\backslash}$$

The equation stated above expresses the quantitative relationship between the object distance (do), the image distance (di), and the focal length (f).

$$M = \frac{hi}{ho} + \frac{di}{do\backslash}$$

The magnification equation expressed above relates the ratio of the image and object distance to the ratio of image height (hi) and object height (ho).

Note:
These two equations can be combined to yield information about the image distance and image height, if the object distance, object height, and focal length are known.

Use this equation to determine the magnification of an eyepiece:

$$M = \frac{\text{F.L. Telescope}(\text{mm})}{\text{F.L. Eyepiece}(\text{mm})}$$

$$\text{Example: } \frac{1{,}200\text{mm}}{25\text{mm}} = 48\text{M}$$

A "Periscope-Style" Right-Angle Finder

The importance of finders and their use on telescopes of all kinds cannot be underestimated. Finders serve a purpose and the average astronomical observer would be lost without one. Most if not all of the finders found on telescopes today are refractors with a low power eyepiece and a cross hair (illuminated or not) that yields a fairly large field of view. Most telescope finders utilize two separate support rings and adjusting screws that can center the finder on the object that is the object of observing interest in the main telescope. The traditional finder design consists of an achromatic objective lens and a low power eyepiece and cross hair. Some are illuminated and some are not. A telescope's finder is usually positioned and secured on the telescope's main tube parallel to the telescope's optical axis. While some employ a right-angle prism diagonal for viewing, others are sometimes viewing through the eyepiece with no diagonal at all. Finders can come in all sizes with objectives up to 80 mm in diameter. Some are as small as 20 mm. Some finders possess a single cross hair, while others have a double cross hair. The Telrad finder is probably one of the more popular types of finders that one can find on some of homemade amateur and commercial telescopes (Fig. 7.24). It utilizes a familiar illuminated two and four degree circle FOV "Heads Up" type of display that can make finding celestial easier than using some of the more traditional finders. It has been many years since a different type of finder has been seen on amateur telescopes. The "Periscope-Style" right-angle finder made its first appearance on the homemade 10-in. f/4 "Dob Buster" telescope in 2011 (see Figs. 6.14, 6.15, and 7.25).

Fig. 7.24 (Cartoon credit: the Author)

Fig. 7.25 A "Periscope-Style" right-angle 30-mm finder on the 10′ f/4 homemade "Dob Buster" Newtonian telescope (Image credit: The Author)

The "Periscope-Style" right-angle finder has a couple of distinct advantages over a typical standard type of finder that one finds on the majority of amateur telescopes. The first important advantage is that there is no set of support rings needed to support the "Periscope-Style" right-angle finder. The finder's optical system is housed within its wooden case and can be aligned on an object by adjusting the three screws that move the small first surface mirror about in the alignment process. Once aligned on an object, just a small twist of the lock nuts secures the mirror in place. Another advantage is that it has a very low profile and can be attached to the primary telescope in a more permanent way or used as magnetic base to attach it and/or remove it when the observer decides to use his telescope without a finder. If the "Periscope-Style" right-angle finder is placed in a horizontal fashion near the eyepiece focusing system as in Fig. 7.25, one only has to move his eye only a few inches to view through the finder, making it more convenient to use (Fig. 7.26).

Making a "Periscope-Style" right-angle finder is a fun project. Not only is it easy to make, but one can use the majority of parts from a standard finder to make the "Periscope-Style" right-angle finder. The only thing needed would be to obtain a small first surface mirror that is at least 1/8 wave or better. The finder objective is from a standard 30-mm Celestron finder and the mirror is a small 32 mm × 32 mm first sur-face mirror that had been acquired years ago. It's the author's opinion that it would be easier

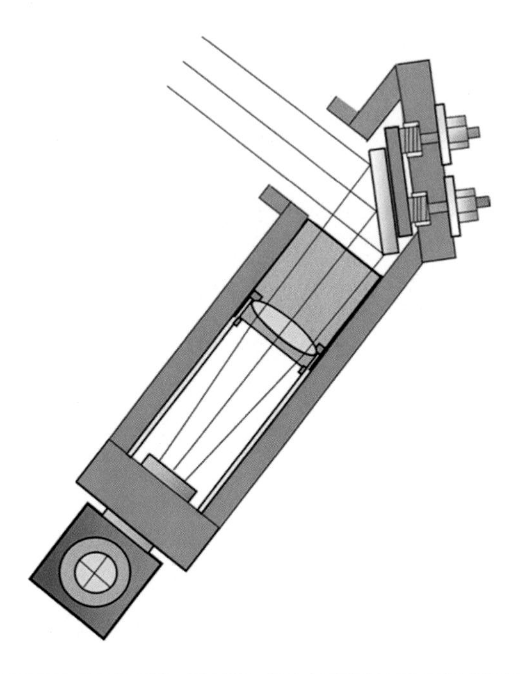

Fig. 7.26 A “Periscope-Style” right-angle finder diagram (Image credit: the Author)

to build and assemble the “Periscope-Style” right-angle finder around the original standard finder tube assembly with the objective in its cell and connected to the original finder tube as was done with the 30-mm “Periscope-Style” right-angle finder.

There are times when one doesn’t need to use a finder, and having the option of removing it when desired without disturbing its original alignment is really a time saving advantage when it comes to observing. Building this type of finder with a magnetic base made this option possible, and if you’re going to build a big binocular telescope, then perhaps you might consider a “Periscope-Style” rightangle finder to use on your binocular telescope.

Multiple Mirror Newtonian

We’ve seen some pretty interesting binocular telescopes that are pictured and described in the book with lots of neat and clever ideas that amateur telescope makers have designed and built into their binocular telescope. Just about any telescope design has its strengths as well as limitations. As an exercise, let’s take a look at a multiple mirror Newtonian telescope. What are the advantages a multiple mirror Newtonian telescope has over a typical binocular telescope of the same aperture? The answer is that it has more light gathering power and resolution. Can it out perform a binoscope or binocular telescope of the same aperture? A multiple mirror telescope has the potential (depending on its focal ratio) to outperform a binocular telescope of the same aperture!

How difficult would it be to build? It will take a little more time to construct, but it would not be any more difficult to build than a binocular telescope. So let's first take a closer look at a multiple mirror Newtonian telescope with three 8-in. primary mirrors. The multiple mirror Newtonian telescope has a total of three 8-in. f/8 mirrors that have a total of 192 square inches of light gathering power compared to an 8-in. binocular telescope that has 128 square inches of light gathering power. The 8-in. f/8 multiple mirror Newtonian has 1.5 times the light gathering power.

Collimation and the alignment of a Newtonian optical system with a combined total of nine mirrors would be challenging to say the least. To help accomplish this, a possible solution is the use of three adjustable (the key word here is adjustable) secondary mirrors that have the appropriate size. Each is attached to the front of an "adjustable" aluminum holder. Three elliptical-shaped secondaries reflect the incoming cone of light at a 45° angle to each of the adjustable individual triangle- shaped secondaries. Figure 7.29 shows in some detail the optical arrangement of the three primaries and the three elliptical secondaries along with three center-mounted elliptical secondary mirrors and their aluminum holder in a three-mirror Newtonian telescope. In my future book about multiple mirror telescopes, I will describe in detail how this kind of optical system would work using a total of six elliptical secondary mirrors with three primary mirrors in an 8-in. f/8 Newtonian optical system, along with the attributes and limitations.

Note:
It could be a little challenging to design an "adjustable" center-mounted elliptical mirror holder system that supports three elliptical secondary mirrors that would divert the individual cones of light coming from the three Newtonian OTAs secondary mirrors. But I believe a multiple mirror system like the one illustrated in Fig. 7.29 can be built and, once completed and tested, would be a very rewarding and successful project and will have an appropriate place in amateur telescope making history.

Homemade and commercial refractor binoscopes and binocular telescopes are rapidly becoming more and more popular for big binocular astronomy. In another 5 years or so, binocular telescopes and refractor binoscopes will be fairly common instruments to be seen at local star parties, amateur telescope maker's conferences, and national and international astronomy conventions. Today's amateur telescope makers have access to more high-tech materials, bigger and lighter weight primary mirrors, better coatings, as well as a multitude of accessories to increase their viewing pleasure with. With the rise in popularity of binoscopes, then it's only a question of time before we start to see amateur telescope makers experimenting with multiple mirror telescopes, perhaps similar in design as the multiple mirror telescope in Fig. 7.29. Once one is successfully built, then others will follow. One of the main reasons why a multiple mirror telescope is going to be a popular amateur telescope making project someday is because the images perceived in it will be far superior in brightness and resolution than those seen in a single telescope of the same aperture. Once observers get a chance to view images through a multiple mirror telescope, it should be enough incentive to "light the fire" to start building multiple mirror telescopes just like Lee Cain's 17½-in. binocular telescope did when it was first introduced to amateur telescope makers at the Texas Star Party back in 1984. In reality, when three images are superimposed (stacked) on top of each other and focused in an eyepiece of a multiple mirror telescope, the result is going to be an image that will be very intense in its brightness and resolution and possibly even show more depth than a binoscope of the same aperture. A simple multiple mirror telescope design could have three 8-in. f/8 primary mirrors, three secondaries, and three center-mounted secondaries and be mounted on an Altazimuth GOTO mount with a Cassegrain-style rear focus.

One of the more interesting questions that sometimes brought up in a forum discussion about multiple mirror telescopes is: "Why haven't we seen an amateur made multiple mirror telescopes at star parties or telescope makers conferences on occasion?" Good question. The evolution and introduction of new, unique and potentially good telescope designs has always been a little slow to catch on and, in some cases, even slower to be accepted by the amateur astronomy community. It's very possible that a working multiple mirror telescope design has already been made by some creative amateur telescope maker. But a complex multiple mirror design and other problematic issues may have slowed the telescope's progress. The more mirrors you have in your design, the more difficult the collimation and overall optical alignment. Consider the complicated design of a ten mirror Dall-Kirkham Cassegrain binocular telescope. It's no wonder why binocular telescopes and binoscopes have taken such a long time to evolve to the point where they are today. With the availability of new materials and bigger and lighter mirrors, we should see some new big binocular telescopes being made in the next 5 years or so. Europe, Asia, Australia, and New Zealand may make significant contribution to the work amateur telescope makers have done in Northern America.

Another common question that amateur telescope makers have asked is "What is the most challenging thing about making a multiple mirror telescope?" In my opinion, it's designing and building an efficient center-mounted adjustable secondary mirror system. The centrally-mounted adjustable secondary mirror system has to divert the light coming from the initial three individual Newtonian secondaries at a 45-° plus angle and superimpose each individual cone of light on top of each other (more commonly called stacking) and bring each of them to a focus at the same exact distance below the primary mirrors (see Fig. 7.29). Once all three images are superimposed with each other, expect a very bright intense image that has great depth and resolution. That is why a multiple mirror telescope will be superior when it comes to imaging, more so than a binocular telescope of the same aperture.

The multiple mirror telescope pictured in Figs. 7.27 and 7.28 was designed and made by Harvard model maker Mr. Hal Robinson and Mr. Ken Launie of Massachusetts. It's surprising to think that we haven't seen more amateur-made multiple mirror Newtonian or Cassegrain telescopes since their first homemade multiple mirror telescope appeared on the scene at Stellafane in 1979. However, here is a recent update from Mr. Ken Launie on the progress of this interesting amateur multiple mirror telescope.

A Work in Progress

It's correct that the scope was never finished.

We were inspired by a late 1970s talk given at a meeting of the Boston ATMs at Harvard by Nat Carleton about the original "MMT," which he was helping to lead.

Fig. 7.27 Hal Robinson and his unique multiple mirror telescope at the 1979 Stellafane convention (Image credit: Jack Welch and the Springfield Telescope Makers)

Fig. 7.28 Another view of the very unique multiple mirror telescope designed and built by Ken Launie and Hal Robinson at the 1979 Stellafane convention (Image credit: Denis Milon/Ken Launie)

Hal happened to already have the very light and strong satellite frame that reminded us very much of the MMT's optical support structure, something he'd found in a California aerospace junkyard. After the meeting, we went straight to his house and measured the openings, seeing that we could tuck four 6″ mirrors in the sides, and a central 8″ (rather than waste that aperture), and just like that, the project was born.

The Optical Design

The optical design was for the five separate Cassegrain systems to have the same net focal length, with a very fast primary and a high secondary magnification in order to make very long and shallow light cones that would be combined through mirrors to a single focus point. The four off-axis beams were all slightly tilted, of course, so the usable field would be quite narrow; photometry was how we justified the project to ourselves, though we still wanted to look at planets to see how it'd do on-axis. To help combine and fine align the systems, we had four 1″ thick plane parallel quartz windows in the side optical paths that could be tip-tilted to shift each beam slightly to line up with the central one. Alignment was done using a laser at a significant distance, and I remember how "high tech" we felt in the 1970s to have our own Honeywell laser system, with a base unit and a cord to the separate "head."

Making the Quartz Mirrors

Hal made the primaries and I did the secondaries (of quartz). One of the things that made it seem a little more easily doable was that I was working at Polaroid, and instead of a Hindle test, I was able to bring the convex secondaries to Polaroid's Henry Street optical shop/laboratory, where they could be measured by "Clyde," Polaroid's massive in-house-designed optical measurement profilometer. That was interferometrically controlled and had been built primarily to measure the injection mold cavity inserts for plastic lenses and the viewfinder optics of the SX-70. Those often had very complex higher-order aspheres, some of which, as in the case of the SX-70, were not even rotationally symmetric. "Bonnie" was the similarly controlled grinder to make the shapes, with human opticians to do final polishing as needed (while repeatedly measuring with Clyde). At the time, I think they were unique, though 35 years later the technology is now much more common, being used for diamond turning. I would drop off a secondary on my way in to work in the morning and pick up the previous secondary and bring it (and its printout) to my office, where I had a very small DC motor spindle, and I'd spend my lunch hour polishing/correcting. On my way home, I'd drop it back off (or another if I didn't think I'd made enough progress to justify the retest) and repeat the cycle the next day. The quartz didn't polish as quickly as the Pyrex I was more familiar with, and we only got far enough to get two beams to combine, something we played with at its second Stellafane appearance.

The Alignment

Alignment WAS a bear, and we thought that for practicality (as if anything about the project was practical) what was really needed was some sort of on-board laser setup for each optical system. Now tiny lasers are everywhere, but at the time we only had a couple of very expensive big lasers found on the surplus market, and frankly, after getting to see beams combine from two systems, we decided we'd spent enough time over the 2 years and needed to get back to everything else that had been neglected during our joint obsession.

An "Unfinished Project">But still an impressive project nonetheless (the Author) It sits in a very big box in my basement, with the yoke nearby, and the space it occupies has me thinking about eventual destinations for it. David Devorkin is potentially interested if we can figure out what satellite project the frame was meant for. Once something is in a museum, it drops off one's "unfinished projects" list as far as I'm concerned, as it's then an artifact that I'm not allowed to modify (Fig. 7.28).

Ken Launie

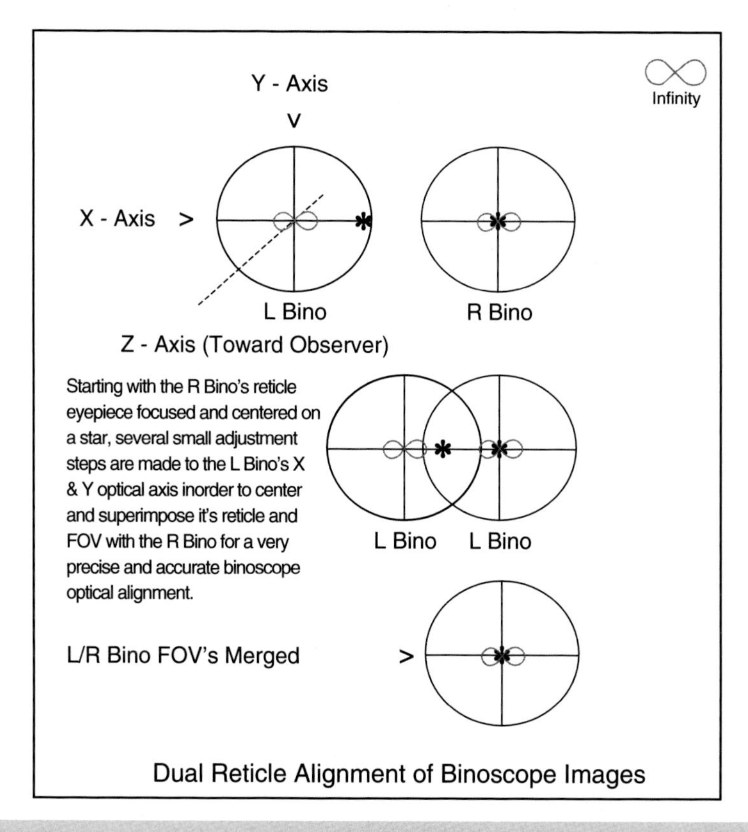

Fig. 7.29 The use of a reticle eyepiece (12mm to 20mm for example) combined with a 2x Barlow lens will help the observer achieve an accurate level of binoscope optical alignment compared to using a singular non-reticle eyepiece. (Image credit: the Author)

Coatings

It's always good to know what kind of coatings you have on your telescope's optics. But at the same time, it can be also very confusing trying to understand what kind of benefits can be derived from the various optical coatings that are so common on today's modern optics, especially if you're shopping around for a refracting telescope, a good pair of binoculars, or spotting scope. Unless you know for sure the types and kinds of coatings that are on your optics when you purchased it, then it would be practically impossible to tell just by looking at them. Anti-reflection coatings, multiple coatings, dielectric coatings, and high-reflective coatings all come in different colors. When you're walking around in a store that sells telescopes and binoculars, one look around and you can immediately see the variety of different colors that coatings have on the lens surfaces. So in order to get a better understanding of the types of coatings that are available, let's start with mirror coatings first.

The commonly known process of depositing vapor-thin layers of metallic coatings, such as aluminum or silver, on glass and other kinds of substrates to make mirror reflective surfaces is known as silvering. The actual metal coating used determines the reflectance characteristics of the mirror's surface. Aluminum (Al) is the cheapest metallic coating and it is the most commonly used metal coating, especially when it comes to telescope mirrors. For example, the aluminum coating thickness for a mirror with a 15-in. radius would have an average coating thickness of 1.53 waves. Aluminum has a reflective yield of around 88–92% over the entire range of the visible spectrum. Silver (Ag) as a precious metal is more expensive and has a nominal reflectance value that can range from 95 to 99% and even extend far into the infrared regions of the visible spectrum, but there is a noticeable large drop in its reflective ability (<90%) in the blue and ultraviolet parts of the spectrum. The more (most) expensive of the precious metals is gold (Au) which has the highest (98–99%) reflectance range throughout the infrared spectrum. It's reflectivity is somewhat limited at spectral wavelengths that are lower than 550 nm, and the typical gold color is the result that one would see in a gold metallic coated mirror.

When it comes to the selection of metallic coatings for your primary, secondary, and diagonal mirrors, then the following information will be important, especially if you want to get the most reflectivity in the visible spectrum. Contrary to popular belief, standard aluminum coatings have much better reflectivity for certain wavelengths of light than silver coatings. When it comes to aluminum coatings for your telescope's primary and secondary mirrors, you can expect them to have much better reflectivity from about 1 nm (nanometer) wavelength of light to around 500 nm and have borderline reflectivity from about 500 nm through the rest of the visible spectrum. Aluminum coatings can have a standard reflectance range from 87% to as high as 93% maximum reflectivity of visible light and the reflectance level of a silver coated mirror can be brought up to over 95–99% with an enhanced coating. Keep in mind that aluminum has the potential to oxidize more rapidly if it's not over-coated with, for example, silicon monoxide or silicon dioxide/silicon monoxide for enhanced reflectivity.

Visible Spectrum

Your eyes are normally sensitive in the visible spectrum to wavelengths that typically have a range from 400 to 700 nm (430–790 THz). Beyond 700 nm, your eye's sensitivity to the visible spectrum drops off rapidly as it is seen in the black in Fig. 7.30. For an average person with healthy eyes, one can expect their eyes maximum sensitivity to be in the nominal range of 500–550 nm. Wavelengths that

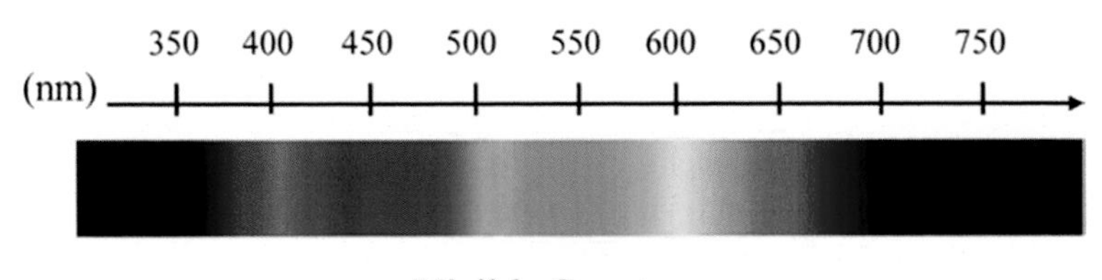

Fig. 7.30 The visible spectrum (Image credit: the Author)

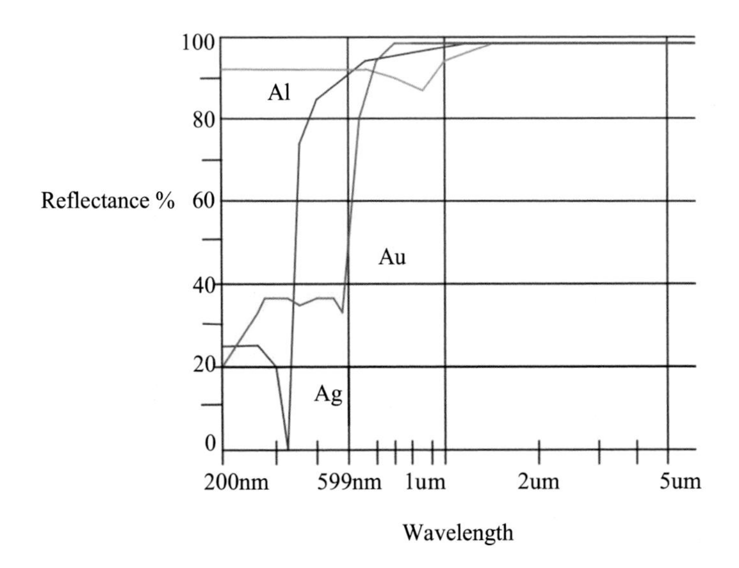

Fig. 7.31 Reflectance versus wavelength chart for aluminum (Al), silver (Ag), and gold (Au) metal optical coated mirrors at normal incidence (Image credit: the Author)

are longer than 750 nm are, in fact, part of the infrared (IR) spectrum and are usually for the most part, not visible for most people's eyes (Fig. 7.31).

Dielectric coatings are commonly referred to as thin-film coatings or interference coatings. They are in fact one of the most familiar types of optical coating (also known as a Bragg mirror) and are typically found on the optical surfaces of lenses used in binoculars, cameras, spotting scopes, eyepieces, microscopes, etc. It is a type of an optical coating that is comprised of very thin multiple layers (typically in the submicron range) of dielectric material, likened to a vapor deposited on the surface of a substrate of glass or some other optical material. Thanks to today's modern, commercial coating techniques, it allows us to create an optical coating by a specific process design, choice and selection of the type, and thickness with a certain number of the dielectric layers that can and by it's initial design, produce a specific level of reflectance at different spectral wavelengths of light. Using special optical coating processes, dielectric coated mirrors are also used to produce ultrahigh (UH) reflective mirror coatings with a range up to 99.999% and even higher over a narrow range of wavelengths. Also, these special optical coatings can be made to reflect a broad spectral range of wavelengths of light, such as for example, the entire visible range of the spectrum. Very high quality dielectric coated mirrors (1/20 wave or better) are commonly used in optical laboratories for experiments that use highly specialized optical lasers. For example, such as the use of a femtosecond optical laser. Anti-reflection (AR) coatings (commonly called thin-film coatings) are specifically designed for the reduction and cancellation of reflections from optical surfaces. For instance, when a ray of light moves from one media (mirror) to another, a percentage of the light is reflected from the mirror's surface (commonly known as the interface) between the two mirror surfaces. While simple single-layer optical coatings are frequently used as anti-reflection coatings (AR), dielectric mirrors normally use several dozens (multilayers) of thin-film layers, and even sometimes exceeding 100 layers or more.

Today's standard single-layer dielectric coatings are traditionally so designed to work specifically within a strict and limited bandwidth or wavelength range and can be made as to have virtually "no" reflectivity at only "one" particular wavelength, most commonly at or near the middle of the visible spectral range. Anti-reflection coatings that are single layer are typically selected for a mid-range wavelength like 550 nm (green). If one wants to choose a single coating thickness of a quarter wavelength in the medium, the desired reflectivity can be selected by using what is called "normal incident reflectance coefficients." Thin-film multilayer coatings are generally more effective over the entire visible spectral range. Sometimes it may be difficult if not impossible to find a suitable coating material with a desirable refractive index. A specific example would be where the greater part of the medium has a relatively lower refractive index. For instance, the following is a commonly known characterized effect that can be used to reduce reflection by using a single-layer coating at the substrate's surface interface with an index of refraction that is suitable between those of the two media that have been chosen. The standard equation that is most commonly used for determining the index of refraction of a thin film for the two beams to totally cancel each other is as follows:

$$\mathbf{n}_f = (\mathbf{n}_0 \mathbf{n}_s)\frac{1}{2}$$

$\mathbf{n}_f$ is the index of refraction of the thin film.
$\mathbf{n}_0$ is the index of refraction of air (or the incident material). $\mathbf{n}_s$ is the index of refraction of the substrate.

Such coatings can reduce the reflection for typical ordinary glass from 4% per surface to around 2%. Lord Rayleigh in 1896 actually discovered (perhaps by accident) the first anti-reflection coating known. He found that old, slightly tarnished pieces of glass seem to transmit considerably more light than pieces of glass that were new due to this tarnishing effect. The more common anti-reflection (AR) coatings not only rely on an intermediate layer for its direct reduction of reflection coefficient but also use the interference effect of a thin layer. If the intermediate layer is closely monitored and controlled to be exactly one quarter wavelength coating, the number of reflections from the front and back sides of the thin layer will destructively oppose and effectively annul and cancel out each other. In reality, the performance of a single-layer interference coating is somewhat limited for the reason that the reflections will only exactly cancel out for one wavelength of light at one angle and also by "not" being able to find and use materials that are considered suitable for this dielectric coating process.

Coating materials with layers of specific thicknesses are carefully chosen to produce maximum destructive interface in the beams of light that are reflected from the interface and, in particular, examples where the majority or the greater part of the materials have a relatively lower refractive index. For typical ordinary glass ($n \approx 1.5$), the optimum coating index is $n \approx 1.23$. Generally speaking, there are only a few useful substances have the desired refractive index. Magnesium fluoride (MgF_2) is commonly used, since it is very durable and hard wearing and can be easily applied to the surfaces of glass substrates using a physical vapor disposition process even though its refractive index is a bit higher than what is considered desirable ($n \approx 1.38$). With such near perfect coatings, reflections as low as 1% can be reached on common glass substrates and even more desirable results can be achieved on media with a higher refractive index. It's also possible to see an even greater reflection reduction by using multi-coated layers that are designed such that reflections from the surfaces are exposed to interference. The coating is designed so that the relative phase shift between the reflected beam at the top and bottom border of the thin film is approximately 180°. When destructive interference occurs between two reflected beams of light, both beams will annul each other before they leave the surface of the media or substrate. The thickness of the optical coating must be an "odd" number of quarter wavelengths ($\lambda/4$, where λ is the specific design wavelength or wavelength that is being "carefully chosen" for desired optimal performance in order to reach the preferred path difference of one half wavelength between the reflected beams and thus leads to their annulment. The use of two or more multilayer, anti-reflection (AR) broadband coatings that cover the visible spectral range (400–700 nm) with maximum reflectance typically results in levels of less than 0.5%. In wavelength bands that are a little narrower, reflections can be as low as 0.1%. Also, a series of multiple layers with small differences in their refractive indexes can be used to create what is called a broadband anti-reflective (AR) coating by a process of what is called a refractive index gradient.

When one looks closely at high-reflection (HR) coatings compared to anti-reflection (AR) coatings, it turns out that they actually work by opposing each other. Typically, the overall concept is based on a system of periodic layers that are made up from two completely separate and different distinct materials, with one that has a high index of refraction, such as zinc sulfide ($n \approx 2.32$) or titanium dioxide ($n \approx 2.4$), and low refractive index coating material, known as magnesium fluoride ($n \approx 1.38$) or silicon dioxide ($n \approx 1.49$). This periodic layer system raises the reflectance level of the surface of the substrate in a specific wavelength range commonly referred to as band stop, whose initial width is determined by the ratio of only the two selected indices (for quarter-wave system), while the maximum reflectance is increased to almost 100% with multiple layers in the stack. The nominal thicknesses of the layers are usually found in the range of a quarter-wave that give way to the broadest high reflectance band in comparison to the nonquarter-wave systems composed from the same (non-similar) materials. Reflected beams are designed such, where they can produce interference in a constructive and a reciprocal way with each other in order to achieve a desired maximum reflection and minimum transmission result. Those coatings that are considered to be extremely good high reflectance (HR) coatings are made up from dielectric lossless materials that are deposited on media and substrates with a surface that is perfectly smooth that can yield reflectance values that can exceed 99.999% (over a fairly narrow range of wavelengths). Typically it's common for high-reflection (HR)) coatings can reach reflectance levels of 99.9% in the visible spectrum that covers a very broad nanometer wavelength range.

An incoming beam of light's incidence angle can affect high-reflection (HR) coatings when it is used in a way that's in direct contrast to a line or path that is a normal incident angle. As a result, the range of reflectance moves in the direction of wavelengths that are typically somewhat shorter and thus becomes dependent on polarization. This particular kind of effect can be made to produce coatings that can in effect, polarize a beam of light. By a special technique of manipulation, the exact thickness, and composition of the multilayer reflective stack, the reflectivity characteristics can be micro-tuned or "honed" to a specific application and may include both high-reflective (HR) and anti-reflective (AR) wavelength regions. The coating can be designed as a long- or short-pass filter, a band pass, or a mirror with a specific reflectivity (often useful in lasers). For example, the dichroic prism assembly used in some commercial cameras requires two distinctive types of dielectric coatings, one is a long-wavelength pass filter that reflects light below 500 nm (to separate the blue component aspect of the light) and the other is a short-pass filter used to reflect red light, above the 600 nm wavelength range. The remainder of transmitted light is classified as the green component (Fig. 7.32)—(Adapted from: Optical Coatings—Wikipedia)

Thickness of the coating is ¼ of the wavelength of green light in the coating material. Light reflected from the bottom surface travels an extra distance of half the wavelength of green light. Green light reflected from the top surface interferes

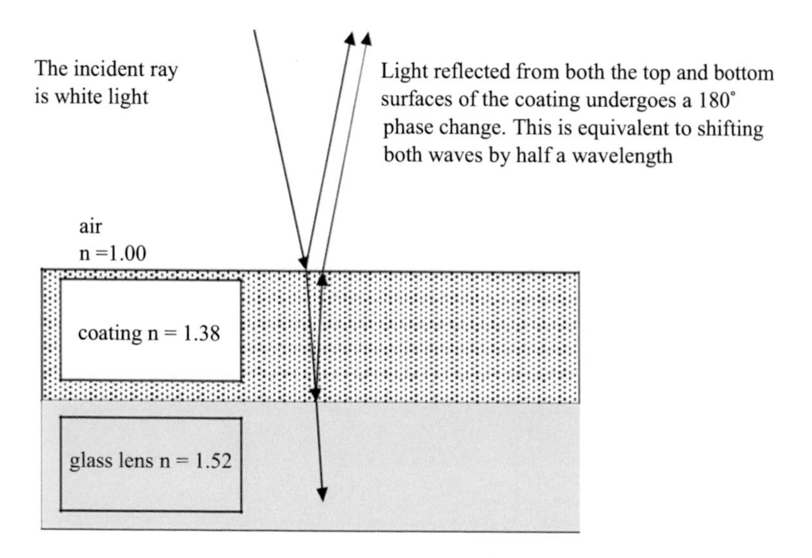

Fig. 7.32 Example how destructive interference coating works (Image credit: the Author)

Fig. 7.33 Special reflective coating on large naval binoculars (Image credit: Public Domain—Wikimedia Commons)

Fig. 7.34 Binoculars with red colored multi-coated objective lenses (Image credit: Dfrg.msc—CC BY-SA 3.0—Wikimedia Commons)

Fig. 7.35 IC 1848, often called the "Soul Nebula," can be found in the direction of the constellation Cassiopeia (Image credit: Sara Wager—www.swagastro.com)

Fig. 7.36 (Cartoon Credit: Jack Kramer)

Fig. 7.37 Looking down the business end of Keith Wilson's splendid looking dual homemade 10-in. binocular telescope (Image credit: Keith Wilson)

Fig. 7.38 Keith's binocular telescope tubes are made of carbon fiber. Check out his elegant dual eyepiece arrangement (Image credit: Keith Wilson)

Fig. 7.39 An aluminum adjustable mount and pier setup with triangular-shaped tripod legs (Image credit: Keith Wilson)

Fig. 7.40 Keith's outstanding Altazimuth mount (Image credit: Keith Wilson)

Fig. 7.41 A great example of a small, nicely-built backyard observatory, which is home to Keith's big binoscope (Image credit: Keith Wilson)

destructively with green light from the bottom surface. In other words, the green light and most of the light in the middle of the visible spectrum is transmitted. Note that some red and violet light is reflected, so it resembles a purple coating (Figs. 7.33, 7.34, 7.35, 7.36, 7.37, 7.38, 7.39, 7.40, and 7.41).

Further Reading

Eyepieces

Amateur Telescope Making, Vol. 1 (1931).
Astronomy Technology Today (monthly periodical) 1374 North West Dr. Stafford, MO 65757. Astro-Tom.com – Huygens (4, 10).
Eyepiece Evolution, Brayebrook Observatory.org
EyepiecesEtc.com. http://www.eyepiecesetc.com
Harrington, P. S. Star Ware, pp. 181, 183, (3, 8).
Nagler, A. United States Patent US4286844.
Nagler, A. United States Patent US4747675.
Nagler, A. United States Patent US4525035.
Nagler, A. Finder scope for use with astronomical telescopes. Nagler. TeleVue: A historical perspective.
Scientific American. http://www.scientificamerican.com
Stellafane, 1980 Convention. Norm Fredrick, Springfield Telescope Makers. http://stellafane.org
Wikimedia Commons. http://commons.wikimedia.org

Coatings

Acmite Market Intelligence. Market report: Global optical coatings market.
Clark, et al. (2001). Two-color Mach 3 IR coating for TAMD systems. In *Proceedings of the SPIE* (Vol. 4375, pp. 307–314).
Hecht, E. (1990). Chapter 9: Optics (2nd ed.). Addison Wesley. ISBN 0-201- 11609-X.
MIT researchers create a 'perfect mirror'. MIT press release. 26 Nov 1998. Retrieved 17 Jan 2007. Moreno, et al. (2005). Thin-film spatial filters. *Optics Letters, 30*, 914–916.
Thin-film spatial filters, (PDF). Retrieved 30 May 2007. Wikimedia Commons. http://www.wikimedia.org
RP Photonics, http://www.rp-photonics.com

References for Optical Coatings

Wikipedia Commons, http://www.wikipedia.org (1) (2) (4)
Wikispaces.com. http://www.nothingnerdy.wikispaces.com (3)

Chapter 8

So what is an equatorial drive platform? Some prefer to call it a Poncet drive system, while others just call it an equatorial drive platform for Dobsonian tele-scopes. We'll start with the most common name first. The Poncet platform was invented by Adrien Poncet in the 1970s. Poncet's original platform design was a type of equatorial platform that was simple in design and function that uses a single pivot as one support and inclined along a plane that is aligned with the Earth's equator along which two other supports slides. It was publicized in the January 1977 issue of *Sky & Telescope* magazine. The type of equatorial platform that Poncet constructed was made out of simple plywood, with a pivot made from a nail. Poncet also used Formica to cover the top of his incline plane along with the use of plastic 35-mm film canisters as bearing feet for the platform, on top of which he mounted a Newtonian telescope of a 6-in. aperture. Simply built, the Poncet equatorial platform and mount in its basic design and fabrication requires just the use of regular hand tools and common construction materials and hardware to build. The only real accurate and precise setting required is the angle at which the inclined plane of the Poncet platform is set at in order to match the angle of the Celestial equator (e.g., 39°). The Poncet equatorial platform with its simple design has been used to track in the equatorial mode, a multitude of instrumentation, including just about everything from small cameras, optical instruments, telescopes to an entire observatory. Its simple design and low profile has made it a very useful "retrofit" for Altazimuth mounted telescopes, such as the popular Dobsonian telescope. Those who use a Poncet equatorial platform simply place their telescope on top of its mount in a secure fashion to get the added feature of tracking in the direction of right ascension that's accurate enough for visual observing at higher magnifications or even astro-imaging.

N. Butler, *Building and Using Binoscopes*, The Patrick Moore Practical Astronomy Series, DOI 10.1007/978-3-319-46789-4_8

When it comes to the subject of driving your binocular telescope, then a Poncet equatorial platform is probably going to be your first consideration. For example, if you're planning to build a 10-in. binocular telescope, then you would need to scale up a standard equatorial platform design that would be suitable to drive your dual 10-in. binocular telescope. This would mean scaling up the size of your equatorial platform's base by at least a factor of two compared to an equatorial platform that was designed for a typical 10-in. Dobsonian telescope. Remember your equatorial platform will be carrying at least twice the mass and size of a regular 10-in. Dobsonian equatorial platform. That's why it's important to choose a good solid equatorial platform design and build it with the "times two" amount of mass in mind that will ultimately be sitting on top of it and tracking with it. Obviously the equatorial platform will be about twice the size as the base of your 10-in. binocular telescope, so everything you design and build for it will be twice as big and should be built with a low profile in mind.

As simple and elegant as a Poncet equatorial platform is, it does have some limitations. It is typically designed to track in the equatorial mode for up to 1 h and perhaps slightly longer (15° of tilt) since longer tracking could possibly cause the instrument on top of the Poncet platform to fall off. After an hour or so of tracking, the platform's mount has to be reset back to the original equatorial tracking mode position before restarting the mount's clock drive again. Since the Poncet equatorial platform has no roller bearing surfaces that can be driven, the clock drive mechanism itself has posed some mechanical design challenges for telescope makers. For instance, equatorial platform drives that operate in the horizontal plane that have threaded nut/bolt drive mechanisms demonstrate a different drive rate when changed to a circular motion. It also takes time to reset the plat-form's drive mechanism by spinning the nut back to the mount's original starting point. There have been some platform drive mechanism designs that use a half-nut lever mechanism to help accomplish this. Amateur telescope makers have seriously taken on this problem in a very creative way, with some using circular bolt designs and even mechanical cams with specific shapes to change the horizontal line drive motion to one that even has a variable speed controller. Overloading the mount with instrumentation and heavy telescopes and/or using them at lower geographic latitudes can cause the platform's mount to potentially bind or even seize up. This requires making improvements and design changes to bearing surfaces to try and overcome this kind of problem. Improvements such as making mechanical design variations are based on the original Poncet platform design that includes equatorial platform mounts using a more complicated cylindrical bearing design in place of the inclined plane and equatorial platform designs which use a sophisticated conical bearing system. Building an equatorial platform can be as much fun as building a Dobsonian binocular telescope itself. If you want to spend time looking at each of the great celestial objects your binocular telescope can provide without always having to try and keep the object in view by hand, then using an equatorial platform to track it is the only way to go. And adding an equatorial table is a good solution.

An equatorial platform is relatively easy and inexpensive to build and no modification to the telescope itself is necessary, no field rotation at the eyepiece, and no computer needed. An equatorial or Poncet platform has very few draw-backs and is best used within a latitude band of ±5°, polar alignment takes time, and a heavy scope requires the platform to be designed to rotate. The scope around its center of gravity, guiding a scope on a platform, will induce field rotation unless you are imaging near the meridian, the platform must be rewound every hour or so, and generally speaking, an equatorial platform, unless it was computerized, does not have motorized "GOTO" functionality. When it comes to field rotation, a field derotator should be considered; however, it is only needed for long exposure astrophotography on an Altazimuth mount. For most areas of the sky using an Altazimuth mount, one can image with a CCD camera for a few minutes and expose film for a couple of minutes before field rotation shows in the extreme corners.

A few excellent examples have been highlighted of Poncet platforms and equatorial platforms in this chapter that were made by some very creative individuals, which describe in detail in their own words the design and construction for the benefit of those who wish to construct one of their own someday. This will give the reader an opportunity to better understand the mechanical complexities that went into building each of these following splendidly crafted equatorial platforms.

Heijkoop's Equatorial Platform

By Andre Heijkoop

I based my design for my platform mainly on Warren Peter's design. I want to thank him for getting me started and for his excellent excel spreadsheet in which you can calculate the dimensions of the platform. The platform I made is for a 14″ f6 truss Dobson telescope. With my truss Dobson on my equatorial platform, I can track the planets or other objects in the sky for 45 min.

Main Dimensions

The main dimensions (in cm and degrees) for my platform are:

Telescope's central gravity	CG	50
Length of baseboard		70
Width of platform		70
Altitude		52°
Rotation angle		11.25°

Materials

18-mm plywood

40 × 90-mm Meranti wood

30 × 2-mm aluminum

Inline skate wheels ø 72 mm, hardness Durometer 82A, and ABEC 3 bearings Hurst stepper motor AS3004-001SP2702

Unipolar stepper motor driver CK1404 from Carl's Electronics Inc. Ø22mm Nylon stepper motor drive wheel

Dimensions in cm

Radius North bearing	Rnb	64.4
Radius South bearing	Rsb	19.12
Height baseboard	D+B	9.8

I used John Reagan's idea of the axial South bearing conversion of Warren Peter's design. This way you have only one bearing at the center of the South radius (1.97 cm above the baseboard).

The Woodwork

The North Bearing Supports

To make a good radius on the North bearing sector, I made a simple wooden tool, and I attached it with two bolts on the ground plate of the router. I made two of these sectors and glued them together to get a sector of 36 mm thickness (Figs. 8.1, 8.2, 8.3, 8.4 and 8.5).

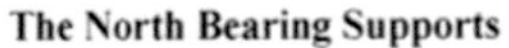

Fig. 8.1 North bearing supports (Image credit: Andre Heijkoop)

Fig. 8.2 Constructing the North sector bearing (Image credit: Andre Heijkoop)

Assembly of the North Sector

Fig. 8.3 For a perfect bearing surface, I used 30 × 2 mm aluminum (Image credit: Andre Heijkoop)

Assembly of the South Sector

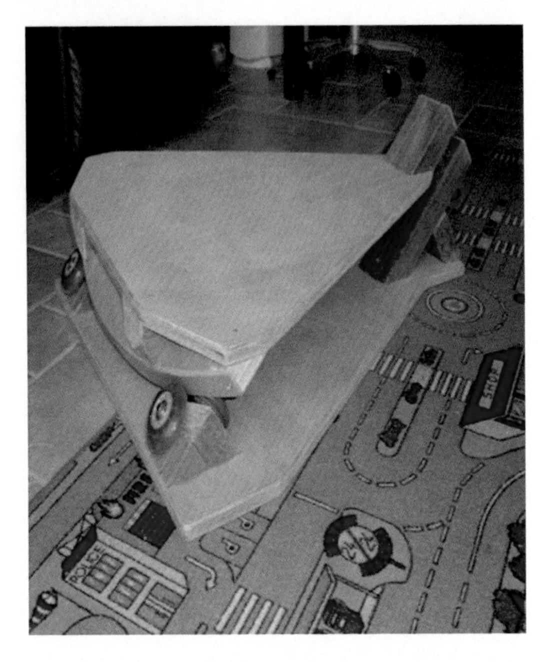

Fig. 8.4 Assembly of the South sector (Image credit: Andre Heijkoop)

Fig. 8.5 I also used an inline skate wheel for the South bearing (Image credit: Andre Heijkoop)

The Direct Drive

The Hurst Stepper Motor

The nylon wheel drives an inline skate wheel at the North sector; the inline skate wheel drives the platform. With this setup, the stepper motor makes an approximately 19 steps per second (Fig. 8.6).

Fig. 8.6 The Nylon wheel drive system (Image credit: Andre Heijkoop)

I attached a spring to get a quick release of the drive wheel during Polar alignment. The spring is also useful for a fast rewind (Fig. 8.7).

The Inline Skate Wheel Drive

Fig. 8.7 I replaced the little black box with a bigger blue box with all the electronics in it (Image credit: Andre Heijkoop)

Electronics

Blue Box

The electronics to drive the Hurst stepper motor is mounted in the blue box. On the foreground, you see the Optimate III to charge the 12 V 5 Ah battery (Figs. 8.8 and 8.9).

Ten Revolutions Linear Pot with Digital Readout

On top of the blue box, you can see the end switch, which cuts of the power when the platform is at its end of his travel. On the front plate of the blue box is a 10 revolutions linear pot with digital readout mounted; next to the linear pot, you can see the main on/off switch.

Fig. 8.8 The blue box contains the electronics to drive the Stepper Motor (Image credit: Andre Heijkoop)

Fig. 8.9 The blue box contains the platform power cutoff travel switch (Image credit: Andre Heijkoop)

Inside the Blue Box

Inside the blue box, you see (Fig. 8.10):

- On the left the 12 V 5 Ah gel battery
- In the middle a fuse box
- On the right the 10 revolutions linear pot
- Above the linear pot the main switch
- At the bottom the unipolar stepper motor driver CK1404

Notes

I replaced the standard logarithmic1M pot of the CK1404 kit with a 47 K resistor and a 100 K linear pot. With the logarithmic 1 M pot, it is very difficult to get the speed right. The 47 K resistor gives the Hurst motor a basic speed which I can easily fine-tune with the 10 revolutions 100 K linear pot.

2. As you can see in Fig. 8.11, I mounted all the electronics in one blue box:

(a) Pros: No loose controller boxes on long cables.
(b) Cons: Sometimes I have to get out of my observing chair to make small adjustments to the speed (Fig. 8.11).

Inside the blue box

Fig. 8.10 The electronics contained inside the blue box (Image credit: Andre Heijkoop)

Fig. 8.11 Close-up of the power cutoff electronics in the blue box (Image credit: Andre Heijkoop)

My Pride and Joy (Finished)

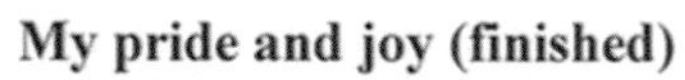

Fig. 8.12 The completed equatorial platform (Image credit: Andre Heijkoop)

The Complete Setup

The complete setup

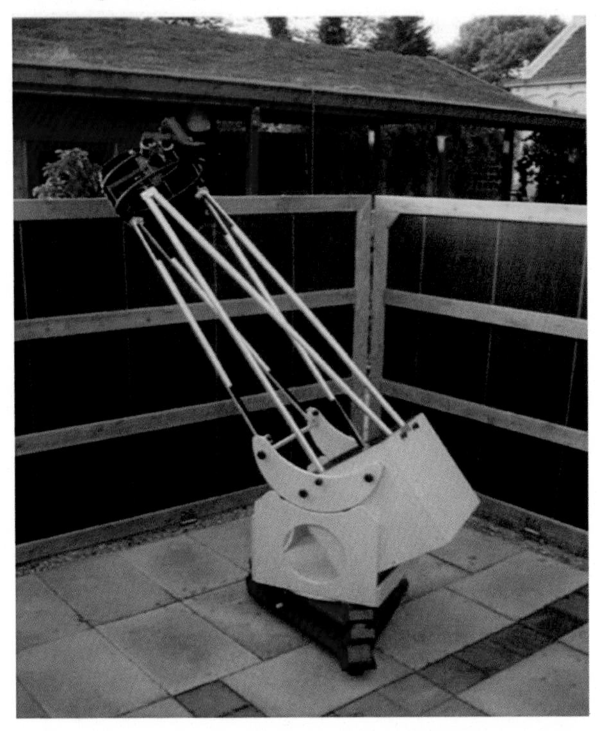

Fig. 8.13 My Dobsonian telescope and completed equatorial platform (Image credit: Andre Heijkoop)

Handle Bars and Wheels

Above you can see the pictures of my homemade handle bars and wheels for my equatorial platform. I leave the handle bars (Figs. 8.12, 8.13, and 8.14) almost always attached to the equatorial platform; they are on the North side of the platform so in the field hardly ever in the way. The wheels (B) are attached in no time. I just lift the South side of the platform and roll them underneath (A). I let the aluminum U-profile underneath the platform (C) rest on the threaded rod between the wheels, gravity does the rest.

Andre Heijkoop

Adapted from "Heijkoop's Equatorial Platform" with permission from Andre Heijkoop (http://www.astrosurf.com/aheijkoop/)

Fig. 8.14 Ready to roll (Image credit: Andre Heijkoop)

Building an Equatorial Platform for My Telescope

By Warren Philpot

I finally finished my platform and had a chance to test it out last night with my 10″ Dob. And I must say it works great. I will be testing out the platform with a webcam in the next few weeks.

I don't want to duplicate the fine efforts of so many contributors at the Yahoo Equatorial Platforms group and elsewhere on the web, but I did want to add the details of my construction so others can use my approach as a base. Nothing I did was new, but of course my final design was a combination of several approaches. Rather typical, I think.

My platform is designed for 38° and most of it is 1/2″ plywood and other woods. I have a fairly complete wood shop, so working with the wood was easy. At first, my bearing supports and drive unit were going to be made of aluminum angle, but I didn't really like all of the metal bits that were being thrown around my basement shop. So, the aluminum was abandoned in favor of hobby plywood and dense wood for bearing supports.

General Design

My criteria for the design were:

Minimized cost
Light in weight
Avoidance of complexity
Low height

Using my wood shop tools consisting of a table saw, band saw (optional), and router (most important tool), I was able to meet the criteria.

Plywood in 1/2″ thickness was chosen for the platform top, bottom, and braces. 3/4″ plywood was used for the sectors to match the 3/4″ aluminum strips used for the bearings (Fig. 8.15).

Sector Design

The original design used a circular North sector, but when I began the drive implementation, it became too complicated for the tools and materials that were available. After a bit of research and consideration of the trade-offs, I decided to convert my design to a VNS approach. The only change was to make a new North sector that was vertical instead of angled. One of my goals was to keep things simple, so I measured the radius on my drawing for the vertical sector and cut a new one using my router. I did not make any special jigs…I just cut a circular sector

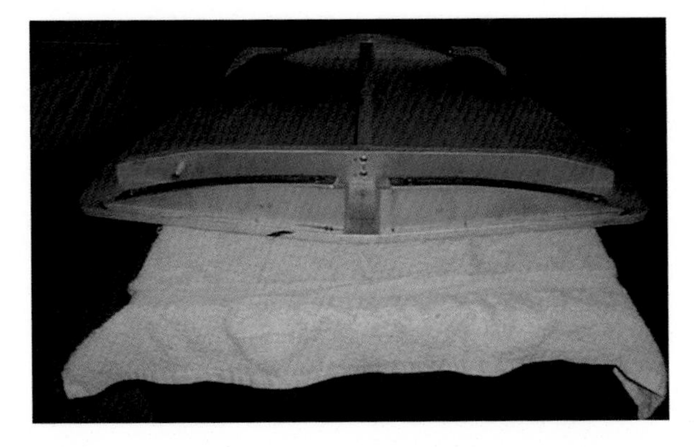

Fig. 8.15 Equatorial platform North sector (Image credit: Warren Philpot)

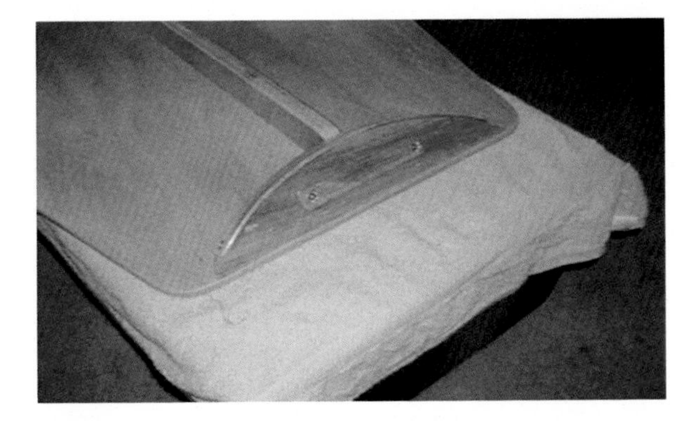

Fig. 8.16 Equatorial platform rear bearing support (Image credit: Warren Philpot)

using a simple router circle jig. I figured that if it didn't work, then I could either go back to the slanted circular sector or make a jig for the parabolic sector (So far, it seems to work fine) (Fig. 8.16).

Support Assembly

I had a box of roller skate wheels with bearings from a yard sale, which were going to be used as the supports. When I drew the design up, it was obvious my platform was going to be rather high, so I pulled the ball bearings out of the wheels and used them as the support bearings. The sectors are covered with aluminum strip, mostly because it is easy to work and form them onto the wood sectors. Time will tell if they will have enough durability.

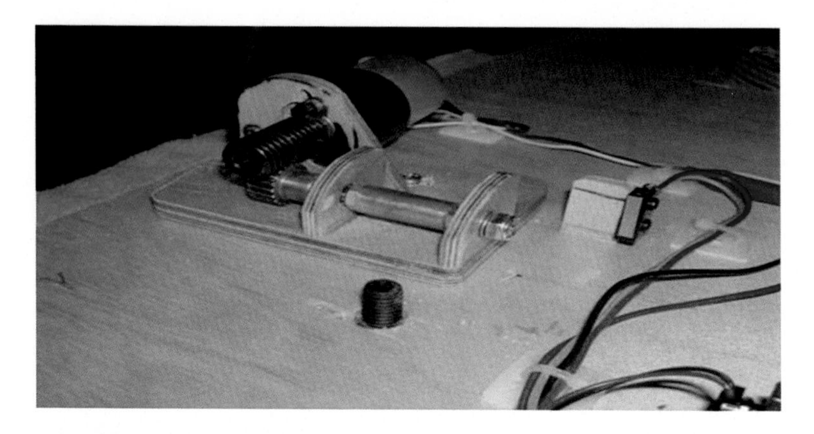

Fig. 8.17 Equatorial platform motor drive unit (Image credit: Warren Philpot)

My original plan was to use aluminum angle for the bearing supports, but as stated above, the mess from working them turned out to be a bit of a problem. I decided to use some old pine and they seem to be holding up just fine. If they turn out to be too soft, I will redo them using a harder wood. The pine was cut at the right angle and the bearings bolted directly to them using 1/4″ machine bolts (Fig. 8.17).

Drive Assembly

This was one of the toughest decisions and I have a fairly technical background. I considered using a stepper motor, but the cost of the controller and motor seemed to be a bit much for my budget. After a bit of searching on the web, I found a geared DC motor for about $30 and a controller for about $25. It turns out that after buying the gears, it may have been about as cheap to buy the stepper motor. However, I didn't go to any real pains for precision measurements on my platform, usually being within 1/8″ was fine for me, so the stepper motor seemed like overkill. We shall see as time goes on if the DC motor will be sufficient.

Designing the gear assembly with a worm gear turned out to be a challenge. I ended up using the worm gear on a 5-mm screw with nuts securing each of the components. It is supported by small ball bearings that are pressed into the 1/4″ plywood. It is amazing how well the hobby plywood and fast setting glues work for this project. And if you need to make adjustments to the motor and bearing mounts (which I did several times), it is a piece of cake. Just make sure you put a solid urethane or epoxy finish on the wood to protect them from the elements (Fig. 8.18).

The drive shaft (aka 5-mm screw) which provides the second bearing support for the platform took some ingenuity. Wouldn't it be great if there were a single hobby hardware supply where we could get all of these pieces? I ended up using a 1/4″ × 1″ aluminum support post and covered it with a piece of vinyl tubing.

Fig. 8.18 Equatorial platform drive assembly (Image credit: Warren Philpot)

Fig. 8.19 Equatorial platform motor control unit (Image credit: Warren Philpot)

It seems to work well, but time will tell if it has the durability for the long haul. But hey, if it needs to be replaced, it is cheap and readily available.

(Side note: Do you get strange looks when you wander through a hardware store looking for that special part but not knowing where you might find it? Not to mention you are only going to spend 35 cents. Try explaining to the clerk you need a 1/4″ drive shaft for an equatorial platform.)

After a bit of experimentation, the drive unit works well (Fig. 8.19).

Motor Control Unit

Using a PWM (pulse-width modulation) circuit to control the speed of the motor is one of the critical decisions. That allows the motor to operate at its rated voltage and lowers the risk of stalling the motor. This also helps keep the motor at a constant speed. This is not a complex circuit, but making a circuit board does add a bit of complexity. So, I decided to purchase an inexpensive prebuilt circuit. The circuit I chose had a built-in potentiometer to control the circuit. This worked out great until I tested the circuit board by using jumper wires and setting it in the driveway next to my platform. It worked GREAT for about 5 min, until I stepped on the controller in the dark and broke the control potentiometer. Ah well, I had planned on replacing the potentiometer with a remote handheld speed control. The only other major change to the circuit was to add a 1 k ohm resistor in series to the circuit to offset the speed control settings. (Note: The 1 k ohm resistor is connected in series with the white wire. It is installed in the remote box.) Otherwise, the motor would not start turning until it was set to the halfway point (Fig. 8.20).

A power switch was added to the control box and a limit switch wired in series with the battery was added so the platform would not drive itself off the support bearings or wear out the drive wheel.

Operation

So, how does it work in practice? For visual purposes, it seems to work great. I have only had it out for one night, but using an 8-mm eyepiece on my 10″ Dobsonian along with a 2× Barlow, it was rock solid while at my local astro group's visitor nights. It will be nice to crank up the magnification a bit and not have to tell the visitors to "LOOK QUICK!" Or to view the planets at high magnification for an hour at a time.

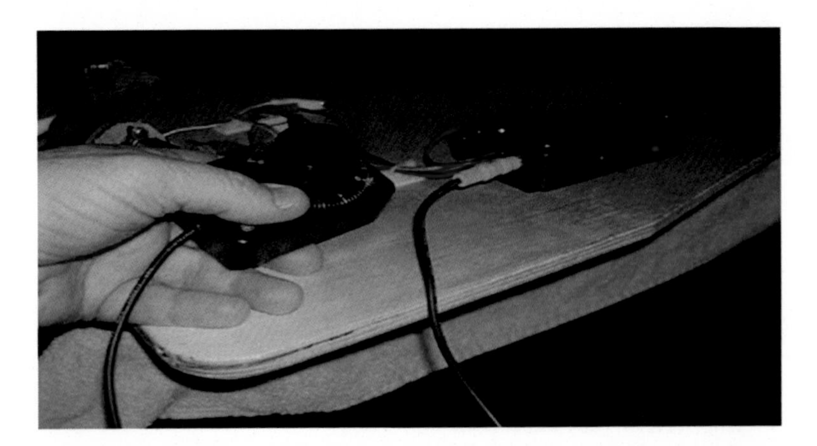

Fig. 8.20 Equatorial platform motor speed control unit (Image credit: Warren Philpot)

Summary

I really enjoy projects where many different skills are required to solve problems. The web is amazing in its ability to allow us to share our experiences and create our own variations of different projects. And it doesn't have to be a technical project… it might be learning a new craft or athletic sport. And I would like to thank the members of the Yahoo Group equatorial platforms for all of the information that proved invaluable for my project. And thanks to all of the other web bloggers that posted about their experiences while designing and building their equatorial platforms. In particular, a design on the website of Reiner Vogel inspired my own design and gave me the confidence that I could build the project. There were numerous others who gave me direction and I thank them all. And one of the most important things is that if you are thinking of building an equatorial platform, but not sure if you have the skills to build a working model…give it a shot! What do you have to lose? It is much easier than I would have thought and the design of the platform is much more forgiving than you might expect. Give it a shot and have fun (Fig. 8.21).

Warren Philpot

Adapted from "Building an Equatorial Platform for My Telescope" with permission from Warren Philpot (http://www.warrenthewizard.com)

Fig. 8.21 Warren Philpot and his Dobsonian telescope mounted on the equatorial platform (Image credit: Warren Philpot)

Equatorial Cradle

By Chris Luton

The following images show a basic platform for a Dobsonian telescope which allows the scope to track an object simply by rotating the platform about its axis. It was easy to build and the parts cost under $40 (Fig. 8.22).

The platform swings between the two bolts which attach it to the base. The angle between the bolts is set at 35° (my latitude). The virtual axis formed between the two bolts needs to be aimed at the celestial pole. The platform has approximately 15° range of motion—the range is inherently limited by the crossbeam of the swinging section hitting the central beam of the base.

The swinging section is braced with two pieces of ply. This bracing seems to be the key to the strength of the whole platform. Initially, I thought the base would also need to be braced lengthways, but the rigidity of the swinging part also strengthens the base because it is attached rigidly to the base. The platform is surprisingly sturdy—it easily supports 65 kg (me), so the 21 kg of my Dob is not an issue (Fig. 8.23).

Ideally, the center of gravity of the scope and the swinging section combined should lie close to the axis. Most of the weight of the base section of my Dob falls below the axis so it's only really the tube and a small part of the base which is above the axis. The mass of the swinging part of the platform offsets much of the weight of the tube (Fig. 8.24).

The motion of the platform is reasonably smooth. Initially, the motion was a bit jerky as there was some friction between the wood and the bolts, but a little oil on the bolts has mostly eliminated the "sticking." Unfortunately, there is some wobble

Fig. 8.22 The equatorial cradle. A very simple but effective inexpensive equatorial cradle design (Image credit: Chris Luton)

Fig. 8.23 The equatorial cradle with a small Dobsonian telescope mounted on board (Image credit: Chris Luton)

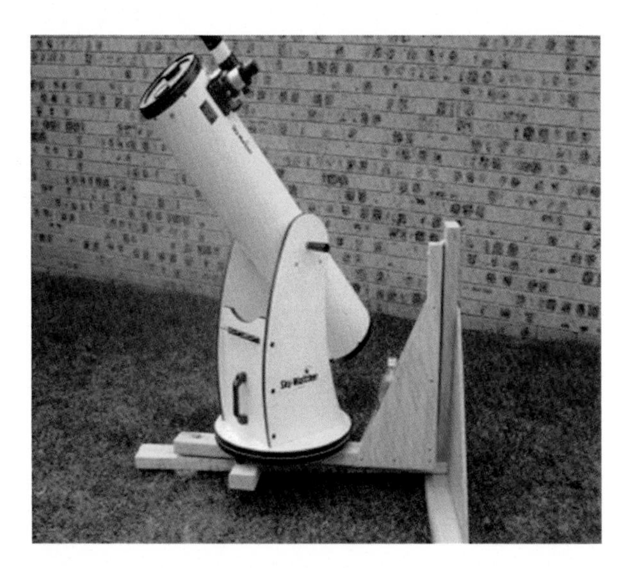

Fig. 8.24 The equatorial cradle can reach any part of the sky dome (Image credit: Chris Luton)

when moving the platform. However, without any wind, I was still able to use a 10-mm eyepiece with a 2× Barlow to view Saturn. I suspect the wobble would be reduced by using something other than bolts as the bearings—maybe too short, fat axels sitting on Teflon pads a bit like a Dobs altitude bearing.

One of my early ideas was to use ball joints at each end—this would make construction quite easy, but they might swing a little too freely (Figs. 8.25 and 8.26).

Fig. 8.25 The equatorial cradle showing the top polar pivot end (Image credit: Chris Luton)

Fig. 8.26 The equatorial cradle showing the bottom polar pivot end (Image credit: Chris Luton)

The motion of the platform is reasonably smooth. Initially, perhaps a ball joint at one end and something like a Dob's altitude bearing at the other? Maybe a piece of threaded pipe screwed into a tee as in the seesaw example below. (Update: I think a tangent arm attached toward the bottom of the swinging part would eliminate the wobble—it works well with ball joints. The whole platform weighs about 12 kg, which is about the same as the base section of my Dob. It isn't exactly portable, so it's more of a backyard fixture. The total cost of parts was under $40. It uses approximately 5 m of 90×45 mm structural pine ($3/m), 1 sheet of 12-mm ply ($14), 2 bolts, 2 nuts, around 25 washers, and around 25 screws.

Chris Luton

Adapted from "Equatorial Cradle" with permission from Chris Luton (http://eq.brindabella.org)

The McCreary Mount

By Dann McCreary

The McCreary Mount is the first truly novel equatorial mount design in this generation. It is intended to provide superior smoothness and accuracy in tracking when compared to any previous design (Fig. 8.27).

Why Use a McCreary Mount?

Benefits

- Dramatically inexpensive to build or purchase.
- Silent in operation, allowing you to listen to "the music of the spheres."
- Inherently strong—supports far greater weight than other comparable designs.

Fig. 8.27 The McCreary Mount with a Dobsonian telescope sitting on the equatorial platform (Image credit: Dann McCreary)

Fig. 8.28 The wooden base and the platform for the McCreary Mount (Image credit: Dann McCreary)

- Precise—subarcsecond accuracy possible in a well-designed and well constructed implementation.
- You can build it yourself with simple tools.
- Produces superior astrophotography results.
- Observatories can realize tremendous cost savings and significantly improved accuracy when compared with other designs (Fig. 8.28).

The Prototype: Some Preliminary Specifications

Two Major Components

- Base (equilateral triangle 4′×4′×4′)
- Platform—handles rocker box up to 22.5″ square
- Each component easily carried by one person (Figs. 8.29 and 8.30)

Tracking angle—20°
Tracking time—1 h, 20
mini Periodic error—None
Tracking accuracy—<1 arcsecond during 80 min
Tracking adjustment—RA only (i.e., rate of motion) Maximum instrument weight—200 lb
Polar alignment—Altitude and East/West bearings are adjustable with 10–32 pitch adjustment screws for adjustments in arcseconds rather than arcminutes (Figs. 8.31 and 8.32)

Fig. 8.29 The platform resting on the base of the McCreary Mount (Image credit: Dann McCreary)

Fig. 8.30 Close-up of the upper bearing support (Image credit: Dann McCreary)

Fig. 8.31 Close-up of the lower bearing support (Image credit: Dann McCreary)

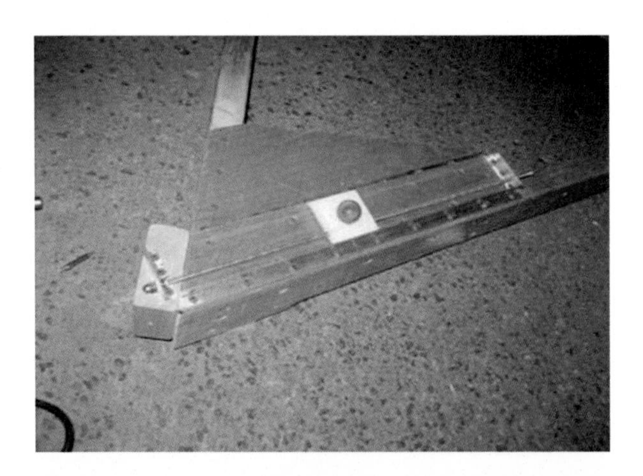

Fig. 8.32 This adjustment allows for a latitude range roughly between San Diego and Boston (32–42°) as well as "fine-tuning" the polar axis altitude (Image credit: Dann McCreary)

McCreary Mount Fundamentally Different and Why Is It Better Than Anything That Came Before?

The fundamentals of the McCreary Mount are extremely simple. This simplicity is the secret of the mount's basic strength and accuracy.

Two Points

In simple geometric terms, two points exactly specify a line. In the McCreary Mount, these are two literal points—in the initial concept and "proof of concept" implementation, these were two steep, sharpened points of two metal bolts, mounted on a base or on piers to establish a polar axis. In newer and more refined implementations, these points are the centers of two extremely precise bearing balls.

Each of these two bearings is finely adjustable in position along a single axis, resulting in the ability to produce extremely accurate polar alignment.

Gravity-Driven Fluid Motion

An incompressible fluid (typically water) is used to regulate and control the motion of the McCreary Mount. The weight of the platform and the instrument drive the fluid through a valve that precisely, smoothly, and silently regulates the mount's motion. There are no motors, gears, roller bearings, or other sources of mechanical "noise."

Design Variations

There are many variations possible, limited only by your imagination. The design inherently lends itself to great strength in construction, making it suitable for supporting very heavy observatory instruments much more economically than earlier existing technologies (Fig. 8.33).

Fig. 8.33 Preliminary fluid drive uses a five gallon water bag for test purposes (Image credit: Dann McCreary)

Claiming as Original?

I claim the following as being, in their combination and synergy, the essential elements of my invention. To the best of my knowledge and belief, this unique combination of elements has not been previously invented or disclosed.

The McCreary Mount is a telescope mount that falls in the general category of "equatorial mount" in that it counteracts the motion of the earth's rotation in order to keep a telescope or other astronomical instrument accurately pointed at a celestial object. However, it differs radically from all prior devices in that it brings together for the first time the following combination of design features (Fig. 8.34).

Specific Combined Elements of the Invention

The McCreary Mount employs a base comprising two literal, firmly fixed but finely adjustable geometric points which uniquely define a polar axis of rotation that can be precisely adjusted to specify a line exactly parallel to the earth's axis of rotation. In the preferred embodiment, these points are the geometric centers of two extremely precise spherical bearing balls mounted in any suitably stable supporting structure with their upper surfaces exposed to bearing sockets in the associated instrument platform.

It incorporates a platform or mounting for the telescope or other instrument or instruments, having two bearing sockets arranged so as to mate with the two support bearing points of the base. The shape and design of the platform may be infinitely variable and adaptable to the needs of the device that it supports.

The bulk of the platform is situated on the west side of the two supporting point bearings. This platform need only be sufficiently strong and rigid for the particular application and to prevent changes in the amount of sag or flexure during the range of motion desired for any particular implementation. The placement of the platform on the west is deliberate for good and sufficient engineering reasons.

Fig. 8.34 Extra strong. The McCreary Mount platform can hold a lot of weight (Image credit: Dann McCreary)

Toward the west end of the McCreary Mount, a hydraulic means of regulating the motion of the platform is employed. In its preferred embodiment, gravity is used as the motive force during tracking, and the flow of an incompressible fluid such as water may be finely adjusted through any suitably precise valving mechanism to regulate the motion.

Obviously, a curved fluid cylinder with a constant flow valve will provide the most regular platform motion, free of tangent error. Also, in applications at great distances from Latitude Zero (the equator), modified versions of this hydraulic means may be employed while still retaining the basic identity of the McCreary Mount.

Variants

An infinite set of variations of and enhancements to the above described essentials are possible, but any system incorporating these elements in combination is fundamentally a "McCreary Mount."

It should be obvious to those skilled in the art of telescope making that such a device can be adapted to a wide range of applications and can also benefit from devices and methods previously developed for existing equatorial mounts, such as setting circles and closed-loop guiding feedback mechanisms.

The inventor explicitly discloses and releases this invention into the public domain with the express desire that the use of his invention might further enable and facilitate the revelation of the Glory of God, who through Jesus Christ created the heavens that we as astronomers all enjoy.

McCreary Mount Eliminates Virtually All Sources of Periodic Error

All the usual error generators are eliminated, such as:

- Worm
- Worm gears
- Worm wheel
- Spur gears
- Auxiliary gear trains
- Motors
- Polar shaft
- Roller bearings
- Clutches
- Counterweights

What are the principles of engineering design that have been overlooked in prior art, but are implemented in the McCreary Mount?

The following is a set of notes and outlines that feature some of the principles employed:

Seriously Engineer Only What Is Most Essential

- Shun conventional wisdom.
- Put strength in the supporting piers and adjustments.
- Pay attention to platform rigidity, not shape.
- Require ultimate precision only where most needed.
- Bearing balls Grade 24 (1/6th arcsecond).
- Bearing balls Grade 10 (1/20th arcsecond)
- A smooth, accurate hydraulic cylinder.
- Eliminate tolerance stack-up.
- Use gravity, don't fight it!

The Ultimate Simplicity of the Polar Axis

A line is defined by two points:

- Mount bearings are reduced *literally* to two points!
- The center of a sphere is a point!
- Most important, most solid, most precisely adjustable!
- Bearings should have one-dimensional, tight adjustments.
- Adjust elevation with one bearing point.
- Adjust East/West with the other bearing point.
- Polar alignment becomes very easy because when performing adjustments, you only need to move the two bearings, not an entire assembly consisting of a base, shaft, bearings, etc. The base remains stationary and the platform and instrument are coupled to the bearings that you are adjusting only with a bearing surface (Fig. 8.35).

Geometry Is Far More Important Than Weight!

- Don't think that making your device heavy will make it better!
- Use triangles and trusses, resulting in low flexure.
- Use larger rather than smaller dimensions wherever possible.
- The wider apart the bearing points, the smaller the alignment change with a given motion and the more precise the resulting alignment.
- A curved hydraulic cylinder gives a true and steady rate.
- No tangent-arm errors.
- Pressure compensation may be necessary.

Fig. 8.35 The McCreary Mount with a Dobsonian installed on top of the platform with assistant (Image credit: Dann McCreary)

Avoid Balance! Use Pre-stressing Instead!

- A balanced mount is susceptible to "noise"... every little stress will move it off target.
- An off-balance design helps to prestress all structures into a strong, stable, steady form as well as to preload bearings and eliminate any possibility of backlash.

Eliminate Tolerance Stack-Up

- Be sure all stresses are "unidirectional."
- Stability and accuracy actually improve to an extent with instrument weight. The weight of the telescope removes all slack, stresses all bearing materials in one direction, and prestresses the platform structure.
- Flexure self-eliminates during modest length exposures.

The McCreary Relaxation Drive

Gravity is a steady, stable, reliable force.

- It drives the motion of the mount.
- The weight of the instrument helps and stabilizes.
- It prestresses the structure and bearings.
- Gravity supplies both power and "transmission," i.e., there are no gears, etc., needed!

Swamp the Noise!

- Weight far overwhelms miniscule cylinder and support point friction.
- Overcome resistance (resistance is futile!).
- Be sure driving forces far overpower the friction of any slip/stick forces.

Limitations

While superior in many ways, the McCreary Mount admittedly has a few drawbacks.

- The mount is (deliberately) "larger than your average mount"—a price I believe to be well spent for simplicity, accuracy, and great strength and stability.
- Portable versions are also "tall," often requiring larger ladders to reach the eyepiece.
- The direct gravity-driven fluidic drive is best used at latitudes of ±50°. There are ways to adapt the design for use outside of this range, involving alterations to the hydraulic drive design.
- The bearing design also works best at lesser latitudes, but there are also ways to stabilize the polar axis bearings for use outside of this range.

Problems Shared with Other Designs

Like other "platform type" equatorial drives:

- There is no "convenient" declination tracking adjustment.
- A dec adjustment could be implemented in other ways. Like ALL equatorial drives:
- Does not directly correct for refraction
- Requires the telescope itself to not flex or bend
- Requires reasonable care in design and construction

You Want This Mount!

Once you understand the principles of its operation, and recognize the potential specifications of a well-built version of the McCreary Mount, you will want one for your serious astrophotographic efforts.

The McCreary Mount is sure to become the "sine qua non" for all serious ground-based astrophotography.

Dann McCreary

Adapted from "The McCreary Mount" with permission from Dann McCreary (http://www.subarcsec.com)

A Large Equatorial Platform for a Big Dob

By Michael Davis

Here is a photo of the 17.5-in. Dobsonian telescope on my new equatorial platform. Here the scope is set up at the Orange Blossom Special Star Party at Alafia River State Park in February 2006. It was the first real test of the platform (outside my back yard). Although I had built it specifically for use with the big Dob, I had not yet had a chance to actually test it with anything as heavy as the big Dob. It was only finished a few days before the star party and I had only tested it with one of my much smaller scopes. I was a little worried about whether it would really work with all that weight sitting on it. If anything, it seems to work even better and more smoothly with the larger load of the big Dob (Figs. 8.36 and 8.37).

Fig. 8.36 Michael Davis's big 17.5-in. Dob sits on his newly constructed equatorial platform (Image credit: Michael Davis)

Fig. 8.37 Another view of Michael Davis's big 17.5-in Dob at a public star party (Image credit: Michael Davis)

Here's another view of the scope sitting on the equatorial platform. Having an equatorial platform is a big help when I am at a public observing event and I have a long line of people at the telescope wanting to see the wonders of the universe. I can put the scope on an object (even at really high power) and just walk away. The platform will do the work for me. Dozens of people can get a good long look through the scope without me having to jump in after every person or two and re-aim the telescope as the Earth rotates the object out of interest out of view. Below are some close-up photos of the platform (Fig. 8.38).

Here is a view of the platform without the telescope sitting on it. The top is 24 inches square. The three pads of carpet help prevent vibration from the drive motor exciting the telescope. Here it is in the middle of its range of travel and the top is flat. This is the best position for setting up the telescope on top of it. After the scope is set up, the platform is reset to its extreme Easterly position. The platform was built to work at 30° North latitude. This is a good compromise since I will be using it both in Florida and Arizona. In Florida, the North end needs to be tilted down a couple of degrees. In Arizona, it needs to be tilted up a few degrees. How did I design this platform? Here are a series of drawings that show the mental process I went through to come up with this design. I need to warn anyone thinking of copying this platform that the design must be modified for your particular latitude. My platform was designed to work in Central Florida. The same design won't work for someone significantly North or South of 28° North latitude, without rework. So I am going to show you how to design your own platform for your particular area. Note: These instructions should work just as well in the southern hemisphere if North and South are reversed in the diagrams, and the motor drive direction is reversed.

Fig. 8.38 Michael Davis's equatorial platform for his big 17.5-in. Dobsonian telescope (Image credit: Michael Davis)

I spent a long time brainstorming how to do this. One mental image I "kept" coming back to over and over again was the idea of a cone on rollers. Imagine a giant cone lying on the ground with the point facing due South. If the half-angle of the cone is equal to your latitude, then the axis of the cone should point to the North Celestial Pole (Fig. 8.39).

Now put the cone on a set of roller wheels that support it with the tip still pointing due South and the bottom side parallel to the ground. If the cone is rotated clockwise (as seen from the North side) at the sidereal rate of one revolution per day, the Earth's rotation would be canceled out, from the point of view of the cone, and any camera or telescope attached to the cone should follow the stars. If the platform is designed in such a way that the axis of the (now virtual) cone passes through the center of gravity of your telescope, the system will be very nearly balanced, and very little torque will be required to drive the platform, even with outrageously heavy telescopes like my massive 17.5-in. Dob. Note that I made a mistake in these drawings and have the axis passing through the CG of the tube assembly. The actual center of gravity of the whole telescope, including rocker box and ground board, is really lower than this. Keep that in mind (Fig. 8.40).

The cone shape is pretty inconvenient for mounting cameras and telescopes to. So I next imagined slicing off most of the cone parallel to the ground to provide a flat surface to mount a telescope on. The cone can no longer rotate a full 360°. It now can only rotate through a few degrees on either side of center before we run off the rollers. So in use, the platform will be run clockwise for some period of time (about 40 min in my case); then it must be reset back to the fully counterclockwise position. The drive motor can be run much faster than the sidereal rate when resetting to make the wait until the platform can be used again quite short.

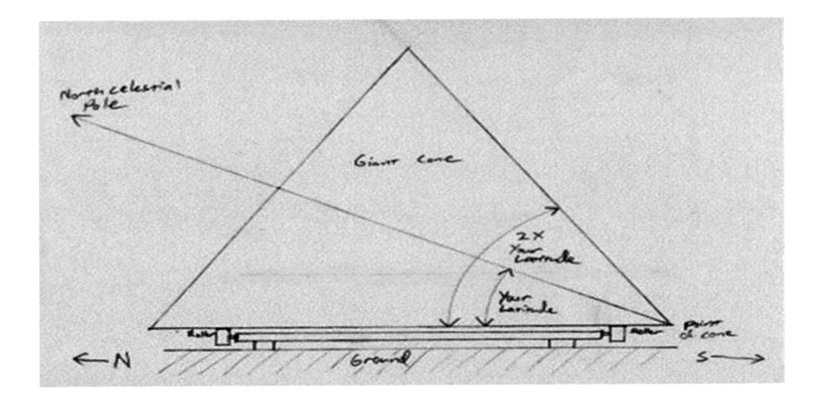

Fig. 8.39 The cone on rollers sketch for the preliminary equatorial platform design showing polar North (Image credit: Michael Davis)

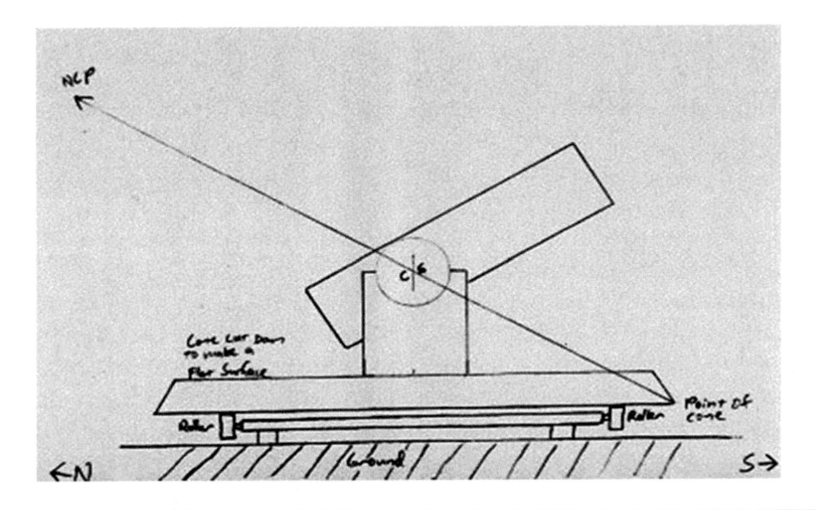

Fig. 8.40 Another sketch of the preliminary equatorial platform design (Image credit: Michael Davis)

My next imaginary leap was to realize that only the parts of the cone riding against the rollers needed to follow the shape of the cone. The entire rest of the cone could be cut away. Now only two slices or sectors out of the original cone are left to ride against the rollers and the rest of the platform is just a flat plate. The physical point of the cone is now gone and only exists virtually out in space South of the platform (Fig. 8.41).

The final revelation was changing my vision of the cone from a perfect straight-sided cone to a stack of short cylinders of different diameters, like in the Towers of Hanoi game. At this point, I was about to start building. I had a design that I was

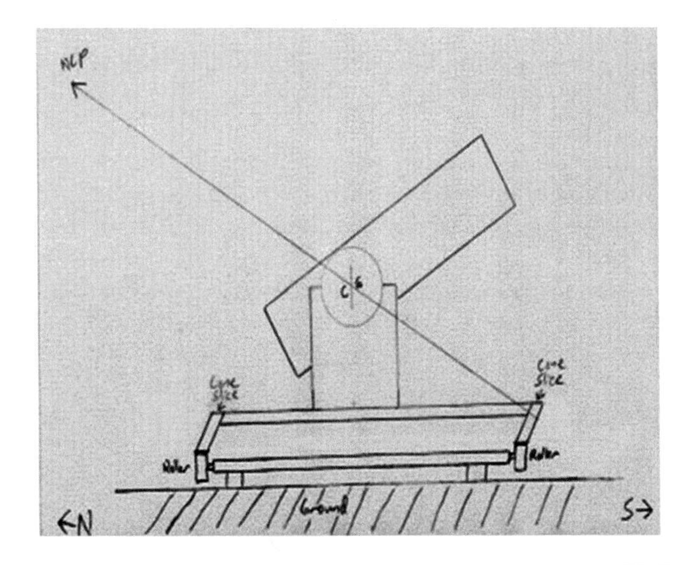

Fig. 8.41 Getting closer to the final design (Image credit: Michael Davis)

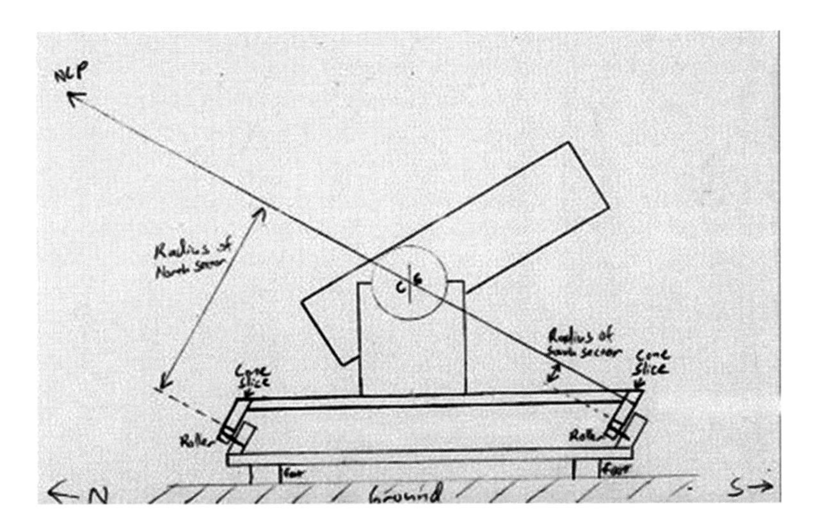

Fig. 8.42 The final design sketch of the equatorial platform (Image credit: Michael Davis)

sure would work and that I felt sure I could build. The only fly in the ointment was the cutting the correct curves and angles on the sectors. I was already pretty good at cutting straight-sided circles out of plywood. However, cutting the sectors at the angle of the cone was a new twist. I realized I was going to have to design and build some sort of special jig to do it. This prompted me to think about how to simplify the design a little more (Fig. 8.42).

Now each sector is a slice out of a straight-sided cylinder with a radius equal to its distance from the axis of the virtual cone. The rollers would be mounted on supports tilted at the angle of latitude. Not only does this simplify making the sectors, but the design change also provides constraint on the motion of the platform and prevents it from running off the rollers to the North or South. A simple way to size everything is to find the height of the center of gravity of your telescope and make a drawing to scale with the line pointing to the NGP passing through the centerline of the top plate at that height. Then just take measurements off the drawing for the radii of the sectors. Note that these particular drawings are just for illustration and are not to scale, and my own sector radii and heights are different. On my platform, the North sector has a radius of 35 in. and the South sector is 24 in. Both have a height of 4 in.

Now I was ready to build.

While still on the subject of initial design considerations, I should point out that there are several possible methods of driving the platform. The two that immediately come to mind are using a driven wheel in place of one of the idler rollers or using a linear rack system with a pin that engages a slot attached to the top plate. I chose the latter system, even though it is considerably more complicated. I was concerned that a drive wheel might not be able to transmit enough torque in an out of balance situation and might slip, especially when wet with dew. A driven wheel might work just fine in your application, food for thought (Fig. 8.43).

Here is an overview of the equatorial platform as seen from the North side with the top plate removed. I'll go into some of the details about how it works below. Here is an overview. A stepper motor turns a worm gear which turns a piece of half-inch threaded rod. There is a nut with a steel pin welded to it that travels left to right on the threaded rod as it rotates. The pin engages a slotted tab on the underside of the top plate and drags it along. The top plate sits on sectors which ride on tilted bearings which point through the North Celestial Pole. The platform will

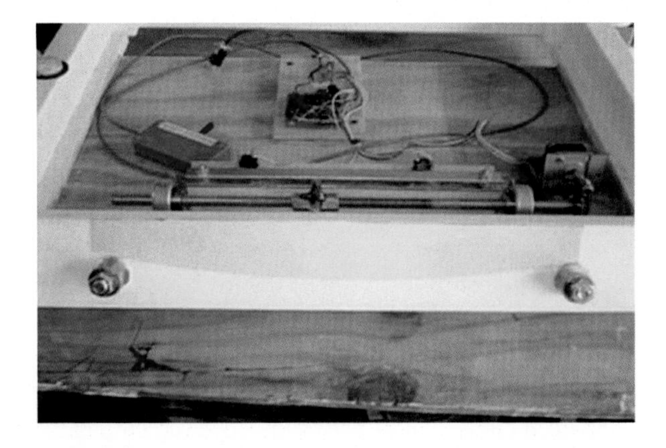

Fig. 8.43 Another view of the equatorial platform from the North side with the top plate removed (Image credit: Michael Davis)

Fig. 8.44 Another view of the equatorial platform from the side with the top plate removed (Image credit: Michael Davis)

run for 40 min before the pin on the nut hits the West limit switch and shuts the motor down. Another limit switch on the East side shuts the motor down at the end of a reset cycle (Fig. 8.44).

Here is another overview of the platform from the side with the top plate removed. The North side is on the left and the South side is on the right. As you can see, the North and South sides are tilted up at 30°. The bearings on the South side are inside the base. The bearings on the North side are on the outside of the base. The frame of the base unit is made from 2×4 s. The 2×4 s are screwed together with long wood screws. The corners are reinforced with wood glue blocks. A plywood panel screwed onto the bottom provides a platform for the drive system and further reinforces the frame by preventing it from skewing out of square. The bearings are 1 in. diameter sealed ball bearing units with 1/2-in. bores. They are mounted on 1/2-in. carriage bolts which pass through the 2×4 frame on the North and South sides. Formica crescents are glued on the North and South faces between the ball bearings. GoreTex pads on the backs of the plywood sectors of the top plate ride against the Formica, creating another bearing surface (Fig. 8.45).

Here's a close-up of the drive motor and gearing. A 200 step/rev stepper motor salvaged from an old disk drive is half stepped at 375 steps per second. Half stepping reduces vibration. The motor is mounted on rubber pads to further damp out vibration. The motor turns a 30:1 ratio worm gear, which is attached to the end of the main threaded drive rod. Everything about this layout is about reducing and damping out vibration from the motor. The half stepping and the 30:1 ratio mean that the motor must be stepped very fast. Fast steps mean a high-frequency vibration that damps out much more easily than slower clunk-clunk-clunk type steps if the

Fig. 8.45 A close-up view of the motor drive and gearing (Image credit: Michael Davis)

Fig. 8.46 Here's a view of the threaded rod and its coupling nut assembly (Image credit: Michael Davis)

motor were coupled directly to the threaded rod. The 30:1 gear ratio also means that a much smaller motor can be used to drive the platform. A smaller motor means—you guessed it—less vibration. If I had it to do over, I'd probably try to use something other than a stepper motor to drive it. It's just so much easier to make stepper motors move at an arbitrary rate of speed as compared to the alternatives like servos and ac motors. I didn't have the time to get elaborate with the drive electronics, and I have a lot of experience with steppers, so that's the rout I took. The down side is getting rid of the vibration from the stepper motor. It took a little trial and error, but in the end I managed to totally eliminate the vibration at the telescope eyepiece (Fig. 8.46).

Here is a view of the threaded rod. It has 1/2 in. diameter, 13 threads to the inch. The long nut rides on the rod. This type of nut is called a coupling nut. It is ordinarily used for coupling two pieces of threaded rod together. That's why it's so long. I used a coupling nut here because it wobbles less on the rod than an ordinary nut because it has more threads in contact with the rod. You can find coupling nuts in just about any hardware or home center store right next to the threaded rod. A steel pin is welded to the nut. One side of the pin rides between two steel plates which prevent the nut from rotating. The other end of the pin engages a slot in a tab (visible in the next photo) which sticks down from the underside of the top plate. You can also see one of the limit switches on the left. Everything is thoroughly greased for smooth operation. The slotted steel tab and the steel pin welded to the drive nut are the only metal pieces that had to be custom fabricated. A machinist friend of mine made them for me. Everything else is made from standard hardware and fasteners available in most well-stocked hardware or home center stores (Fig. 8.47).

Here is a view of the underside of the top plate removed from the platform. The two sectors were cut from hardwood plywood using my router and a circle cut-ting jig. Each sector has a different radius corresponding to slices of a tilted cone.

The North sector has a radius of 35 in. and the South sector has a radius of 24 in. Both are 4 in. tall. Each sector has a 1/8-in. thick band of aluminum attached to its outer rim to provide a smooth surface to ride against the bearings on the base unit. Each sector also has two squares of super slippery GoreTex which rides against Formica crescents on the base unit reducing friction. You can see the bearings and Formica backing on the North side of the base in the fourth photo from the top. The inside of the South side of the base is similar. Here you can also see the slotted steel tab sticking up which engages the drive pin on the base unit (Fig. 8.48).

Fig. 8.47 A view of the underside of the top plate removed from the equatorial platform (Image credit: Michael Davis)

Fig. 8.48 A side view of the threaded rod and its coupling nut assembly (Image credit: Michael Davis)

Fig. 8.49 The North bearing (Image credit: Michael Davis)

This picture was taken by squeezing the camera through the gap between the upper plate and the base unit. It shows how the slotted tab on the upper plate engages the steel drive pin on the base section. I designed the platform this way to keep as much of the drive machinery as possible covered up and protected from the elements, dust, and grit. I've seen a lot of platforms that had all the drive hardware stuck on the outside of the unit and exposed to everything Mother Nature and careless astronomers could throw at it. I do think I made one mistake here though. I now think the pin should engage the slot at a 90° angle. I think that might reduce the amount of torque required to move the platform, reduce binding at the extremes of travel, and perhaps allow it to move a little further in either direction. Once summer arrives and the weather goes to hell here in Florida, I will probably rework the platform so that the pin engages at 90° and see if it really works any better. If so, I'll have an even better platform ready when the weather dries out, the mosquitoes die off, and the nights get long again next fall (Fig. 8.49).

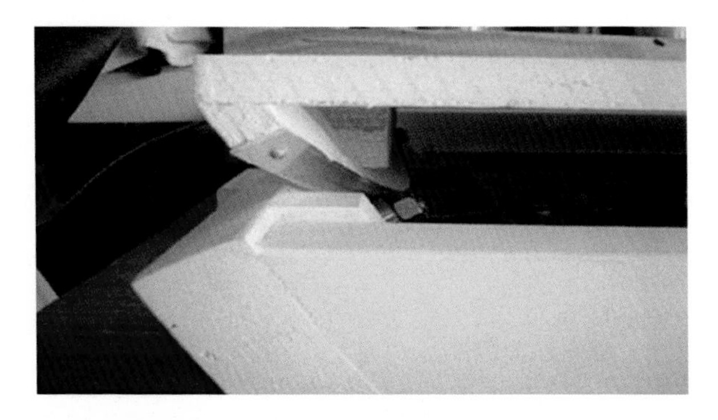

Fig. 8.50 The South bearing (Image credit: Michael Davis)

Fig. 8.51 The South side (Image credit: Michael Davis)

Here is a close-up side view of the North bearings from the side with the top plate in place. The plywood sector with its metal banding on the rim rides against the sealed ball bearings. Between the plywood sector and the 2×4 base is the GoreTex on Formica bearing (not visible) (Fig. 8.50).

Here is a view of the South bearings from the side with the top plate in place. This is identical to the North side except that the bearings are mounted inside the base unit (Fig. 8.51).

Here is a view of the south side showing the single South leveling foot. There are two leveling feet on the North side. You only need three feet since every object

Fig. 8.52 The underside (Image credit: Michael Davis)

only rests on three points. Now you know why all four-legged tables rock. One leg will always be off the ground. With three legs there is no rocking. Just spread the three points out into the widest possible triangle for the most stability. When setting up in grass or dirt, I place small squares of plywood under each foot to prevent them from sinking into the soft ground. On a hard surface like concrete or asphalt, the plywood isn't needed (Fig. 8.52).

Here is a shot of the underside of the platform showing the three leveling feet. I should also explain why it is painted white. I get a lot of questions about that for some reason. I wanted to protect all the wooden surfaces that would be subject to having dew condense on them, have dew running off the telescope drip on them, or be in contact with dewy grass. I chose white paint so that the platform is as visible as possible in the dark so I don't trip over it or accidentally kick it while working around the telescope. This is not much of a problem when the 17.5-in. Dob is sitting on the platform since it pretty much completely covers it. However, in my testing of the platform with smaller scopes, it was a problem. It's also nice to be able to easily see exactly where the platform is when setting up or tearing down at night without having to turn on a light, thus preserving my precious night vision for hunting down faint galaxies and such (Fig. 8.53).

Here is a photo of the electronic circuitry that drives the stepper motor. It is based on the UCN5804B single chip stepper motor driver. A 555 timer circuit is tuned with a trimpot to provide about 375 pulses per second to the driver chip to move the motor at the sidereal rate. The motor is a 200 step/rev model that is half stepped. This means I get not quite one revolution per second out of the motor at the sidereal rate. For resetting the platform, the motor direction is reversed, the step mode is changed from half to full, and a reed relay cuts a second trim-pot into the

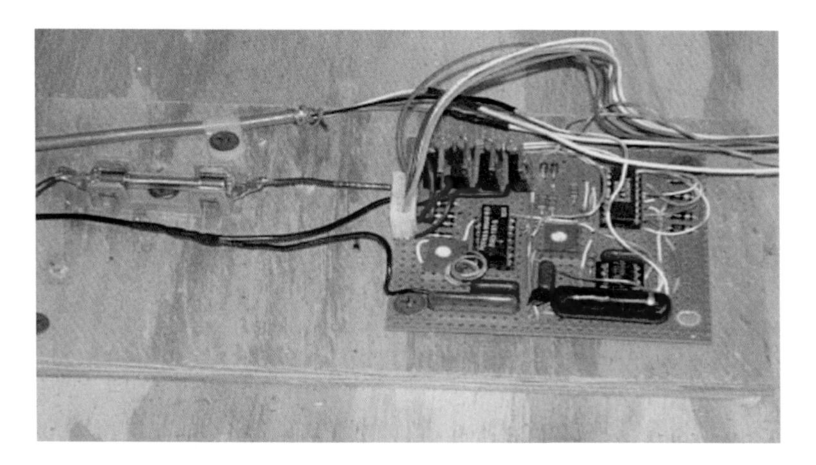

Fig. 8.53 The electronic circuitry (Image credit: Michael Davis)

555 circuit to speed up the pulse rate. The UCN5804B is supposedly (according to its spec sheet) capable of driving the motor I am using directly. However, I initially used a much bigger stepper motor in this project. So I added four big driver transistors to the board and used the outputs of the UCN5804B to switch them. I probably would have done it anyway since I tend to over-engineer things. The stuff I build never dies. For visual observing, the platform works great. For photography though, drift becomes a problem. In the future, I want to add a multi-turn trimpot to the hand controller and wire it in series with the main timing pot on the circuit board so I can fine-tune the speed of the platform for photography (Fig. 8.54).

Here is a schematic of the drive circuitry. Sorry it took so long to all who have been clamoring for it. I somehow lost my notes on this project when I moved some time ago. I couldn't find the schematic I drew up for it. Finally, I decided to bite the bullet and reverse-engineer the thing and recreate the schematic. Click on it for a larger version. I used the UCN5804B stepper motor driver IC by Allegro. Unfortunately, this chip is now obsolete. I wish I'd bought a boatload of them before they went obsolete. You may be able to still find some NOS out there. This circuit will drive a small stepper motor as is. To drive a big stepper that needs an Amp or more per phase, you should probably add some hefty driver transistors. To calibrate the electronics, start by tuning the 500 K pot to the speed to give you the sidereal tracking rate. This will take a lot of fiddling and drift tests. Tune the 50 K pot to the highest speed your stepper motor can handle for resetting the platform without stalling out. Both calibrations should be done with the normal weight load on the platform that it would see in use. Setting up and tearing down the big Dob for all testing and calibration the platform needed was too much of a hassle, so I used my "little" 6-in. Dob for drift testing and piled a lot of scrap iron on the platform to simulate the weight load of the big scope (Fig. 8.55).

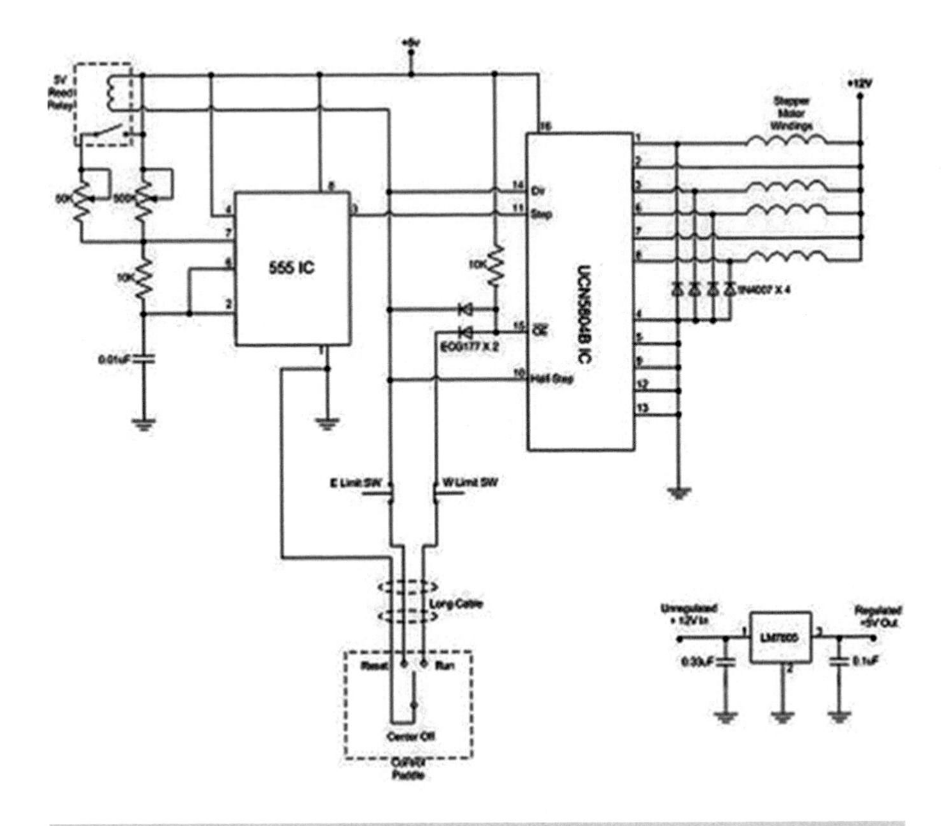

Fig. 8.54 The drive circuitry (Image credit: Michael Davis)

Fig. 8.55 The hand controller (Image credit: Michael Davis)

Fig. 8.56 Saturn taken through the big Dob (Image credit: Michael Davis)

Here is a photo of the hand controller for the platform. It is on a long cable and has Velcro on the back which allows me to stick it to a convenient spot on the side of the telescope. The controller only has three functions: run, stop, and reset. In the run position, the platform rotates from East to West at the sidereal rate counteracting the Earth's rotation. The platform will run for 40 min before needing to be reset. In the reset position, the platform will run at high speed from West back to East. The platform stops on its own at either end of travel. I can also stop it anytime by putting the switch in the stop position, which I do at times when the telescope doesn't need to be driven so as to minimize the number of reset cycles during an observing session. A reset cycle takes about 10 min. The little motor just doesn't have enough torque at higher speeds to move the mass of the platform with a big telescope on it. It just stalls out if I try to speed up the reset cycle anymore. That's ok though. The 10-min reset cycle gives me a chance to sit down for a while, relax and consult my star charts, and decide which objects to observe next (Fig. 8.56).

Here's a photo of Saturn taken through the big Dob on the platform, using a ToUcam. Astrophotography with a Dob! What a concept! With improvements, I hope to do a little deep-sky photography with the platform too (Figs. 8.57 and 8.58).

Michael Davis

Adapted from "A Large Equatorial Platform for a Big Dob" with permission from Michael Davis (http://www.mdpub.com)

Fig. 8.57 John Bauer (Asteroid 4525johnbauer) who was a noted Astronomy Professor at San Diego City College for almost 40 years is shown here in the photo observing the heavens with his homemade 14-inch Newtonian telescope in his Encinitas, California observatory. (Image credit: Vicky (Bauer) Melanson)

Fig. 8.58 Omega Centauri is a beautiful globular cluster in Centaurus (Image credit: Public Domain—Wikimedia Commons)

Further Reading

Davis, M. A large equatorial platform for a big Dob. http://www.mdpub.com
Heijkoop, A. Heijkoop's equatorial platform. http://www.astrosurf.com/aheijkoop/
Luton, C. Equatorial cradle. http://eq.brindabella.org
McCreary, D. The McCreary Mount. http://www.subarcsec.com
Philpot, W. Building an equatorial platform for my telescope. http://www.warrenthewizard.com

Chapter 9

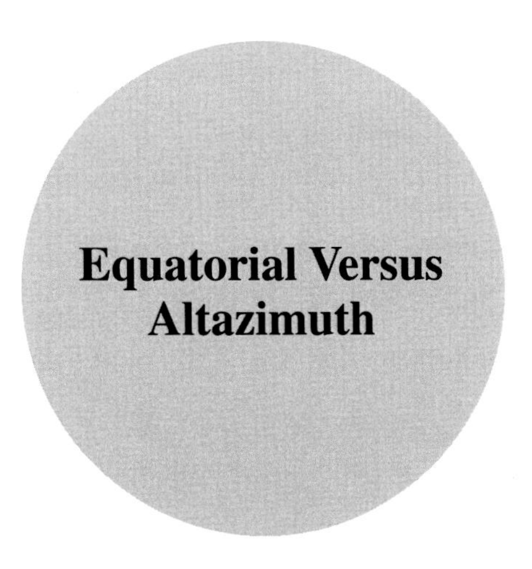

Equatorial Versus Altazimuth

The two most common popular types of telescope mounts in use today are the equatorial and the Altazimuth mountings. Both types of telescope mounting systems have been used by amateur and professional astronomers alike to mount equipment ranging from telescopes and cameras to an extremely large assortment of scientific instrumentation. For astronomers, a telescope once mounted needs to have some way to track celestial objects with it. An equatorial mount can be made to track quite easily in its equatorial axis (sometimes called the polar axis) with the addition of appropriate drive gears and a motor. With an Altazimuth mounting, the ease of use with its two axes (azimuth and altitude) makes it a simple process to maneuver a small or even large telescope manually about the sky for observing. However, when it comes to tracking celestial objects with it, things become a bit more complicated.

The use of an equatorial table or Poncet platform then becomes a primary consideration and a simple solution to remedy the tracking problem on an Altazimuth mount. With the use of an equatorial table or Poncet platform, tracking celestial objects becomes relatively easy. An equatorial table or Poncet platform is relatively inexpensive to build, no modifications to the telescope are needed nor necessary, and no field rotation at the eyepiece and no computer-assisted dual-axis drives are needed for tracking. Even one can do astrophotography for short periods of expo-sure time to over an hour and 15 min on their Altazimuth equatorial platform. One of the drawbacks is that an equatorial table or Poncet platform is optimally used within a latitude band of ±5°. Also, precise and even accurate polar alignment can take time to adjust. A heavy telescope requires the platform to be arranged to rotate the telescope about its center of

N. Butler, *Building and Using Binoscopes*, The Patrick Moore Practical Astronomy Series, DOI 10.1007/978-3-319-46789-4_9

gravity, and guiding a telescope on an equatorial platform will produce undesirable and undesirable field rotation unless you are imaging very close to the meridian. The equatorial platform or Poncet table must be reset approximately every hour or so, which will limit the exposure time for astrophotography to about 15° of tracking movement (approx. 1 h) across the sky. There has always been a great deal of debate about which type of mounting is the best one to use for doing basic observational astronomy. Some of the initial debate or issues can be attributed to incorrect facts and some obvious misinformation from some good intended individual opinion. But there are some real misconceptions about the equatorial and Altazimuth mounting types. So perhaps it is an opportune time in this chapter to discuss and even dispel some beliefs about their obvious positive advantages and negative disadvantages. One thing to bear in mind in this following discussion on telescope mounts is: There is no perfect mounting that can accommodate for all of the tracking eccentricities. Below are some of the positives and negatives of the two most popular telescope mounting systems.

Altazimuth Mount

The Altazimuth mount or more commonly referred to as a Dobsonian mount has gained considerable popularity over the past 70 years having been first seen at earlier popular telescope making conferences across the country, such as, Stellafane which is held annually in Vermont, actually evolved from the old "pillar and claw" system originally used only in the smaller inexpensive telescopes. Many cheap telescopes found in stores have small metal tripods with flimsy thin legs and small screw-tightened pivot points for bearings. We call these the "pillar and claw" mounts and they are the cheapest way to mount a telescope of just about any aperture. For example, in a typical "pillar and claw" mount, the single tiny support bearings are so small and of poor quality that once it is bumped or shaken, the telescope will often vibrate making for an uncomfortable viewing session. Most standard camera tripods in the past were considered almost as bad, but they too have also evolved in a positive way over the past 40 years or so. With advances in modern mount design and composite materials, computerized Altazimuth mounts can boast of supporting some of the largest apertures in amateur astronomy today.

Altazimuth Mount (the Positives)

Traditional Altazimuth mounts are stable telescope mounting systems, which are simple in design that do not require the use of an on/off-axis counterweight system that can create an unwanted vibration or mount flexure problem. Altazimuth mountings with their simple design only need two opposing vertical trunnion bearings and one horizontal bearings (e.g., a Lazy Susan bearing) is needed for horizontal rotation support and lateral movement.

1. Most Altazimuth mounts are relatively simple in design and construction and are a user-friendly mount for beginners and novices to learn and develop their skills on.
2. They are extremely portable with a faster setup time compared to most types of equatorial mountings, especially for telescopes with apertures over 8 in.
3. Easy to design, build, and assemble and often allows for a much simpler overall telescope primary mirror-cell and support system.
4. Altazimuth mounts are generally cheaper to purchase, setup and maintain compared to an equatorial mount, especially for the larger apertures.
5. An Altazimuth mount can take more abuse than an equatorial mount. Easier to replace or repair if damaged.

Altazimuth Mount (the Negatives)

1. Traditional Altazimuth mounts are unable to track objects with a single axis motor drive system compared to a clock drive German equatorial mount for long term tracking capability, and must be computerized with either dual-axis servomotors, stepper motors, or supported on an equatorial table or Poncet platform (approximately 1 h 15 min maximum tracking time on an equatorial platform). If it proceeds longer, there is the potential danger of the telescope falling off the platform when it reaches the critical angle, thus creating damage to the telescope.
2. Altazimuth mounts cannot track objects entirely through the zenith in dual-axis driven mode (e.g., being "Dobson's Hole" at the zenith, the "bane" (meaning > woe) of Altazimuth mounted telescopes).
3. With the absence of fixed field orientation makes the use of "star hopping" the main method of trying to locate faint celestial objects on non-computerized Altazimuth mounts. Celestial objects that are not located in readily familiar recognizable star fields can be more difficult to locate visually. Note a right-angle sweep cannot be used with an Altazimuth mount to find objects for visual observing.
4. Changing altitude and azimuth coordinates can make finding celestial objects more challenging with only the use of setting circles. This often requires using computerized digital readouts or the nearly continuous manual movement and manual calculations in order to keep track of pointing coordinates for finding desired celestial objects.
5. Field rotation limits the ability for the observer to do astrophotography with traditional 35-mm cameras and CCD cameras to shorter exposures (unless expensive field rotators are used). Using a guide scope for long exposures can be very difficult, since corrections for drift are sometimes more difficult to determine for the less experienced novice observer or beginning amateur.

Equatorial Mount

The typical equatorial mount is aligned using the celestial polar coordinate system (e.g., San Francisco, California, Latitude = 37° 46 min 30 s N), and it has been the primary standard of serious amateur, commercial, and professional astronomical telescopes for over a century. Equatorial mounts come in a variety of interesting and efficient designs that can support a multitude of telescopic equipment, accessories, and instrumentation. The most commonly recognized types used for astronomy are German equatorial mount (GEM), open fork mount, English yoke, English cross-axis, polar disk, fork, split ring, etc.

Equatorial Mount (the Positives)

1. An equatorial mount can be used manually or equipped with a clock-driven motor to drive the telescope in right ascension (RA) for equatorial tracking of celestial objects.
2. The majority of equatorial mounted designs (except for the English yoke mount) can be pointed and track celestial objects through all areas of the sky up to and including the zenith.
3. The requirement for "no" field rotation allows the observer to do astrophotography using longer exposure times with a variety of sophisticated guiding equipment where manual non-in-depth guiding corrections are held to a minimum. With a telescope on an equatorial mount, planetary observing becomes a bit easier, since the object in the field does not rotate.
4. Using well-known techniques such as the time honored "right-angle sweep" and the traditional "star drift" method can also be used in locating faint and sometimes elusive astronomical objects in a somewhat easier and faster way, even with telescopes that have no clock drive (only a reference star in close proximity in the field of view (FOV) is needed).
5. Right ascension (RA)/declination (Dec.) setting circles (both digital and analog) can be used for systematically locating the majority of celestial objects that are not readily visible. Digital readouts installed in both right ascension (RA) and declination (Dec.) for equatorial mounted telescopes can be simplified, since there is no need to perform real-time guiding calculations.

Equatorial Mount (the Negatives)

1. Equatorial telescope mounts of higher quality tend to be very stable but also somewhat bulkier, heavier, and less portable than its Altazimuth mount counterpart. Equatorial mounts often have to be separated and broken down into several smaller parts for transporting, which ultimately results in longer setup

times. In plain and simple terms, a typical equatorial mount generally speaking has more knobs to twist, turn, and tighten compared to an Altazimuth mounting.
2. Typical German equatorial mounts require the use of several heavy counterweights secured in an exact or precise location on a long shaft to balance the telescope and additional equipment properly. Balance of the telescope and its accessories on an equatorial mount takes time to accomplish and may require the addition of more counterweights which sometimes may or may not be readily available for precise balance. An off-balance telescope may cause the telescope to not track properly and may be difficult to maneuver.
3. German equatorial mounts with a telescope equipped with instrumentation have the potential problem of running into the telescope tripod or pier on celestial objects that are located on or near to the zenith. This usually requires a "heads up" on the observer's part, especially during astrophotography or serious observing, inorder to reverse the telescope's motion for continued sidereal tracking.
4. Moment and tangent arms on some equatorial mounts (depending on their length and thickness) have a tendency for flexure and even vibration, thus becoming potential problems unless the mount and its tines (an extended moment/tangent arm) are designed to be extremely massive. Equatorial fork mount moment and tangent arms also have a tendency to flex, making for minor errors in tracking and periodic lower-frequency (or jumping-jack) oscillation vibration with heavy telescopes with longer (tines) moment arms.
5. High-quality commercial or amateur-made equatorial mounts usually have a minimum of four bearings (two bearings on each axis). Some equatorial mounts even have brass or bronze bushings too. When it comes to purchasing a good equatorial mount, they are often more expensive than a standard quality manufactured Altazimuth mounting.
6. Precise polar alignment is fundamental for accurate sidereal tracking. This applies to equatorial and Poncet platforms too. Remember when it comes to polar aligning your telescope's equatorial mount, it's important to learn the field of view (FOV) around Polaris, the pole star in order to properly align your equatorially mounted telescope to the North Celestial Pole (NCP).
7. Equatorial mounts for novices and beginning amateurs require a little more in-depth learning skills, although once the amateur or novice telescope user gets familiar enough with them, they typically can often locate and track celestial objects faster and more easily than with Altazimuth mountings.

In terms of selecting the type of telescope mounting system, one would want to use for their own observing program, none of these "negatives" described above should "not" make one want to disqualify a particular equatorial or Altazimuth mount design from use by beginning and experienced amateur astronomers. The Altazimuth mount for the novice amateur astronomer can be recommended. The experienced amateur who has already enjoyed the service of both mounts needs no introduction to their positive and negative attributes. When it comes to long exposure photography, the popular equatorial equipped mount is often the mount of choice.

Very large telescopes that are designed for easy portability and setup time, the Altazimuth mount will often be the one chosen for the job. For example, the compact ever-popular split-ring equatorial mount design can also enjoy its clever engineered portability even with telescopes as large as 18–20 in. Nowadays, amateurs wanting a computerized telescope GOTO drive systems have the choice between the equatorial and Altazimuth computerized mounts. For obvious reasons, the increase in technology comes at a higher price, but happily without changing their basic design characteristics, except perhaps making them more user-friendly. As history has shown, both the equatorial and the Altazimuth equipped mounts will continue to evolve and to remain ever popular with the amateur astronomy and telescope making community well into the future and rightfully so. Adapted from: Telescope Buyers FAQ, What is the Best Mount? Dennis Bishop (C) 2002 (http://mss.mhn.de).

Using a Finderscope for Polar Alignment

If you are going to set up out telescope for a night of observing, you want to be certain that the telescope's equatorial polar axis is accurately aligned with the Earth's rotational axis. To accomplish this, we must first be able to check if the telescope's finderscope is aligned with the telescope's equatorial polar axis. The finderscope will be used to help us align our telescope. A typical telescope's finderscope is an optical pointing device that is attached to the top or the side of the main telescope tube (see Fig. 9.1). A telescope's finderscope is used to help point the main telescope at a specific area of the sky that you want to observe. Knowing how to use your finderscope is an important part of learning how to locate celestial objects in the night sky with your telescope.

Fig. 9.1 Typical right-angle 8× power finderscope attached to a telescope (Image credit: Halfblue-Wikimedia Commons)

Why do we need a finderscope? Finderscopes are generally very low in magnification. Usually between 6× and 9× compared to the naked eye and with a relatively wide field of view. And there are some finderscopes that have no magnification at all. For example, a Telrad uses a heads up display with an illuminated circle for locating the object. Without the finderscope, locating objects simply by looking in the main telescope would be very difficult. Even at a telescope's slowest magnification, it is still far too much magnification for locating objects easily.

A finderscope is an instrument for bringing objects into your telescope's field of view. Finderscopes when used properly will see the same objects that your main telescope sees. In order to do this, they need to be properly aligned. Finderscopes are usually not delivered attached to your telescope, so learning to set up, use, and remove the finderscope for storage is very important if you plan to use your telescope frequently.

Polar Alignment

By the Fort Worth Astronomical Society and Tom Koonce—Astro-Tom

If you're like most new amateur astronomers, the first thing you probably do when you get your new telescope properly assembled is to put in an eyepiece and point it up to look at the moon. Just the excitement of seeing the lunar landscape up close is enough to keep you entertained for days. But eventually, as you progress to finding more difficult objects, such as planets and faint deep-sky objects, you will want to utilize all the features of your equatorial mount, such as the setting circles or perhaps even a motor drive. A mount is said to be "equatorial" if one of its two axes can be made parallel with the Earth's axis of rotation. Aligning the telescope to the Earth's axis can be a simple or rather involved procedure depending on the level of precision needed for what you want to do. For casual observing, only a rough polar alignment is needed. Better alignment is needed for tracking objects across the sky (either manually or with a motor drive) at high magnifications. Still greater precision is needed in order to use setting circles to locate those hard-to-find objects. Finally, astrophotography will require the most accurate polar alignment of all.

Theory

The polar alignment procedure works on one simple principle: The polar axis of the telescope must be made parallel to the Earth's axis of rotation, called the North Celestial Pole (NCP). When this is accomplished, the sky's motion can be canceled out simply by turning the axis (either by hand or with a motor drive) at the same rate as the rotation of the Earth, but in the opposite direction. Although residents of the northern hemisphere are accommodated with a bright star (Polaris) less than a degree from the Earth's rotational axis, the NCP can still be a somewhat elusive place to locate (Fig. 9.2).

Fig. 9.2 Pointing toward the North Celestial Pole with the telescopes polar axis (Image credit: Fort worth Astronomical Society/Tom Koonce—Astro-Tom)

The North Celestial Pole (NCP) is the point in the sky around which all the stars appear to rotate. The star Polaris lies less than a degree from the NCP, and it can be used to roughly polar align a telescope. However, for accurate polar alignment, the polar axis of the telescope's mount needs to be aligned to the true NCP.

Rough Polar Alignment

By matching the latitude angle of the telescope mount with the latitude of your observing site, you can easily approximate the position of the North Celestial Pole (NCP) (Fig. 9.3).

For ordinary visual observing, the telescope's polar axis must be aligned to the Earth's pole. This simply means positioning the telescope so that the polar axis is aimed up at Polaris. The easiest way to accomplish this is to rotate the telescope tube to read 90° in declination. In this position the telescope will be parallel to the polar axis. Now, move the telescope, tripod, and all, until the polar axis and telescope tube are pointed toward Polaris. Finally, match the angle of your telescope's polar axis to the latitude of your observing location. Most telescopes have a latitude scale on the side of the mount that tells you how far to angle the mount for a given latitude (see your telescope owner's manual for instructions on how to make this adjustment). This adjustment determines how high the polar axis will point above the horizon. For example, if you live at 40° latitude, the position of Polaris will be 40° above the

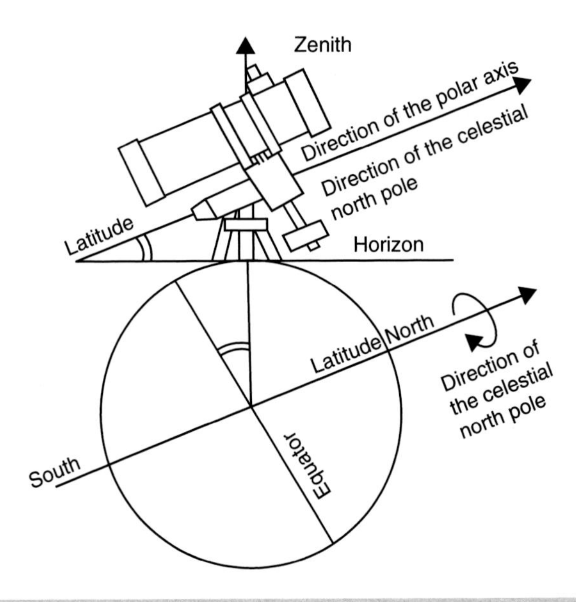

Fig. 9.3 Adjusting your telescope's polar latitude to the North Celestial Pole (Image credit: Fort Worth Astronomical Society/Tom Koonce-Astro-Tom)

northern horizon. Remember your latitude measurement need only be approximate; in order to change your latitude by 1°, you would have to move your observing position by 70 miles! Polaris should now be in the field of view of an aligned finderscope. Continue making minor adjustments in latitude and azimuth (side to side), centering Polaris in the finder's cross hairs or low power eyepiece. This is all that is required for a polar alignment good enough to use your telescope's slow motion controls to easily track a star or planet across the sky.

However, in order to take full advantage of the many features of your telescope (such as setting circle and astrophotography capability), a more precise polar alignment will be necessary.

Accurate Polar Alignment

Before we can be certain that the telescope's polar axis is accurately aligned with the rotational axis of the Earth, we must first be certain that the finderscope (which will actually be used to polar align the mount) is aligned with the telescope's polar axis.

For polar alignment purposes, the finderscope itself can be used to accurately align the mount's polar axis by adjusting the finder inside its bracket. This is quite simple since the finder is easily adjusted using the screws that hold it inside the bracket. Also, the finderscope's wide field of view will be necessary for locating the position of the North Celestial Pole relative to Polaris. Here's how it's done:

Set up your mount as you would for polar alignment. The DEC setting circle should read 90°. Rotate the telescope in right ascension so that the finderscope is positioned on the side of the telescope tube. Adjust the mount in altitude and azimuth until Polaris is in the field of view of the finder and centered in the cross hairs.

Now, while looking through the finderscope, slowly rotate the telescope 180° around the polar axis (i.e., 12 h in right ascension) until the finder is on the opposite side of the telescope. If the optical axis of the finder is parallel to the polar axis of the mount, then Polaris will not have moved, but remain centered in the cross hairs. If, on the other hand, Polaris has moved off of the cross hairs, then the optical axis of the finder is skewed slightly from the polar axis of the mount. If this is the case, you will notice that Polaris will scribe a semicircle around the point where the polar axis is pointing. Take notice how far and in what direction Polaris has moved (Fig. 9.4).

Even with the telescope positioned 180° around the mount, the telescope (and finderscope) should still be pointing at the same object in the sky.

Using the screws on the finder bracket, make adjustments to the finderscope and move the cross hairs halfway toward Polaris' current position (indicated by the "X" in Fig. 9.5). Once this is done, adjust the mount itself in altitude and azimuth so that Polaris is once again centered in the cross hairs. Repeat the process by rotating the mount back 180°, and adjusting the finder bracket screws until the cross hairs are halfway between their current position and where Polaris is located, and then centering Polaris in the cross hairs by adjusting the mount in altitude and azimuth. With each successive adjustment, the distance that Polaris moves away from the center will decrease.

Continue this process until Polaris remains stationary in the cross hairs when the mount is rotated 180°. When this is done, the optical axis of the finderscope is

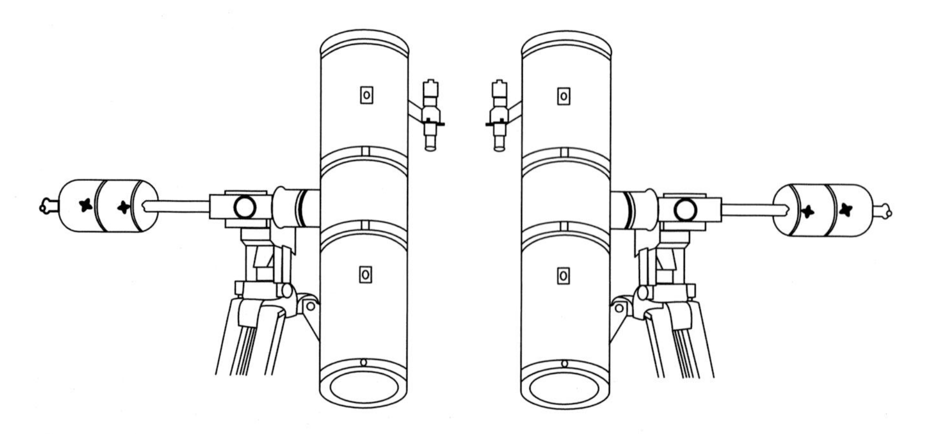

Fig. 9.4 Aligning your telescope on Polaris (Image credit: Fort Worth Astronomical society/ Tom Koonce—Astro-Tom)

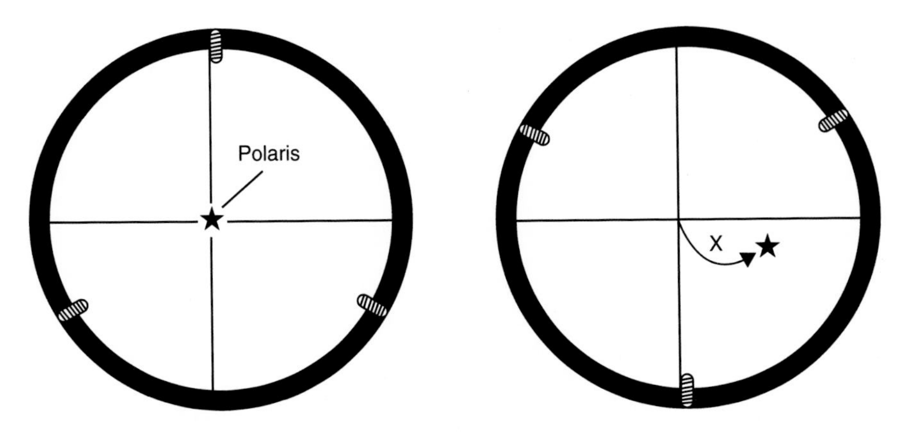

Fig. 9.5 Aligning your telescope on Polaris using your finderscope (Image credit: Fort Worth Astronomical society/Tom Koonce—Astro-Tom)

perfectly aligned with the polar axis of the mount. Now the finder can be used to polar align the mount (Fig. 9.5).

When rotating the finderscope 180° around the polar axis, the cross hairs will rotate around the point which the polar axis is pointing (indicated by the "X" in Fig. 9.5). Adjusting the finderscope and the equatorial mount until an object remains centered in the cross hairs indicates that the finderscope is aligned with the telescope's polar axis.

So far we have accomplished aligning the polar axis of the telescope with the North Star (Polaris), but as any star atlas will reveal, the true North Celestial Pole (NCP) lies about 3/4° away from Polaris, toward the last star in the Big Dipper (Alkaid). To make this final adjustment, the telescope mount (not the telescope tube) will also need to be moved away from Polaris toward the actual NCP. But the question is: "Since Polaris makes a complete rotation around the Celestial Pole once a day, how far should the mount be moved and in what direction?" Let's take an example: Suppose you are out observing on August 1st at 8:00 p.m... a quick inspection of the northern sky will reveal that the last star in the handle of the Big Dipper, Alkaid, lies above and to the left of Polaris in the 10 o'clock position. Now, while looking through the finderscope (with Polaris still centered in the cross hairs), adjust the latitude and azimuth of the mount up and to the left until Polaris also moves up and to the left in your straight through finderscope. (Remember a straight through finder inverts the image, so Polaris will appear to move in the same direction as the mount is moved.) How far to move Polaris will depend on the field of view of the finderscope. If using a finderscope with a 6° field of view, Polaris should be offset approximately 1/3 of the way from center to edge in the finder's view (i.e., half of the field of view, from center to edge, equals 3° and 1/3 of that equals 1°). This calculation can be approximated for any finderscope with a known field of view (Figs. 9.5).

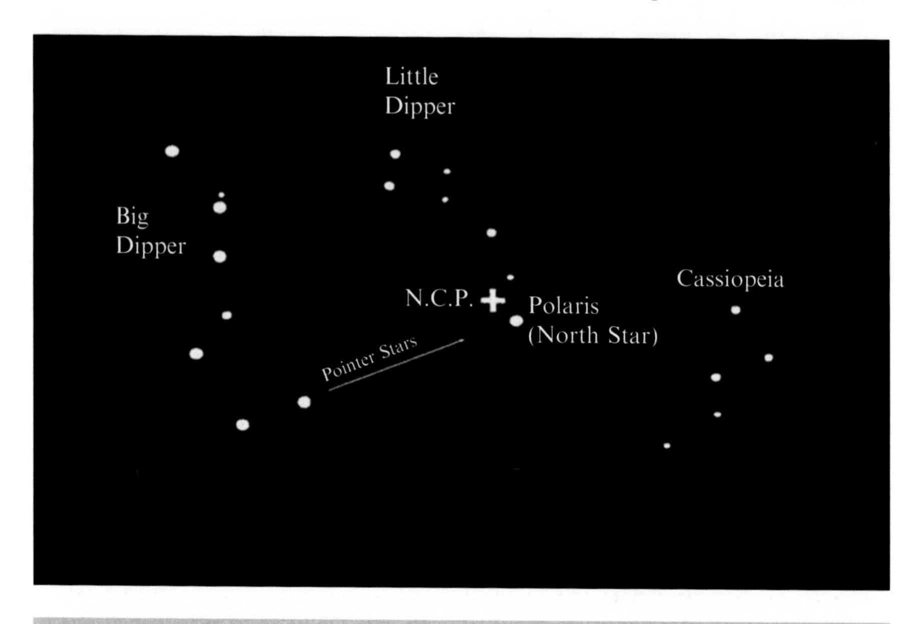

Fig. 9.6 The position of Polaris to the North Celestial Pole (NCP) (Image credit: Forth Worth Astronomical Society/Tom Koonce—Astro-Tom)

The true North Celestial Pole (NCP) lies less than a degree away from Polaris in the direction of the last star in the handle of the Big Dipper (Ursa Major).

The mount's setting circles can now be used to determine just how close the polar axis is to the NCP. First, aim the telescope tube (be careful not to move the mount or tripod legs) at a bright star of known right ascension near the celestial equator. Turn the right ascension setting circle to match that of the bright star. Now, rotate the telescope tube until it reads 2 h 30 min (the right ascension of Polaris) and +89¼° declination.

Polaris should fall in the center of the finder's cross hairs. If it doesn't, once again move the mount in latitude and azimuth to center Polaris.

This procedure aligns the telescope mount to within a fraction of a degree of the NCP, good enough to track a star or planet in a medium power eyepiece without any noticeable drift. However, long-exposure astrophotography is far less forgiving and film will easily reveal even the smallest amount of motion. At this point, you may be wondering why bother polar aligning any more accurately if you can use the slow motion controls or drive corrector to keep a guide star centered in the cross hairs of an eyepiece. Unfortunately, keeping the guide star centered in the cross hairs is only half the battle. Since, the polar axis is not perfectly in line with the Earth's axis, the stars in the field of view will slowly rotate as you guide. You will get a sharp image of the guide star, but the other stars on the photograph will appear to rotate around the guide star. This is also why you cannot accurately do guided photography with an altitude-azimuth (Altazimuth) style mount (Figs. 9.4 and 9.5).

Fig. 9.7 Dr. Chris Sorensen (blue cap, black shirt—front right) and his Kansas State University Astronomy class getting ready to do some H-Alpha and White Light solar observing in Manhattan, Kansas. (Image credit: Dr. Chris Sorensen—Dept. of Physics—Kansas State University)

Fig. 9.8 The Pleiades (M-45), more commonly called the "Seven Sisters", is located in the constellation Taurus (Image credit: Public Domain—Wikimedia Commons)

Fig. 9.9 (Cartoon credit: Jack Kramer)

Precise Polar Alignment

Precise polar alignment is limited by the accuracy of your telescope's setting circles and how well the telescope is aligned with the mount. The following method of polar alignment is independent of these factors and should only be undertaken if long-exposure, guided photography is your ultimate goal. The declination drift method requires that you monitor the drift of selected stars. The drift of each star tells you how far away the polar axis is pointing from the true celestial pole and in what direction. Although declination drift is simple and straightforward, it requires a great deal of time and patience to complete when first attempted. The declination drift method should be done after the previously mentioned polar alignment steps have been completed.

To perform the declination drift method, you need to choose two bright stars. One should be near the eastern horizon and one due south near the meridian. Both stars should be near the celestial equator (i.e., 0° declination). You will monitor the drift of each star one at a time and in declination only. While monitoring a star on the meridian, any misalignment in the east–west direction is revealed. While monitoring a star near the east horizon, any misalignment in the north–south direction is revealed. As for hardware, you will need an illuminated reticle ocular to help you recognize any drift. For very close alignment, a Barlow lens is also recommended since it increases the magnification and reveals any drift faster. When looking due south, insert the diagonal so the eyepiece points straight up. Insert the cross hair ocular and rotate the cross hairs so that one is parallel to the declination axis and the other is parallel to the right ascension axis. Move your telescope manually in RA and Dec. to check parallelism.

First, choose your star near where the celestial equator (i.e., at or about 0° in declination) and the meridian meet. The star should be approximately 1/2 h of right ascension from the meridian and within 5° in declination of the celestial equator. Center the star in the field of your telescope and monitor the drift in declination.

If the star drifts south, the polar axis is too far east.
If the star drifts north, the polar axis is too far west.

Using the telescope's azimuth adjustment knobs, make the appropriate adjustments to the polar axis to eliminate any drift. Once you have eliminated all the drift, move to the star near the eastern horizon. The star should be 20° above the horizon and within 5° of the celestial equator.

If the star drifts south, the polar axis is too low.
If the star drifts north, the polar axis is too high.

This time, make the appropriate adjustments to the polar axis in altitude to eliminate any drift. Unfortunately, the latter adjustments interact with the prior adjustments ever so slightly. So, repeat the process again to improve the accuracy, checking both axes for minimal drift. Once the drift has been eliminated, the telescope is very accurately aligned. You can now do prime focus deep-sky astrophotography for long periods.
Note:

If the eastern horizon is blocked, you may choose a star near the western horizon, but you must reverse the polar high/low error directions. Also, if using this method in the southern hemisphere, the direction of drift is reversed for both RA and Dec.

Even with a telescope with a clock drive and a nearly perfect alignment, most beginners are surprised to find out that manual guiding may still be needed to achieve pinpoint star images in photographs. Unfortunately, there are uncontrollable factors such as periodic error in the drive gears, flexure of the telescope tube and mount as the telescope changes positions in the sky, and atmospheric refraction that will slightly alter the apparent position of any object.

Polar alignment, as performed by many amateurs, can be very time consuming if you spend a lot of time getting it more precise than is needed for what you intended to do with the telescope. As one becomes more experienced with practice, the polar alignment process will become second nature and will take only a fraction of the time as it did the first time. But remember that when setting up your telescope's equatorial mount, you only need to align it well enough to do the job you want.

Using the Star Drift Method

In the following article, differing directions for the southern hemisphere are indicated by braces [example]. If you have an equatorial telescope, you may have experienced the pain of tracking error. If you plan on doing long-exposure astrophotography, you must be accurately aligned on the North [or South] Celestial Pole. Having a drive corrector isn't enough. Even if you keep a star right on the crosshair in your reticle, if the mount isn't polar aligned, you'll get some error from field rotation. Polar finderscopes are useful tools and will get you very close to the mark, but imagine having your scope track "dead on" with no declination corrections for 10–20 min! If you have period error correction (PEC), you'll be free from right ascension corrections as well. Polar alignment by the "star drift" method takes more time than using a polar scope, but a simple procedure using two stars will get your mount RIGHT ON the Celestial Pole. It's a simple method that does not require setting circles or the knowledge of date and time. Another valuable aspect of the two-star method is that Polaris does not need to be visible while aligning.

Start by doing a quick and dirty eye-ball polar alignment. The closer this initial alignment is, the faster you will complete the task. Point the telescope at a star close to the celestial equator and near the eastern horizon. Not too close to the horizon now… you don't want diffraction entering the equation. Track on the star and make corrections in right ascension only; don't correct in declination yet. Watch the star drift. An illuminated reticle makes this a lot easier. If the star drifts south in the field of view, then the polar axis is too low (pointed below the pole—toward the horizon). If the star drifts north, the axis is too high (pointed above the pole toward zenith). Make your adjustments to the altitude of the polar axis until the drift becomes negligible.

Now go to a star near the celestial equator again but this time near the meridian. Guide only in right ascension as before. If the star drifts south, then the polar axis is pointing east [west] of north [south]. If it drifts north, then the axis is west [east] of north [south]. This time adjust the azimuth of the mount until the declination drift goes away. Do this a few more times until there's no drift in declination from stars near the horizon and meridian.

There's an even easier way to remember which direction to adjust. Forget all about north and south in the eyepiece. Just remember that it's the altitude and azimuth of the mount that you are adjusting. Astrophotographer Barry Gordon calls it the "altitudE is Easy and aziMuth is Mad" method. The altitude adjustment is easy to remember—adjust it so the star moves back toward the center of the field. The azimuth adjustment is maddening because you move the star further away from the center, in the direction it's drifting.

It really is that simple—it just takes longer than any of the other methods. But if you are taking shots over 10 min in length, it gets pretty boring constantly staring into an eyepiece. Wouldn't you rather look every minute or 2 and make a tiny tweak rather than being hunched over with your eyes straining and watering?

Adapted from "Polar Alignment" with permission from Fort Worth Astronomical Society and Tom Koonce—Astro-Tom (http://www.astro-tom.com)

Further Reading

Celestron website. Polar alignment. http://www.celestron.com/polar.htm

Coco, M. J. (1993, February). Dialing for deep-sky objects. *Astronomy Magazine*.

Michael Porcellino, M. (1992, May). Polar aligning your telescope. *Astronomy Magazine*.

Mihalas, D., & Binney, J. (1981). *Galactic astronomy: Structure and kinematics* (2nd ed.). San Francisco: W.H. Freeman. ISBN 0-7167-1280-6.

Polar alignment—Adapted from Fort Worth Astronomical Society and Tom Koonce/Astro-Tom. http://www.astro-tom.com

Schmeidler, F. (1994). Chapter 2: Fundamentals of spherical astronomy. In G. D. Roth (Ed.), *Compendium of practical astronomy*, revised translation of Handbuch für Sternfreunde (4th ed., pp. 9–35). Springer Verlag. ISBN 0-387-53596-9.

Telescope Buyers FAQ What is the Best Mount?, Dennis Bishop © 2002, http://mss.mhn.de

Chapter 10

Binoscopes of the Third Kind

What are "binoscopes of the third kind"? Binoscopes for the most part can be either commercial (first kind) or homemade (second kind). But there are a lot of binoscopes out there in telescope land that are both. The author likes to call these kinds of binoscopes simply "binoscopes of the third kind." They can have a varied combination of parts or accessories that are both homemade and commercial and still can be a lot of fun to build and use for astronomy. These types of binoscopes can have mounts that are commercial but have homemade optical systems. Or they can have mounts that are homemade but carry commercial optical tube assemblies. Either way, "binoscopes of the third kind" can have a little of both built in the way of commercial and homemade qualities (see Fig. 10.13). For the most part, it doesn't have to be necessarily a binoscope either. It can be anything associated with and/or used for binoculars or binocular astronomy in general. A new binoscope ATM astroterm been has probably not been "coined" here. And even if for some reason it never catches on as a new "household word," the author won't be too disappointed.

The following ATM projects are just a few interesting examples of what "binoscopes of the third kind" could look like. Each of the articles is told by the individuals in their own words giving the reader some great insight on how these particular "binoscopes of the third kind" were designed and made.

Eight-Inch Binocular Telescope

By the members of the Wabash Valley Astronomical Society

The *Wabash Valley Astronomical Society* in Lafayette, Indiana, started its discussions for a new project in the fall of 1998. After months of discussions and presentations, we

N. Butler, *Building and Using Binoscopes*, The Patrick Moore Practical Astronomy Series, DOI 10.1007/978-3-319-46789-4_10

Fig. 10.1 Eight-inch binocular telescope (Image credit: Wabash Valley Astronomical Society)

decided to build an 8-in. binocular telescope. Three major issues of design came up: The first was portability. Two 8″ optical tube assemblies mounted permanently together would be quite bulky. Using two separate tubes that would be separated between uses and reattached for use might make the alignment problem (second issue) more difficult. We decided to make the tubes in sections with only the rear parts permanently attached in a box frame (Figs. 10.1 and 10.2).

The secondary cage construction was first. The upper section includes a truss tube section to increase portability and decrease the needed storage space. The second issue was alignment: With a binocular scope, there has to be a way to adjust the distance between the eyepieces. The individual scopes are Newtonian reflectors, so when the light cone comes out of the side of the tube, it is bounced off a star diagonal (the "tertiary") so that the light cones and eyepieces are parallel for viewing. With this setup, the easiest way to vary the interocular distance is to use a focuser at the normal Newtonian position to vary the distance from the telescope tube to the tertiary and eyepiece. The problem with that is that it changes the distance from the primary mirror to the eyepiece, so it changes the focus view from the bottom, showing the reflection of the built-in Barlow in the 2.14″ secondary. The secondary holder was made from a piece of black PVC pipe. Jim machined the helical focuser from two pieces of spare aluminum. The inner piece threads into the diagonal where the original eyepiece holder tube was. The inner piece is threaded on the outside. The outer piece then threads onto the inner piece for fine focusing. The upper inside of the outer piece is not threaded, rather the original eyepiece

Fig. 10.2 Completed optical tube assembly (Image credit: Wabash Valley Astronomical Society)

holder slides into it for coarse focusing. The outer piece has a knurled grip ring. Next, the truss assembly was constructed. Another focuser is placed after the tertiary and holds the eyepiece. We wanted to build a scope that could be used at public events, and it has been our experience that the general public is often very timid about touching the controls on a scope, so having to adjust the first two focusers and then make a significant change with the other two focusers may be too much "fussing" for the general public. Another option is to change the interocular distance by moving one or both telescope tubes. The difficulty with this is that with the magnification, the tubes need to maintain their alignment to within about 1 arc minute (at high power and especially in the vertical direction) for the eyes and brain to combine the two images into one. This is difficult to do, given the size and weight of the full tubes. Several methods to do this have been tried by other people, and it is generally regarded as one of the most difficult parts of building such a telescope. We ultimately decided to suspend the lower sections of the tubes from pivots so the tubes would swing out away from each other to the needed interocular distance (much like conventional binoculars, except that each tube will have its own pivot) (Fig. 10.3a, b).

The third issue involved optics: As mentioned above, an extra diagonal is needed to aim the light cones and eyepieces in a parallel direction. This means more of the light path is past the secondary, so a larger secondary is required, and that will degrade the image. A possible cure for this is to install a Barlow lens in the side of

Fig. 10.3 Completed optical tube assembly (Image credit: Wabash Valley Astronomical Society)

the tube where the light cone exits the tube. This stretches out the remainder of the cone so that a smaller secondary can be placed farther from the primary and still achieve focus at the eyepiece. The drawback is that the scope operates at a high focal ratio, which means higher magnification, a correspondingly smaller field of view, and tighter alignment tolerance. In fact, since the light path bounces through the tertiary star diagonal before going to the eyepiece, a standard 2× Barlow will operate at about 3×. We used a "shorty" Barlow, with the Barlow tube itself acting as the extension tube between the side of the main tube and the tertiary. In our Barlow, the cell holding the Barlow lens itself unscrews from the tube so it can be removed for low power viewing. There may be a bit of vignetting in this case since in choosing the secondary size, we compromised between the two options, but leaned toward the smaller size, leaving open the possibility of getting a larger second set of secondaries later for better low power viewing. The lower tube sections will have two sets of mounting holes for the primary mirror cells since the mirrors will have to be moved forward if the Barlow is not used. It is a common knowledge that using two eyes allows for easier viewing and usually provides a better view, but after reading about the extreme difficulty of the alignment problem, you may be wondering why we didn't just start with a single primary with twice the surface area but use a binoviewer. After all, a larger primary is theoretically capable of producing higher resolution, right? There are several reasons why we didn't take that route: first, the cheapest binoviewers cost about as much as our entire budget for this project. Second, while a larger primary would achieve better resolution under perfect conditions, conditions are rarely good enough to get better resolution than an 8″ mirror provides, and a larger mirror will actually be more affected by atmospheric turbulence on bad nights. Also (and not often recognized), since the two mirrors will be looking through different air masses, they will show two slightly different images, depending on how each is affected by the turbulence. When the brain gets two views of the same object, it is remarkably good at concentrating on the sharper image. Thus, a true binocular scope provides good views almost twice as often as a single tube scope. Finally, another reason we made

Fig. 10.4 Completed optical tube assembly (Image credit: Wabash Valley Astronomical Society)

Fig. 10.5 Optical tube with Barlow attached (Image credit: Wabash Valley Astronomical Society)

this particular type of scope (and not some other type of scope altogether) was that Jim Sattler had two 8″ mirror blanks to donate to the project. Curious about how we built this? Tour the construction (Figs. 10.4 and 10.5).

The first step was to build the box frame to hold the two tubes. Jim Mettler took charge of this part. The frame was made from 3/4″ square aluminum tube (1/8″ wall thickness) held together with machined fittings press-fit into the ends. The fittings were made very slightly oversized, with the corners of the square pegs rounded to avoid splitting the square tubes. After a few hours of cutting aluminum tube lengths and pressing the fittings into them, the basic frame took shape. Then, we added two 1/8″ aluminum sheet plates to hold the altitude bearing, and since one section of the frame has to be left out so that the frame can be moved in altitude when on the pier, more 1/8″ aluminum sheet was used as a gusset in the back to keep the two sides from sagging toward each other (Figs. 10.6 and 10.7).

After the frame was complete, the mount bearing was created. The following parts make up the bearings (Fig. 10.8).

Fig. 10.6 The fitting being pressed into a tube section (Image credit: Wabash Valley Astronomical Society)

Fig. 10.7 A close-up of a corner fitting about to be pressed into a tube section (Image credit: Wabash Valley Astronomical Society)

At the left is a section of 4″ PVC pipe that slides into the pier. Two snug wooden spacer rings made by George Gourko provide a tight fit (only one is shown—the other is already in the pier). At the top of the pipe is a threaded adapter. In the center of the picture is a 4″ PVC "T" sliced in half, that forms the heart of the mount (Jim Sattler's idea) (Figs. 10.9 and 10.10).

Fig. 10.8 Close-up view of the rectangular tube frame section (Image credit: Wabash Valley Astronomical Society)

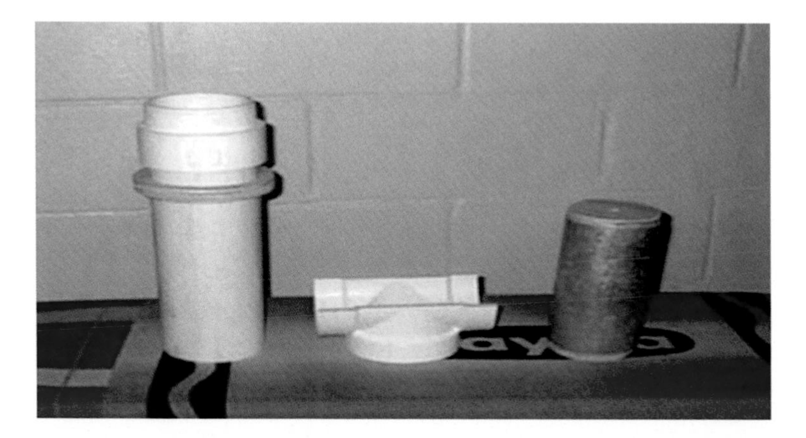

Fig. 10.9 The 4-in. PVC pipe section on the left that slides into the pier (Image credit: Wabash Valley Astronomical Society)

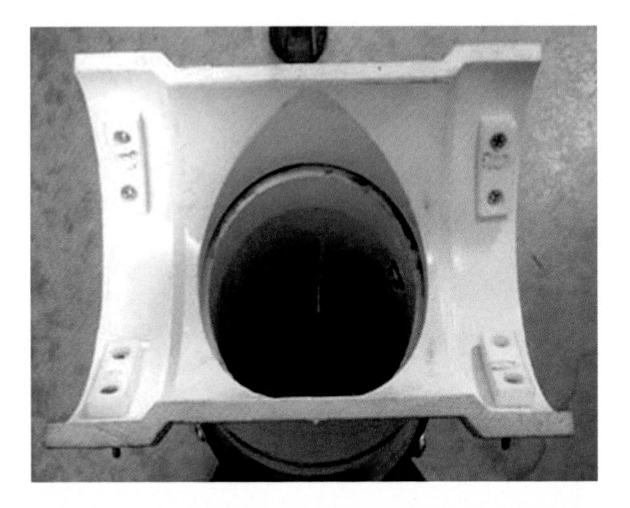

Fig. 10.10 PVC section sliced in half "T" that performs the heart of the mount (Image credit: Wabash Valley Astronomical Society)

In the center of the picture is a 4″ PVC "T," sliced in half, that forms the heart of the mount (Jim Sattler's idea). The third ("stem of the T") opening is threaded and screws on top of the pipe to form the azimuth bearing. Heavy grease is applied for smooth movement. A 4″ diameter aluminum tube that will be cradled on the top of the "T" to form the altitude bearing is shown in Fig. 8.13. This tube will be mounted in the box frame between the optical tubes, so it will be sort of an "inside out" version of a standard Dob mount (Figs. 10.11 and 10.12).

For the altitude bearing, wooden disks made by George Gourko were pressed into the ends of the 4″ diameter aluminum tube. These were drilled off-center (left) to allow a vertical adjustment for balancing the scope. Teflon pads were added to the T (Fig. 10.11) and a stop was added to make sure the mount wasn't accidently "unscrewed." But the off-center holes mean there will be some torque on the tube, so sandpaper disks were glued to the disks (right) for better friction when the bearing is mounted between the side plates in the box frame. To allow for horizontal adjustments, slots rather than holes were cut in the side plates for mounting the bearing (Figs. 10.12 and 10.13).

The pier is from a 16″ Meade Starfinder Newtonian. The pier is on semi-permanent loan from the Prairie Grass Observatory.

Wabash Valley Astronomical Society

Adapted from "Eight Inch Binocular Telescope" with permission from Wabash Valley Astronomical Society (http://www.stargazing.net)

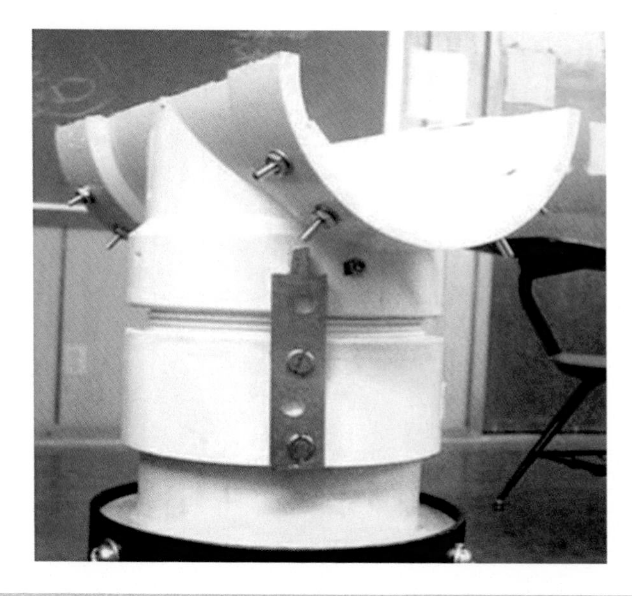

Fig. 10.11 PVC section sliced in half "T" that performs the heart of the mount (Image credit: Wabash Valley Astronomical Society)

Fig. 10.12 Teflon pads were added to the "T" along with a stop (Image credit: Wabash Valley Astronomical Society)

Fig. 10.13 The handle in the middle section rotates an arm that makes the two optical tubes swing apart via a pivot point. (Image credit: Wabash Valley Astronomical Society)

Short Notes on Asymmetric Binocular Telescopes

By Dr. Alex Tat-Sang Choy (Hong Kong)

I've always dreamed of having a pair of big binocular telescopes since teenage. In 2002–2003, I built a pair of asymmetric binocular telescopes with what I believed to be the simplest design, using very simple tools. This design gives essentially the same optical quality for each telescope tube and is very suitable for amateur ATM. Recently, I noticed a new Cloudy Nights article by Bill Zmek showing a binocular telescope he intended to build with a design he discovered independently. Since I already built one and used it for a long time, I think I should share some thought so anyone who tries to build binoculars with this design will have the benefit of my experience (Fig. 10.14).

If we are given a pair of 40-mm refractors with diagonals, it would take 5 min to build simple binoculars. Simply find a piece of wood, drill a few holes, and secure the refractors in parallel and we're set. The separation of the two eyepiece's center would have to be the interocular distance of our eyes, which is usually between 5 and 8 cm. However, if we're given a pair of larger refractors, say 120 mm in diameters, such simple design would not work. The asymmetric binoculars are a design to solve this problem. (Bill Zmek has drawn nice figures and explained the principle clearly, so I would not repeat here.)

Fig. 10.14 The 8-in. binocular telescope pier with altitude bearing attached (Image credit: Wabash Valley Astronomical Society)

When building binoculars with this design, the following details may be helpful:

1. Let me call the uncut OTA as tube A and the cut OTA as tube B. Since tube B is cut and rejoined at 90°, in order to retain the original optical quality, it is crucial to have a mechanism to align the optical axis of tube B or else the optical quality of the final product will not justify the hard work of making the binoculars. I learned this the hard way (Fig. 10.15).
2. There must be mechanism for the alignment of optical axis of A and B. Unlike regular binoculars, which goes up to as much as 40× magnifications, these binocular telescopes are expected to work at 100× or 200×. At these powers, a minute off in optical axis will result in 1–2° off in the visual image. Although the human eyes/brain can adjust for minor mismatch in images between the two eyes, it may only take a few minutes of observation to make one tired and stressed. The result is reducing productivity and enjoyment (Fig. 10.16).

   The design shown in Fig. 10.16 does not have alignment mechanism between tube A and B, giving me a lot of trouble in actual observation. I have since reworked the mounting plate between A and B and added an x-y fine adjustment for tube B, which makes my observation night a lot more enjoyable.
3. Rigidity of the assembly is important. The distribution of weight of this pair of binoculars is strange, parts holding the OTAs in place must be strong, and vibration dampening may be necessary.

Fig. 10.15 The asymmetric 120-mm binocular telescope—front view (Image credit: Dr. Alex Tat-Sang Choy)

Fig. 10.16 The asymmetric 120-mm binocular telescope—diagonal view (Image credit: Dr. Alex Tat-Sang Choy)

4. Center of gravity is important. The center of gravity for this binocular telescope is somewhere strangely placed, both in the up/down and front/back direction. The binocular telescope itself is very heavy and has large moment of inertia; slight mismatch between the mount axis and the center of gravity of the assembly can make the scope hard to use.
5. This is a heavy scope and requires stronger mount. Fortunately, I had a handy LXD55. My LXD55′s "polar" axis can be adjusted to point upward, and the

Fig. 10.17 The asymmetric 120-mm binocular telescope—front rear view (Image credit: Dr. Alex Tat-Sang Choy)

controller can be made to believe it is an LX90. My binoculars on the azimuth LXD55 is usable, although not great. But a 120-mm GOTO binoculars can draw some attention. In any case, strong mount is important (Fig. 10.17).

1. There are three 90° binoculars designs I know of, and they have their disadvantages:

   (a) *Matsumoto design.* It uses two mirrors per tube and gives erect images, which is great for wide angle panning in star fields. I thought it is a brilliant design, but it is probably not easy to build and not cheap to buy.
   (b) *Asymmetric design.* Mirror image could be hard to get used to for scanning and panning in the star field. I used only a few days to build the prototype, but a lot more effort to improve upon it. I had only simple electric drills and saws, if I could build one, it's not difficult.
   (c) *Three mirrors per tube design.* Its interocular distance is adjustable, but buying six mirrors and aligning them could be not favorable for ATMs. The multiple reflection also places strong requirement on the quality of mirrors.

2. Interocular distance adjustment is overrated. I had great trouble finding a simple, strong, and low cost solution for interocular distance adjustment. In the end, I decided to build first, think later. My binoculars have a fixed interocular distance of 5.5 cm. I have since shown them to some members of the Columbus Astronomy Society and even once to the public at the Perkin's Observatory; to my surprise, only about 1 in 10 people complaint about not able to merge the two images. Large percentage of the public grasped at the sight of a 23× moon. (And since they never used my telescope for long observations, I do not know if they would

still feel comfortable in the long run.) Therefore, from an ATM's point of view, NOT building the interocular distance adjustment could be a decision that saves more than 40% of the effort, a nice trade-off in my opinion.

3. The viewing. While I was still in the USA, where dark skies were accessible, this pair of binoculars gave me some of the most memorable views. In a good night using a pair of mid-priced 30-mm 80° eyepieces, I could view the entire M31 filling up most of the 80° apparent field of view, while at the same time clearly see the dust lane and the companions. M42 was 3D like. The whole Veil Nebula could be seen together without any filter. And the Milky Way was simply stunning. Unfortunately, the 120-mm achromat was not good for planet observation. Despite that, the most touching view I had was during a total lunar eclipse, at 23×, when some faint cloud passed in front of the pale orange moon, the color and perception of 3D was breathtaking.

The refractive binocular telescope brings a different dimension of enjoyment to star observing due to its great image clarity and contrast. The asymmetric design is probably the easiest to build for ATMs and in my opinion, worthy of all the hard work and resources.

Dr. Alex Tat-Sang Choy

Adapted from a Cloudy Nights story "Short Notes on Asymmetric Binocular Telescopes" with permission from Dr. Alex Tat-Sang Choy (Hong Kong)/Astronomics (www.astronomics.com)

Easy-to-Build Binocular Chair

By Greg Walton (MPAS/ASV Member)

Most binos are designed to look at the horizon (Terrestrial) and very few come with 45° or 90° eyepieces and these are very costly. So I had to come up with some way of holding the binos and be able to look through them at the zenith (straight up). First, I tried the camera tripod but these were too low and hard to aim at the object, and I would end up with a bad neck trying to look through them. The second thing I tried was the trick of looking down into a mirror laid on a table, which was much easier on the neck, but my arms quickly became tired. Also the sky was upside down, and all the stars were double stars because I used an ordinary mirror that was aluminized on the back instead of the front. The third thing I tried was to lay on my back and rest the binos on my head; this worked best so far, until I wanted to look at an object close to the horizon.

I then thought I should get a reclinable car seat and put it on a swivel base and then motorize it all and mount the binos on a bracket in front of my nose. With a push of a button, I could move around the sky. But it all looked like too much work (Fig. 10.18).

Then I came across a metal-framed reclinable chair in an opportunity shop for $30 and it had a swivel base. I thought to myself, that's it, I can make this work. I would

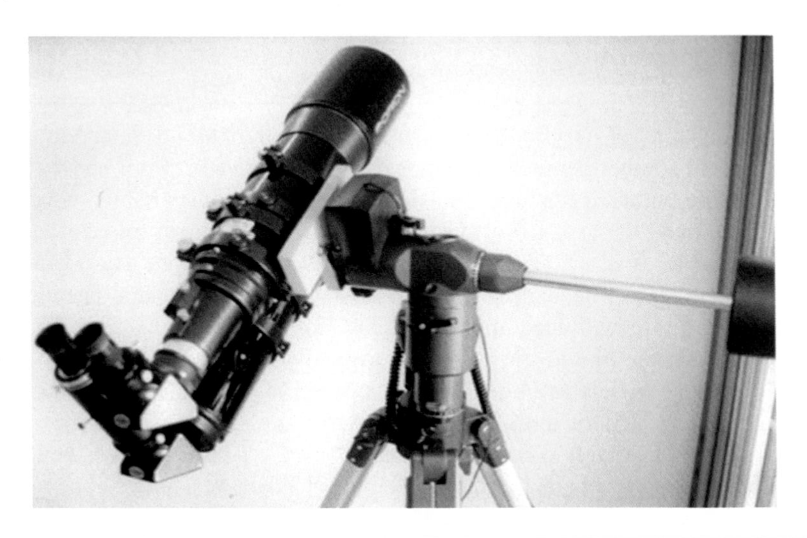

Fig. 10.18 The asymmetric 120-mm binocular telescope—rear view (Image credit: Dr. Alex Tat-Sang Choy)

not need electric motors; my legs can do the job of steering the chair. So I dragged it home and fitted a bracket to hold a pair of 80-mm binos. I also added a 300-mm long spring with a diameter of 25 mm at the rear to balance the weight of the binos. Then I found when I leaned all the way back, I could not lean forward again, because of the extra weight of the binos. So I added a strong spring under the chair to help me lean forward (Fig. 10.18b). I added 4 knobs to the mounting bracket, so I could adjust the angle of the binos quickly. I found as I lent back in the chair, I tended to slipped down lower in the chair. I had to compensate by changing the angle of the binos, so when the chair is all the way back, the binos are looking straight up at the zenith (Fig. 10.18c). The chair is light weight in construction, so it's easy to move around or take to a dark sky site. I thought about adding wheels, maybe when I'm 90.

I have also added a pair of 100-mm binos to one of these chairs and have spent many hours at a time looking at the sky with no ill effects. My only complaint is that everybody who comes along wants to test the chair, and I can't get them out. I have found these types of chairs are quite easy to come by at $50 and have bought 5 in the last year, so I am sure there must be a lot around.

I used 25-mm U bolts to attach the balance bars to the steel frame, at the top of the back rest. By adjusting the tension on the nuts, I can get the right amount of friction when adjusting to the desired angle. I bent the balance bars to a slight S shape, to make it easy to get into and out of the chair. The balance bars are made from 22-mm round steel tube 1.5-mm wall thick and are easily bent or flattened in a vice. I am sure anyone could make this chair and improve upon it.

By Greg Walton

Adapted from "Easy to Build Binocular Chair" with permission from Greg Walton—Mornington Peninsula Astronomical Society/ASV member, Australia (http://www.mpas.asn.au)

EZ BINOC Mount™ KIT

As with almost all of our products, this mount was originally developed to meet the individual's rather demanding personal needs. Its weatherproof and virtually indestructable, so that it can be left outside for instant use. Its portable - it can also be moved around the yard to various locations or broken down for transport. Its solid and stable. It is a 5 axis design that allows viewing from any position. It takes no more than 10 seconds to mount or dismount the binoculars. The mount's inexpensive, and it will accommodate almost any size binocular from 7X50 to 25X100. And some like to travel. While originally designed as a fixed in place mount, it is nice to be able to get away to a darker sky. By unscrewing the three legs, the counterbalance arm, and one of the two short grip arms the mount collapses into a relatively compact 41-inch x 7-inch x 7-inch package. You don't NEED to wrench these parts tight to reassemble at a dark site, but its good to keep a small 12-inch pipe wrench handy as the mount isn't rigid unless at least the legs are tight.

All parts that required machining, welding, drilling or thread tapping. Plus parts that were carefully chosen for this mount and are not commonly available elsewhere. Except for the modified pipe fittings, all metal parts are stainless steel. At final assembly you configure the mount for either a vertical post type mount (typical of 25X100 and larger binoculars) or a horizontal screw mount on pair of 15X70s for example. The kit contains all mounting screws and other hardware for mounting your binocular in either configuration (Fig. 10. 19).

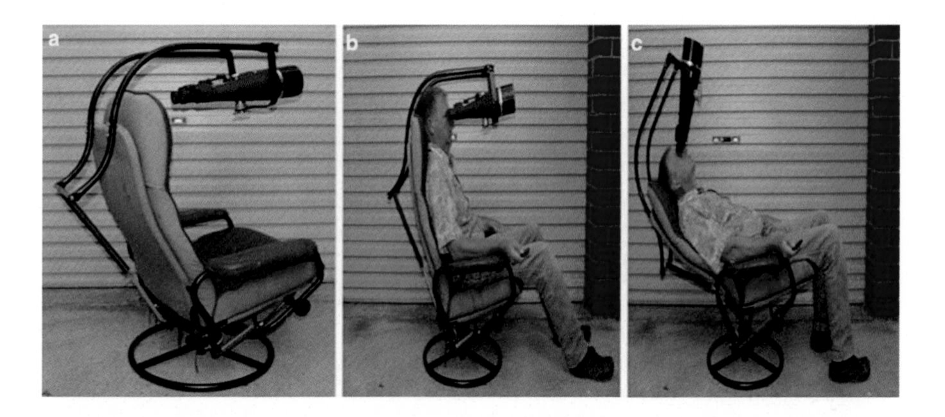

Fig. 10.19 Three photos of Greg Walton's easy to build binocular chair (Image credit: Greg Walton)

You can unscrew the 3 legs and the counterweight arm. At this point the mount becomes very compact - less than 4-feet by 1-foot by 1-foot and will fit almost anywhere. - Peterson Engineering.

Adapted with permission from the website: www.petersonengineering.com.

Dual C-11 Celestron and Borg Binocular Binoscope

Binoscopes

By Jochen Schell

At this website (www.binoscopes.de) I want to tell you about my experiences with binoscopes.

After being an amateur astronomer for over 25 years having owned several small and big telescopes as well as many binoculars I felt that I am most impressed by observing with two eyes using binoculars.

So I had the wish for even more binocular aperture and tried out several big binoculars in the 80- and 100 mm range but unfortunately all have been very disappointing…

I was close to give up this idea when I decided to make a last try in building a binoscope using two Borg 125 mm ED APO refractors and Tatsuro Matsumoto's EMS systems.

The result and experiences with this instrument were way better than expected and motivated by this result I wanted even more.

I thought about building a double dobson but gave up this idea quite soon because of the size, weight and unfriendly usage of these huge instruments.

Finally I ended up with Schmidt-Cassegrain telescopes, also because I already owned a C8 and a C11 and only needed to find matching second tubes for them.

The advantage of these SCT is there big aperture while still being quite small and lightweight instruments.

Also the usage is quite similar to a refractor by looking into them from the back and the possibility to use the Matsumoto EMS

I will talk about my binoscopes, how I did build them and the theories behind them at the following subsites:

Thoughts, theories and experiences about binoscopes
The Erecting Mirror Systems by Tatsuro Matsumoto
My Borg 125 mm ED APO binoscope
My Celestron C8 binoscope
My Celestron C11 binoscope
Building a binoscope

Fig. 10.20 Two photos of the EZ Gazer binocular chair (Image credit: AstroGizmos)

Jochen Schell—www.binoscopes.de (Figs. 10.20, 10.21, 10.22, 10.23, 10.24, 10.25, 10.26, and 10.27).

The genius EMS systems invented and produced by Tatsuro Matsumoto in Japan are a central part of building binoscopes. EMS stands for Erecting Mirror System and these systems are available in varying sizes for different telescope projects. Figure 10.28 displays the model EMS-UXL, which I am using for my binoscopes.

One special detail of this EMS is that they display an upright and right-sided picture, just like binoculars. This makes the orientation of the sky easier to deal with, and these binoscopes can also be used for daylight exploring.

EMS has only two mirrors instead of three. This means that there is less light loss caused by reflections. It also means that light travels a shorter distance compared to the diagonal path with three mirrors and 90Â° reflections. This short light path is important when you are building SCT binoscopes because you can keep the required back focus as low as possible (Fig. 10.29).

Everyone can easily make experiment with a binocular using just one or both eyes to see the huge difference. Firstly, there is the subjective magnification; it is bigger with two eyes than with one eye. You can also see more faint details with two eyes, which are not visible with just one eye.

While looking at the stars, you will see more stars with two eyes than with just one—the ability of collecting light is increased by binocular viewing. Usually, one is talking of a factor 1.4242 or sqrt^2 for the gain of magnification and the ability of light collection. Generally, this factor is right for the increased magnification and for the increased ability to collect light for pinpoint objects like stars. But when we are talking about extended objects, such as nebulae and galaxies with a low

Fig. 10.21 The big Dual Celestrons are mounted and ready for action (Image credit: Jochen Schell—www.binoscopes.de)

brightness of area things are different, we do not only need the big ability to collect light but also contrast. Contrast increases by using two eyes, so such extended objects have an even bigger improvement and the factor is more around 1.5–1.8

In order to discuss contrast, we have to talk about the different types of telescopes. An APO refractor displays higher contrast than a dobsonian and a dobsonian has more contrast than a Schmidt-Cassegrain with its huge central obstruction.

Another factor of binocular addition is the ability of our brain to compensate for bad information coming from one eye. This is very interesting when it comes to poor visibility because the seeing conditions are never the same in front of both telescope tubes. Our brain is able to amplify the good images and to ignore the poor ones.

I did a comparison of different objects with my 12.5″ Portaball dobsonian with a 2″ binoviewer and my Borg 125 mm APO binoscope, as well as with my Celestron C8 Binoscope. Even having the smallest aperture, the Borg 125 mm APO binoscope showed me galaxies and nebulae the best. Although it had a bigger aperture, the C8 binoscope is way behind on observing these objects because of the poorer contrast caused by the huge central obstruction

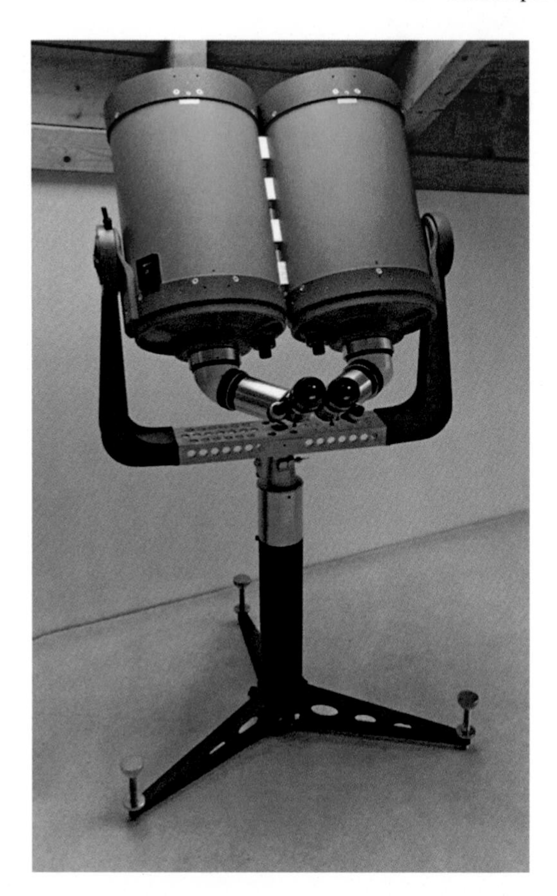

Fig. 10.22 Another view of the dual C-11's in their splendid Altazimuth mount (Image credit: Jochen Schell—www.binoscopes.de)

For brighter objects like planets or globular clusters, things are changing. Here, the C8 binoscope is able to use its aperture and leaves the 125 mm Borg binoscope behind. The 12.5″ dobsonian is at the poorest ranking for all observed objects. Beside weight, size and uncomfortable usage of a double dobsonian, we can also see another disadvantage caused by the long and twice-reflected light path to the eyepiece. The double Newtonian also needs a big secondary mirror which causes an obstruction in the same range of a Schmidt-Cassegrain. The focal ratio and the exit pupil are advantages of the double dobsonian, but they come at a high price.

Unfortunately, we cannot increase the size of an APO double refractor much bigger than 6″ for technical and financial reasons. And even then, we are talking about focal distances of at least 1000 mm, which are not really usable for wide field observations. As a result, for binoscopes, there is no single or universally better telescope.

Fig. 10.23 The dual C-11's setup at the observing site (Image credit: Jochen Schell—www.binoscopes.de)

I prefer the combination of two binoscopes, a double APO refractor with high contrast and a short focal length for wide fields, as well as a double SCT with a bigger aperture and a longer focal length, mainly for planets and globular clusters.

Jochen Schell

(Adapted from his website "Binoscopes" with permission from Jochen Schell—www.binoscopes.de)

Author's Note:

For those ATMs who have built a binocular telescope before, it must have been a rewarding experience to see its "first light". Building a binocular telescope has its challenges and difficulties, but it's worth building one to see the final results. Building a Schmidt Cassegrain binocular telescope is a true challenge, however,

Fig. 10.24 Jochen used three equally spaced aluminum spacers that were machined to match the radius of the C-8 optical tube assemblies (Image credit: Jochen Schell—www.binoscopes.de)

Fig. 10.25 When assembled together notice the close tolerance fit between the aluminum spacers and the C-8 optical tube assemblies (Image credit: Jochen Schell—www.binoscopes.de)

Fig. 10.26 Looking down the double barrel of a dual C-8 photon cannon (Image credit: Jochen Schell—www.binoscopes.de)

Fig. 10.27 Borg 125 mm Binoscope with the 2-inch dual 32 mm eyepiece binoback IPD adjustment system. Notice the interesting vertical dual aluminum knob focusing arrangement on each binoback (Image credit: Jochen Schell—www.binoscopes.de)

Fig. 10.28 Borg 125 mm Binoscope "2nd. edition cover image" with machined adapters and the dual eyepiece binoback IPD adjustment system (Image credit: Jochen Schell—www.binoscopes.de)

Fig. 10.29 The dual eyepiece adapter and the ever popular dual binoback IPD adjustment system (Image credit: Jochen Schell—www.binoscopes.de)

especially if you plan to combine two big Celestrons together into a big Cassegrain binocular telescope on a modified Celestron Altazimuth mount.

Judging by the photos of his excellent, homemade Celestron binocular telescope, Jochen Schell of Germany deserves a lot of credit for being able to successfully combine two C-8 s and a pair of C-11 s together to build his binoscope. His act is a hard one to follow, regardless of how good of an engineer or machinist you are. In a Schmidt Cassegrain binocular project like that you have to know exactly what you're going to do from the start. If not, then the surprises come later. It's no wonder why there are very few homemade Schmidt Cassegrain binocular telescopes.

A Celestron has a primary mirror that is the focusing mechanism in the Celestron Schmidt Cassegrain telescope. It's a beautifully designed optical system, but difficult to make into a binocular telescope. However, once it is in the binoscope, then it tends to become an excellent f/10 binocular telescope. If you decide to invest in two C-8 s to make into your future binocular telescope, prepare to be mechanically challenged. Not only in terms of combining the two the C-8 optical systems, but also modifying the original Altazimuth mounting. One look at the amount of work involved in making a C-8 binocular telescope and you can see why so few are actually made. Nevertheless, it would still be a challenging engineering project if two C-5 s were being used. Obviously, anyone who has the ambition and resources to undertake a Celestron Schmidt Cassegrain binocular telescope project is in for a lot of work regardless of its size. Seeing the results achieved by Jochen Schell after completing his Celestron binocular telescope project should serve as an inspiration for others to attempt the challenging project. There are alternatives to Schell's use of machining aluminum for such an ambitious Celestron binocular telescope project. Wood is a possible option that would be easy to cut and shape into the parts and pieces needed for the same or a similar project.

Building my dual 6 in. f/15 Dall-Kirkham Cassegrain binocular telescope, which I completed in 1980, was a terrific learning experience since I had never attempted any binoscope projects before that. It was a one-of-a-kind binocular telescope and it taught me a lot about building binocular telescopes, Cassegrain optics and optical alignment. Some 36 years later, I still enjoy building my own binocular telescopes, Dobsonians and refractor binoscopes. Someday I would like to build my dream binoscope using a pair of beautifully matched 7-inch Questars Maksutove Cassegrains attached and aligned together on a computerized GOTO mount. that would be my dream binoscope. If you ever had the good fortune to have looked through a 7-inch questar at the moon and planets....then you could imagine what beautiful celestial images two 7-inch questars would render if they were indeed made into a true one-of-a-kind Maksutov Cassegrain binoscope. If by chance there are a couple of interested readers out there have a 7-inch Questar in reasonably good condition and they no longer have any need or interest in them...just give me a hollar. I'll come running.

In all of my astronomical adventures in observing, which date back some 60 plus years, my most memorable ones are of viewing the moon and planets in the mid to late 1950s through scopes of 6–8 in. aperature...Newtonians in particular. An old friend of mine, Jim Starbird of Topeka, owned a beautiful 8 in. f/8 Newtonian that was manufactured by Criterion. That particular telescope, over 60 years ago, I thought was the next best thing to Mt. Palomar. Gotta love those old 1950's Newtonians....an 8-inch Criterian was just one of the many commercial newtonians telescopes that all of us young kids who had an interest in Astronomy would dream about...especially around Christmas time. I alway's hoped Santa would drop one off on Christmas morning....I'm still waiting Santa....ha

The Binochair

By Gary Liming

This is a picture of my binochair after it was completed. The rest of this website is a blog of how it was built and some commentary on using it. At this writing, I've just spent a very pleasant evening with it, and it works great! The large binoculars can be positioned in any direction, and they stay put at the position you leave them. You can cover about 1/5 of the sky before needing to move the chair, so you only have to move the chair five times for a complete sweep of the heavens (Fig. 10.31).

Fig. 10.30 Both Borg OTAs and their accessories are neatly stored in a Rimowa case and ready to travel to dark places and new observing sites. (Image credit: Jochen Schell—www.binoscopes.de)

Fig. 10.31 A photo of Gary Liming's binocular tripod and binochair (Image credit: Gary Liming)

Introduction

This webpage is about making a device to make practical and comfortable use of large binoculars. By "large" I mean the ones I have in mind are 25 × 100 power weighing 10.3 lb. Although that may not sound like much, it's enough to have the following disadvantages:

- Your arms get tired very quickly holding them up to your eyes.
- Holding them up to your eyes renders such a shaky image that it makes them useless and because they are so powerful that the slightest movement—even your heartbeat—makes.
- With that large a weight placed right up against your eyes, a small bump and you get two black eyes quite easily.

Simply stated, they never were meant to be used by just holding them. They do come with a way to place them on top of a standard camera tripod, but those tripods have the problem of being either affordable and rather flimsy or quite expensive for something sturdy enough to hold these. Even if you do get a high-quality tripod, they have three drawbacks.

- It seems the legs of the tripod are always in the way.
- You can't see things directly overhead at all.
- For things high overhead, you neck is bent back constantly, and my neck simply won't handle that anymore.

A common solution among stargazers is to mount a parallelogram type balance beam on a tripod and implement as many degrees of freedom movements as necessary to be able to easily position the binoculars in any orientation and have them counterbalanced. There are many examples of something like this on the net, and I spent some time looking at them.

The design goals of my device are:

- To be used primarily from a lounge type chair. No reason I can't be comfortable while looking up and be able to throw a blanket over me if needed.
- Make the binoculars easily positioned to any spot in the sky.
- Counterbalance the binoculars with just the right amount of friction on the joints so that they seem suspended in front of your eyes—if you let go, they stay put.
- Be able to break it down for easy transport in a regular car.
- Cost is a consideration, which for me means using wood instead of aluminum, so it will probably have a bit of weight, but that I can handle. Besides, wood can look very nice as well.

The actual chair I would like to use, at least to start, is a "zero g" chair I got on sale at Cabela's. It has the advantages of being very comfortable, it is quite easy to lean forward or back with just a slight leg movement, it is fairly portable and lightweight, and best, I already have one! (By the way, they also have a padded version as well as a double seat version that might be great for showing the stars to others (Fig. 10.30).

After looking around the web on various astronomy forums and email groups, I saw pictures of one that I thought would make a good start by Dennis Simmons of Australia. At least initially, I will start with his design until we get to the binocular mount itself.

This device has three main parts, the bottom tripod that provides the azimuth pivot, the parallelogram arm that supports the main up and down weight, and the binocular mount that provides swivel for the binoculars. Accordingly, along the left of this page of the website are the menu selections for the tripod, parallelogram, and mount portions of this build.

Gary Liming

Adapted from "The Bino Chair" with permission from Gary Liming (http://www.liming.org)

The Bolton Group

These binoculars were judged the best at the 2005 Kelling Heath Star Party, the UK's largest and most attended event. This is the first time this award has been made.

I know, I know—we said 8-in. binoculars were the optimum size for both use and transportation but aperture fever has struck again The 12 in. binocular project took nearly 4 years! Gerald got side-tracked into building a new workshop, kitchen, bathroom and driveway. They are similar to the 8-in. except the tubes are 12 sided instead of 6. This was a major project, around 1000 h work, and was completed March 2005. First light was at our dark-sky observing night in March 2005 (Fig. 10.36).

These are just a few minor bits to add in the light of first use but nothing major. The push-pull adjuster for the mirror cell is a new idea by Gerald using a contained (internally and externally) spring. It is adjusted from the top unlike more conventional adjusters. On one telescope they will extend upwards to make them reachable from the eyepieces so final collimation will be a one man job. Unlike the 8-in. binos the eyepieces are not directly between the two telescopes. They are brought out at 45°. This saves a lot of space and makes the binos a lot more compact i.e. not as wide as they might have been.

The finder is a simple red dot type and works excellently. In combination with Pocket Stars on a Pocket PC objects can be quickly located.

Master Optician, Brian Weber Ground and figured both 12-in. (30 cm) f/5 mirrors with their focal lengths differing by less than 1/4-in. (6 mm). The mirrors were made from Suprax blanks (a form of pyrex) and were a bit thicker than desired but by the time both sides were levelle up the thickness was down to 40 mm. As you can see from the ronchigram there is no turned edge! Eagled-eyed might spot a very slight central "hole" but this is of no consequence being in the shadow of the diagonal. The diagonals were also made by Brian and as can be seen by fringes they are smooth and around 1/4 wave—note the slight curve is caused by the camera viewing angle and not the diagonal.

HiLux coated at Orion Optics UK. This enhanced coating claims 97% reflectance—important when there are 6 mirrors! The binos have been our biggest construction job but the effort has been worth it. The following pictures give some idea of the construction. Sorry but there are no drawings available—they just came out of Gerald's head!

The binos were designed to separate into two halves from transport. The bottom cage is attached permanently to the mount and is the heaviest part to lift. Sensibly it requires two people. The top cage is much lighter and is an easy one person lift. The two halves are quickly located and joined with three thumb-screws per tube. These need to be made captive to remove any chance of them falling on the mirror.

Set-up and collimation is also quick and easy. The approximate collimation can be done in daylight with final alignment carried out on an object—usually a star. However, for the Haverthwaite Star Party we couldn't wait and aligned on the crescent moon long before dark. This final tweaking of alignment/collimation is easily done with the push-pull spider adjusters. They keep collimation all night (Fig. 10.32).

Fig. 10.32 A photo of Gary Liming's binochair (Image credit: Gary Liming)

First testing on the very thin crescent moon was breath-taking. During the night we saw all our old favourite deep-sky objects but better than ever before as we were seeing with two eyes! The mount moves effortlessly and is unaffected by the wind.

A common question asked is what about reflections. Sure it looks like the bright ali would produce reflections but of course it doesn't. The necessary parts are blackened such as the spider and inside the focuser but everything else is sparkling ali. There are some black baffles around the mirrors and opposite the focuser.

The binoculars have proved a great success both in use and judged best homemade telescope in the UK's biggest star party at Kelling Heath 2005. They provide a first class tour of the deep-sky, all in the confort of viewing with two eyes and no need for ladders!

215 mm Ultra-Light Binoculars

No sooner were the 12-in. binos finished then Gerald looked around for a new project. The aluminium cut outs from the 12-in. binos were big enough for 8.5 in. ones so the temptation was too great.

The 12-inchers are great for a star party or serious observing but for just carrying out easily for a quick view something lighter is needed. Hence these ultra-lights! The optics are f/6 and all the mirros were made Brian. They are as usual to his very high

Fig. 10.33 Gerald and his almost completed 300 mm binos in his workshop (Image credit: www.deep-sky.co.uk)

standards—if they are not perfect he will not let them out of his workshop! Brian has made many many mirrors of all shapes and sizes. He recently completed a commercial contract for a large test mirror for a company specialising in lasers. To save even more weight the bottom bearing uses a circular rebate in both the upper and lower bearing plates. This is filled with over 100 ball bearings which provides a very wide track for stability plus low weight. It works like a treat. In this new design, the azimuth and altitude axis shafts have been left exposed for the possible option of fitting digital setting circles. I hesitated to show this new idea. Our original 8-in. binos featured crayfords for adjusting the inter-ocular distance. Low and behold when commercial binos appeared a couple of years or so later they too had them! Coincidence? This new solution was not suitable for the 12-inchers as their eyepieces came out at 45°. However, with the smaller size of the 8.5 in. ones we have gone back to inline focusers so some lateral thinking took place. The adjustment only has to be 15 mm maximum—eye separation doesn't vary all that much. Brian devised this method with a single rotating knob (red) to adjust them in synchronisation. The focusing is helical—the black and brass units (Figs. 10.33 and 10.34).

Fig. 10.34 The underside showing the mirror cells (Image credit: www.deep-sky.co.uk)

200 mm Minimalist Binoculars

Our previous light-weight binoculars had weighed in at a very reasonable 65 lbs. But could we go lighter? Gerald had a cunning plan! Going back to basics meant the answer was yes and the these 200 mm Binoculars weigh in at 44 lbs total—an easy one man lift. Yet no stability has been sacrificed and performance will be just as good if not better. To go with these new binos Brian has produced his finest pair of matched mirrors ever. Under autocollimation (double light pass doubling any errors) they exhibit absolutely straight Ronchigrams—no zonal error and no turned edge. We had some great views through them and Luke's adjustable chair made viewing a real pleasure. All design goals have been met, i.e.:

1. Ultra stable
2. Hold collimation
3. Quick inter-ocular adjustment—our simplest yet!
4. Easy one-man assembly
5. Easy one-man move

Note although the side-bearings look like simple Dobsonian bearings they have real bearings—no teflon here. They disassemble into three components for easy transport and storage. What assembly there is, is by thumb-screws—no allen keys required. All the optics were made by Brian i.e. the 2 matching parabolic mirrors, 2 elliptical secondary flat mirrors and 2 rectangular flat tertiary mirrors. The main mirrors are 200 f/5 (1000 mm focal length) and were made from low expansion glass. Their focal lengths are within 4 mm i.e. less than 0.5% difference (Figs. 10.35 and 10.36).

Fig. 10.35 Brian and Gerald at the Haverthwaite Star Party, March 2005 (Image credit: www.deep-sky.co.uk)

Fig. 10.36 The finished 200 mm binoculars (Image credit: www.deep-sky.co.uk)

They were tested using autocollimation with a high quality test flat. In this test the light strikes the mirror under test twice. This is therefore a severe test as any errors present are doubled. In the rochigram they both exhibit straight diffraction lines i.e. no detectable errors or turned edge to the limit of the test. They are certainly better than 1/10 wavelength accuracy probably nearer 1/20 wave.

Author's Note:

It's not often when you come across an excellent group of well designed and splendidly crafted series of binocular telescopes such as the Bolton Group's skeleton tube all-metal binocular telescopes. It's sometimes a rare trait for another active ATM with a healthy ego who can appreciate and applaud the creative talents other ATMs who enjoy the same hobby of telescope making. As an active telescope and binoscope maker myself, I often will become inspired to start another telescope building project when I see a splendid telescope or binoscope that was built by another creative ATM at a local star party or national telescope makers conference. And that's the best part about being an ATM...being able to share your ideas with other ATMs and in doing so, you're helping to advance the science of amateur telescope making...not only on a local level...but internationally too. So let's give three cheers for all of the ATMs out there in the amateur telescope making community who continually surprize us with their creative, innovative and interesting ideas and talents at local star parties and national telescope maker conventions. As an author of telescope and binoscope making books, I'm alway's amazed what the ATMs around the world are creating these days....

Note: The Bolton Group images and website story is presented above with permission from: David Ratledge—The Bolton Group—www.deep-sky.co.uk (Fig. 10.37)

Fig. 10.37 The finished binoculars at Kelling Heath 2012. Gerald is tweaking the collimation. Brian is left and Luke, the new owner, is on the right. We had some great views through them and Luke's adjustable chair made viewing a real pleasure. (Image credit: www.deep-sky.co.uk)

A Comet Discovery?

For those of you who are active comet hunters and are searching the night skies with your binoscope or big binoculars for these elusive small fuzzy snow balls, it's important that you know what to do if you think you've found one you're your telescope. There is a procedure that you can follow in order to confirm your new potential comet discovery.

Author's Note:

A "visual" comet discovery is a remarkable achievement considering the thousands of optical telescopes (amateur or otherwise) around the world that are equipped with the latest CCD camera technology and also looking for comets.

Imagine you're out one night observing with your new big binoculars or your big binocular telescope that you just built, and you were scanning the sky in Ursa Major when you unexpectedly came across a small diffused fuzzy object in your field of view. You immediately become suspicious about the object's small, fuzzy diffused appearance. Because you're suspicious, you reach for your trusty sky atlas and look up the constellation Ursa Major to check for any galaxies, clusters or other objects in that particular area of the constellation that could possibly be mistaken for a comet. After watching the small, fuzzy diffused object in your eyepiece, you think you've detected a slight movement of the object against the background of the stars. Having confirmed using your sky atlas constellation chart that there are no galaxies or globular clusters in the immediate area of your search, you become extremely excited about a possible comet discovery. As you continue watching the object in your field of view, you should check internet Astronomy sources for any known comets that describe their current astronomical locations and visual magnitudes. If you don't find any known comets in the same area from your internet comet source search and you feel confident enough about your potential discovery, then its highly advisable to have it's location confirmed and verified by a professional observatory. After that, you should consider notifying the Central Bureau for Astronomical Telegrams (CBAT) about your possible comet discovery. If you decide to wait another 24 h to see if the small, fuzzy diffused object has moved from its original astronomical location, someone else may spot it and report it. If you are confident in your observation, then notify the CBAT by either telegram or email to cbatiau@eps.harvard.edu along with the following recorded information. CBAT's address: Central Bureau for Astronomical Telegrams, Hoffman Lab 209, Harvard University, 20 Oxford St., Cambridge, MA 02138-2902.

Note:
If for some reason you decide not to notify the Central Bureau for Astronomical Telegrams about your potential comet discovery, someone else probably will.

Example Email or Telegram Template

Dear Central Bureau for Astronomical Telegrams (CBAT),

I would like to report the discovery of a potentially new comet moving slowly across the night sky in the constellation Hercules at the following coordinates with the following physical characteristics:

New Comet Discovery Data:

R.A. (.1"): 17 hr. 30 min.
Dec. (1"): + 42 Deg.
Date: Nov. 1st 2016
Universal Time/GMT (0.001—Time of Day): 10:20:00 GMT—8:20 P.M. P.S.T
I used the following sky atlas to confirm the sighting: Example; Sky and Telescope's "Pocket Sky Atlas" (See attached chart)
Description of Comet: Small, diffused, round
Estimate of Comet's Total Magnitude: 10th. Mag
Comet's Size: 15 min. dia.
Comet's Diffuseness: Diffused
Possible Tail: No visible tail
Degree of Central Condensation 0–9: 2—Diffused coma with definite brightening towards center
Optical Telescope/Instrument used for Discovery: 10" f/4 Binocular Telescope.
Location of Discovery: Boulevard, California 91905

I am looking forward to your confirmation of my new comet discovery.

Sincerely, the Author

Address
Tel./Cell Phone
Email

P.S. See my attached sky chart for tracking the potential new comet's 24 h movement through the constellation Hercules.

So is it difficult to discover a new comet? Even after several hundred hours searching the night skies, some comet hunters may never find a new comet while other comet hunters may find one or two new comets in their lifetime. Discovering a single comet during one's lifetime is a certainly a very worthwhile endeavor and obviously other younger comet hunters will most certainly wish to do the same. However, one comet hunter from Australia deserves naming. With eighteen separate comet discoveries, the late William Bradfield stands out as the legend for all comet hunters to respect and admire (Fig. 10.38).

Brian G. Marsden, the late director emeritus of the IAU's Central Bureau for Astronomical Telegrams had this to say about William Bradfield: "To discover 18 comets visually is an extraordinary accomplishment in any era, but to do so now is truly remarkable, and I think we can be pretty sure nobody will be able to do it again. And it's all the more astounding that in no case did he have to share a discovery with some other independent discoverer. More than any other recipient, Bill Bradfield outstandingly deserves the Edgar Wilson Award" (wikipedia.org) (Fig. 10.39).

For more comet discovery info, visit http://www.nightskyhunter.com/index.html and http://www.cbat.eps.harvard.edu/HowToReportDiscovery.html.

When you're searching for new comets, it really comes down to two or three factors: the longer you search, the better your chances for finding a comet, whether your telescope has a fast focal ratio and a wide field of view, and if it has sufficient light gathering power. However, you also need luck.

Fig. 10.38 They're almost finished with the exception of the un-coated mirrors....It's almost ready to start pumping photons through those dual eyepieces. (Image credit: www.deep-sky.co.uk)

The use of a big and reasonably fast binocular telescope with a wide field of view will give you the biggest advantage for comet hunting. You can also find comets using a CCD camera on a regular telescope, but when it comes to visually searching for a comet, using two eyes and a big binocular telescope is going to be beneficial. I have talked to several comet hunters and some of them have discovered a comet visually after spending more than a hundred and fifty hours searching the skies. If you're not using a big fast binocular telescope, it may take a lot longer. For inspiration, read the story about Comet Macholz 1985e in Chap. 4.

A Few Closing Thoughts from the Author

Building a telescope, binoscope, or binocular telescope can be in itself a very rewarding hobby and one that can be shared with others. Probably not everyone has the same desire or ambition to build a telescope as an amateur telescope maker has. But like any hobby, whether it be making model airplanes, bird houses, or model ships, one always can take pride in his/her finished product. I've always said that if

Fig. 10.39 The 300 mm are finished and waiting for dark to show off it's stuff! (Image credit: www.deep-sky.com)

you have a son or daughter that has an avid interest in astronomy, then building a telescope for the family enjoyment can be a joint venture with everyone becoming involved and a very enjoyable one at that. If you're going to build a telescope, before you start your project, then it would be beneficial to research as many ideas on telescope making as you can find. There are many good telescope makers out there who have shared their telescope making projects and experiences on their websites, and that's a good place to start to gather some ideas and information on what kind of telescope you want to build or even buy someday. It would also be a good idea to join a local Astronomy club in your area, where you enjoy a regular monthly meeting and an observing session afterwards. At the same time, try and pick up some good ideas to incorporate into building your own telescope. It's also good to compare notes with other people interested in making telescopes or binoscopes. A good and sturdy telescope design to build with others helping out would be an all-wood 8 in. or 10 in. f/6 Newtonian telescope on a Dobsonian Altazimuth mount. Building it out of wood gives everyone a chance to cut out the pieces, sand them down and finally gluing them together to make the square wooden Newtonian optical tube. Another fun part about building a telescope is painting and designing the exterior (see Fig. 8.35).

Chances are that once you start your telescope or binoscope making project, you already have a good design in mind, the materials needed, and the approximated cost for everything. There are many good deals on the Internet for telescope mirrors and objective lenses. The author has always started his own particular telescope making projects "after" having purchased the optics. They are, for the most part, the most expensive part of your project and the design and construction of your telescope will always center on the physical and optical characteristics of your mirror or objective lens. Obtaining your optics first before you start your project will, in fact, help reinforce your incentive to start building your telescope project. And with your optics always sitting on the table or shelf in front of you will provide the drive for you to finish it in good fashion (Figs. 10.40 and 10.41).

Unless you're from the old school of amateur telescope making, those who enjoy grinding and polishing their own mirrors, it can be a lot of work but at the same time, it can also be a lot of fun to and very rewarding after you've you've finally finished it. If you're going to make a Newtonian telescope, then grinding, figuring, polishing, and testing your own primary mirror can be, for some, the most rewarding part of building your telescope. The same goes for making a refractor telescope and working anxiously to finish it to so you can see its "first light".

There are a lot of mirror grinding kits available on the Internet and many telescope companies that sell mirror blanks and the grinding and polishing materials that go along with them. For example, Willmann-Bell, Inc. has a fine selection of telescope making supplies to choose from. If you want to get friends or family involved in grinding and polishing a telescope mirror, giving them a chance to walk around the barrel, pushing the glass around will make them a lot more excited about your telescope making project. Plus, everyone gets a chance to contribute in building it. The sight of the beautiful gas cloud in Orion, M-42, in the eyepiece of a telescope will certainly get someone "hooked" on Astronomy. Even the sight of the beautiful Andromeda galaxy (M-31) will thrill those who see it for the first time through a telescope. And seeing these beautiful astronomical objects as part of the "first light" through your new homemade telescope will make this a memorable moment.

Thinking back to the early 1980s, the author originally had thoughts about building a big dobsonian binocular telescope. At the time, Coulter Optical Co. was offering a 10.1 in. f/4.5 Odyssey, and with its overall simple design and optical arrangement would have made it relatively simple to join two 10.1 Odyssey dobsonians together in one large "rocker box" Altazimuth mount, thus making it into a big Newtonian binocular telescope. Once combined together in a large "rocker box" Altazimuth mount, and after making some simple mechanical changes, it would have been an excellent example of an early dobsonian binocular telescope. Had it happened at that particular period of time during the early 1980s, the commercial binocular telescope era would have started much earlier. I believe if only "one" had been built back then, Coulter Optical Co. would probably have offered it as a special product and the binocular telescope era would have started much sooner, especially in terms of amateur telescope making.

In the 1980s, Coulter Optical Co. was selling their large dobsonian telescopes at very reasonable prices (10.1 in. f/4.5 Coulter Optical dobsonian sold for $299), but those days are long gone. Today, a big dobsonian 10-in f/4.5 quality mirror can be priced as high as $1500. So even buying one mirror is a major investment. Buying two 10-in f/4.5 quality mirrors to build a big binocular telescope is almost too

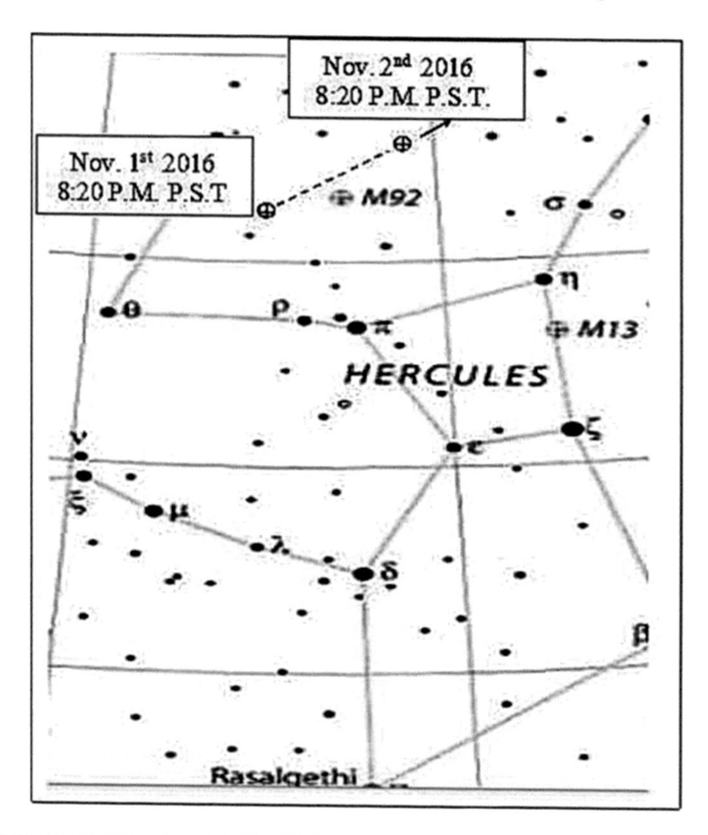

Fig. 10.40 Amateur comet hunter's sky chart showing an example (just for an illustration) of recording a potential new comet's 24-h movement through the constellation Hercules near Messier 92 (Image credit: the Author)

expensive. The cost of secondaries and everything else needed to build a binoscope varies. Binocular telescope and refractor binoscopes have been slow to catch on in the amateur telescope making community because they are very expensive to build.

Even if you only have a humble pair of binoculars or a big dobsonian "light bucket," it really doesn't matter that much at a star party. Everyone is going to get a chance to walk around and check out the great views from all of the telescopes and talk astronomy with all of amateur stargazers and telescope makers at the same time. And that's what makes a star party fun to attend. A star party can offer you an opportunity to check out new telescopes, a variety of different kinds of eyepieces and accessories that you normally might not be able to try out at the telescope store at night if they close before dark. If you buy something online sight unseen, sometimes that can be a disappointing experience too. At least at a star party, you have a chance to compare your telescope and its performance with others and that's a good thing. Your telescope making experience can be shared with others, and once

Fig. 10.41 Comet (Bradfield) from Catus Flat N.E. Colorado, U.S. (Image credit: The Starmon—Wikimedia Commons)

they have the opportunity to look through your own homemade telescope, I'm sure they're going to be impressed.

We have covered several interesting topics on binoscopes and binocular telescopes. All of which are important in their overall design, construction, and optical performance. Obviously some of the topics that are covered in the book are perhaps a little more interesting to read about than others. The important thing is that after reading it, you feel you have learned something valuable about binoscopes and binocular telescopes. In the future, the author aspires to write another book on "The History of Binoscopes" including a chapter in it about building a 3-in. f/10 multiple mirror Newtonian telescope with a Cassegrain style focus that has three primary mirrors (see Fig. 7.29).

As you can probably tell after reading this book, the author is a big proponent of binoscopes. With a 28 incher already having been built in Germany, there is little doubt that an even bigger binocular telescope will be built in the future. As binocular telescopes become more popular with amateur telescope makers and commercial manufacturers, we can expect to see some being built with mirrors exceeding 30 in. or larger in diameter. But in the meantime, still waiting in wings is the multiple mirror telescopes with three or more primary mirrors waiting to evolve into their own special place in telescope making history. From a historical perspective, not much one can find about them, either on the Internet or in telescope making books. However, the concept of using three mirrors to gather light from a single celestial object and focus it into an image that projects three times or more times the light gathering power of a single aperture sounds rather appealing, especially to someone who enjoys the stunning images through a big binocular telescope. One can imagine the stunning view through a multiple mirror with three primary mirrors "pumping" photons into the observer's eye. That is something to really look forward to in the future as amateur telescope makers start to experiment more and more with multiple mirror telescope designs.

What can we expect to see in the next generation of binocular telescopes and refractor binoscopes? Considering the fact that aperture is everything when it comes to observing celestial objects, we're probably going to see some monster binocular telescopes that are being built by amateur telescope makers from around the globe. If someone happens to make a matching pair of 30 or 40-in. f/4.5 primary mirrors, then it's likely that a 30 or 40 in. binocular telescope will be built soon thereafter. In terms of big refractor binoscopes, there is already a big 10-in. APO refractor binoscope that has been made in Japan and an equally impressive 12-in. APO refractor binoscope somewhere in China.

With the increased global popularity of the amateur astronomy community using binoscopes and binocular telescopes over the past 5 or 10 years, more and more ATMs are building their own big binoscopes these days, and more binoscopes are being sold commercially too. Most of us can appreciate the wonderful views that only a big binocular telescope or refractor binoscope can provide. Perhaps a big binoscope or binocular telescope is not for everyone, but everyone can certainly enjoy the views they offer of the starry skies, especially at a local star party or a big astronomical convention. Once you look through both eyepieces of a big binoscope, you may be hooked for life!

Now that we have reached the end of the book, I hope that you have developed a further interest in refractor binoscopes and binocular telescopes. Whether you end up buying or building a binocular telescope or refractor binoscope is really only the beginning of your binoscope adventure. If you do end up building one, when you've finished it and you're ready to try it out, you're going to want to start looking at some really neat deep sky objects with it. I've listed the complete Messier list along with some interesting details about each object that Charles Messier recorded in his famous observational record. You can track them down and observe the sky's wonders with your own binoscope! Once you have observed a few of the Messier objects, you'll soon want to observe them all. Enjoy!

The Constellations

- Andromeda
- Antlia
- Apus
- Aquarius
- Aquila
- Ara
- Aries
- Auriga
- Bootes
- Caelum
- Camelopardalis
- Cancer
- Canes Venatici
- Canis Major
- Canis Minor
- Capricornus
- Carina
- Cassiopeia
- Centaurus
- Cepheus
- Cetus
- Chamaeleon
- Circinus
- Columba
- Coma Berenices
- Corona Australis
- Corona Borealis
- Corvus
- Crater
- Crux
- Cygnus
- Delphinus
- Dorado
- Draco
- Equuleus
- Eridanus
- Fornax
- Gemini
- Grus
- Hercules
- Horologium
- Hydra
- Hydrus
- Indus
- Lacerta
- Leo
- Leo Minor
- Lepus
- Libra
- Lupus
- Lynx
- Lyra
- Mensa
- Microscopium
- Monoceros
- Musca
- Norma
- Octans
- Ophiuchus
- Orion
- Pavo
- Pegasus
- Perseus
- Phoenix
- Pictor
- Pisces
- Piscis Austrinus
- Puppis
- Pyxis
- Reticulum
- Sagitta
- Sagittarius
- Scorpius
- Sculptor
- Scutum
- Serpens
- Sextans
- Taurus
- Telescopium
- Triangulum
- Triangulum Australe
- Tucana
- Ursa Major
- Ursa Minor
- Vela
- Virgo
- Volans
- Vulpecula

Messier List

(M1)—is known as the famous "Crab Nebula" is a supernova remnant in the constellation Taurus. It has an apparent magnitude **(v)** of about 8.4 and is approximately 6300 light years from our solar system.

(M2)—is a globular cluster in the constellation Aquarius. It has a apparent magnitude (v) of about 6.3 and is 36,000 light years in distance from planet Earth.

(M3)—is a globular cluster located in the constellation Canes Venatici. It has an apparent magnitude (v) of about 6.2 and is approximately 31,000 light years from the planet Earth.

(M4)—is a globular cluster in the constellation Scorpius. It has an apparent magnitude (v) of about 5.9 and is approximately 7000 light years in distance from our solar system.

(M5)—is a globular cluster in the constellation Serpens. It has a apparent magnitude (v) of about 6.65 and is about 23,000 light years in distance from planet Earth.

(M6)—is known as the "Butterfly Cluster" is an open cluster in the constellation Scorpius. It has an apparent magnitude (v) of about 4.2 and is approximately 2000 light years in distance from planet Earth.

(M7)—is known as the "Ptolemy Cluster" and is an open cluster in the constellation Scorpius. It has an apparent magnitude (v) of about 3.3 and is about 1000 light years from planet Earth.

(M8)—is known as the famous "Lagoon Nebula" and is a nebula associated with a cluster located in the constellation Sagittarius. It has an apparent magnitude (v) of about 6.0 and is approximately 6500 light years in distance from planet Earth.

(M9)—is a globular cluster in the constellation Ophiuchus. It has an apparent magnitude of (v) about 7.9 and is approximately 26,000 light years in distance from planet Earth.

(M10)—is a globular cluster in the constellation Ophiuchus. It has an apparent magnitude (v) of about 6.4 and is approximately 13,000 light years in distance from our solar system.

(M11)—is known as the "Wild Duck Cluster" and is an open star cluster located in the constellation of Scutum. It has an apparent magnitude (v) of about 6.3 and is approximately 6000 light years in distance from planet Earth.

(M12)—is a globular cluster in the constellation of Ophiuchus. It has an apparent magnitude (v) of about 7. and is about 18,000 light years away from our solar system.

(M13)—is known as the Great Globular Cluster in the constellation of Hercules. It has an apparent magnitude (v) of about 5.8 and is approximately 22,000 light years in distance from planet Earth.

(M14)—is a globular cluster in the constellation of Ophiuchus. It has a apparent (v) magnitude of about 8.3 and is approximately 27,000 light years from planet Earth.

(M15)—is a globular cluster in the constellation of Pegasus. It has an apparent magnitude (v) of about 6.2 and is approximately 33,000 light years in distance from our solar system.

(M16)—is the well-known "Eagle Nebula" and is a nebula with a cluster located in the constellation Serpens. It has an apparent magnitude (v) of about 6.0 and is approximately 7000 light years away from planet Earth.

(M17)—is known as the "Omega Nebula" (or sometimes called "Swan", "Horseshoe" Nebula or even "Lobster Nebula" and is a nebula with a cluster in the constellation of Sagittarius. It has an apparent magnitude (v) of about 6.0 and is approximately 5000 light years in distance from planet Earth.

(M18)—is an open cluster in the constellation of Sagittarius. It has a apparent magnitude (v) of about 7.5 and is approximately 6000 light years from our solar system.

(M19)—is a globular cluster in Ophiuchus. It has an apparent magnitude (v) of about 7.5 and is about 27,000 light years from planet Earth.

(M20)—is known as the popular "Trifid Nebula" and is a nebula with a cluster located in the constellation of Sagittarius. It has an apparent magnitude (v) of about 6.3 and is approximately 5200 light years in distance from planet Earth.

(M21)—is an open cluster in the constellation of Sagittarius. It has a apparent magnitude (v) of about 6.5 and is approximately 3000 light years in distance from planet Earth.

(M22)—is known as the "Sagittarius Cluster" which is a globular cluster in the constellation of Sagittarius. It has an apparent magnitude (v) of about 5.1 and is approximately 10,000 light years in distance from our solar system.

(M23)—is known as an open cluster in the constellation of Sagittarius. It has an apparent magnitude (v) of about 6.9 and is approximately 4500 light years in distance from our solar system.

(M24)—is known as the "Sagittarius Star Cloud" and is a Milky Way star cloud with an apparent magnitude (v) of about 4.6 and is located approximately 10,000 light years in distance from planet Earth.

(M25)—is an open cluster located in the constellation of Sagittarius. It has an apparent magnitude (v) of about 4.6 and is about 2000 light years from our solar system.

(M26)—is an open cluster in the constellation Scutum. It has a apparent magnitude (v) of about 8.0 and is about 5000 light years in distance from our solar system.

(M27)—is the famous "Dumbbell Nebula" and is a planetary nebula in the constellation of Vulpecula. It has an apparent magnitude (v) of about 7.5 and is 1250 light years away from our solar system.

(M28)—is a globular cluster in the constellation of Sagittarius. It has an apparent magnitude (v) of about 7.7 and is approximately 18,000 light years in distance from planet Earth.

(M29)—is an open cluster located in the constellation of Cygnus, otherwise noted as the "Swan". It has a apparent magnitude (v) of about 7.1 and is about 7200 light years in distance from our solar system.

(M30)—is a globular cluster in constellation of Capricornus. It has an apparent magnitude (v) of about 7.7 and is about 25,000 light years from planet Earth.

(M31)—is known as the famous "Andromeda Galaxy" is a spiral galaxy in the constellation of Andromeda. It has an apparent magnitude (v) of about 3.44 and is approximately 2.5 million light years in distance from planet Earth.

(M32)—is a dwarf elliptical galaxy that is located in the constellation of Andromeda. It has an apparent magnitude (v) of about 8.1 and is approximately 2.9 million light years in distance from our solar system.

(M33)—is known as the "Triangulum Galaxy" and is a spiral galaxy located in the constellation of Triangulum. It has an apparent magnitude (v) of 5.7 and is approximately 2.81 million light years in distance from our solar system.

(M34)—is an open cluster located in the constellation of Perseus. It has an apparent magnitude (v) of about 5.5 and is approximately 1400 light years away from planet Earth.

(M35)—is an open cluster in the constellation of Gemini. It has an apparent magnitude (v) of about 5.3 and is approximately 2800 light years in distance from our solar system.

(M36)—is an open cluster in the constellation of Auriga, otherwise known as the "Charioteer". It has an apparent magnitude (v) of about 6.3 and is approximately 4100 light years in distance from our solar system.

(M37)—is an open cluster in the constellation of Auriga. It has a apparent magnitude (v) of about 6.2 and is approximately 4600 light years from planet Earth.

(M38)—is an open star cluster in the constellation of Auriga. It has an apparent magnitude (v) of about 7.4 and is approximately 4200 light years in distance from planet Earth.

(M39)—is an open cluster in the constellation of Cygnus. It has an apparent magnitude (v) of about 5.5 and is approximately 800 light years from planet Earth.

(M40)—is a double star, located some 500 light years away from planet Earth, and it's pointed in the direction of the constellation of Ursa Major. It has an apparent magnitude (v) of about 9.65 + 10.10.

(M41)—is an open cluster in Canis Major. It has an apparent magnitude (v) of 4.5 and is about 2300 light years in distance from planet Earth.

(M42)—the famous "Orion Nebula" and is a nebula located in the constellation of Orion, otherwise known as the "Hunter". It has an apparent magnitude (v) of about 4.0 and is approximately 1600 light years distance from planet Earth.

(M43)—is known as "De Mairan's Nebula" and is part of the Orion Nebula in the constellation of Orion. It has a apparent magnitude (v) of about 9.0 and is approximately 1600 light years distant from our solar system.

(M44)—is known as "Praesepe" or more commonly called the "Beehive Cluster" and is noted as an open cluster located in the constellation of Cancer. It has a apparent magnitude (v) of about 3.7 and is approximately 600 light years in distance from planet Earth.

(M45)—is known as the "Pleiades" or more commonly called the "Seven Sisters" and is noted as an open star cluster located in the constellation of Taurus constellation. It has an apparent magnitude (v) of about 1.6 and is approximately 400 light years from planet Earth.

(M46)—is an open cluster with an apparent magnitude (v) of about 6.1. It is approximately 5400 light years from planet Earth in the direction of the constellation of Puppis.

(M47)—is an open cluster in the constellation of Puppis. It has an apparent magnitude (v) of about 4.2 and it lies approximately 1600 light years from planet Earth.

(M48)—is an open cluster in the constellation of Hydra, also known as the "Water Snake". It has a apparent magnitude (v) of about 5.5 and is approximately 1500 light years in distance from our solar system.

(M49)—is known as an elliptical galaxy located in the constellation of Virgo. It has an apparent magnitude (v) of about 9.4 and is approximately 60 million light years in distance from the planet Earth.

(M50)—is an open star cluster in the constellation of Monceros, also known as "the Unicorn". It has an apparent magnitude (v) of about 5.9 and is approximately 3000 light years from planet Earth..

(M51)—is known as the famous "Whirlpool Galaxy", and it is a spiral galaxy that lies in the constellation of Canes Venatici, also known as "the Hunting Dogs". It has an apparent magnitude (v) of about 8.4 and is approximately 37 million light years from planet Earth.

(M52)—is an open cluster in the constellation of Cassiopeia. It has an apparent magnitude (v) of about 5.0 and is approximately 7000 light years from planet Earth.

(M53)—is a globular star cluster in the constellation of Coma Berenices, and is also known as "Berenice's Hair". The cluster has an apparent magnitude (v) of about 8.3 and is approximately 56,000 light years in distance from our solar system.

(M54)—is a globular cluster with an apparent magnitude (v) of about 8.4. It is located in the constellation of Sagittarius at a distance of about 83,000 light years from planet Earth.

(M55)—is known as a globular star cluster in the constellation of Sagittarius. It has an apparent magnitude (v) of about 7.4 and is approximately 17,000 light years from our solar system.

(M56)—is a globular cluster in the constellation of Lyra. It has an apparent magnitude (v) of about 8.3 and is approximately 32,000 light years in distance from planet Earth.

(M57)—is known as the "Ring Nebula" and is known as a planetary nebula located in the constellation of Lyra. It has an apparent magnitude (v) of about 8.8 and is approximately 2300 light years from the planet Earth.

(M58)—is known as a barred spiral galaxy in the constellation of Virgo. It has an apparent magnitude (v) of about 10.5 and is approximately 60 million light years in distance from our solar system.

(M59)—is known as an elliptical galaxy in the constellation of Virgo. It has an apparent magnitude (v) of about 10.6. It is approximately 60 million light years from planet Earth.

(M60)—is known as an elliptical galaxy in the constellation of Virgo. It has an apparent magnitude (v) of about 9.8 and is approximately 60 million light years from plane Earth.

(M61)—is known as a spiral galaxy in constellation of Virgo. It has an apparent magnitude (v) of about 10.2 and is about 60 million light years from planet Earth.

(M62)—is known as a globular cluster located in the constellation of Ophiuchus, also known as "the Serpent Bearer". It has an apparent magnitude (v) of about 7.4 and is approximately 22,000 light years from planet Earth.

(M63)—is also known as the "Sunflower Galaxy". It is noted as a spiral galaxy in the constellation of Canes Venatici. It has an apparent magnitude (v) of about 9.3 and is approximately 37 million light years from planet Earth.

(M64)—is better known as the "Black Eye Galaxy", and it is a spiral galaxy in the constellation of Coma Berenices. It has n apparent magnitude (v) of about 9.4 and lies approximately 12 million light years from our solar system.

(M65)—is known as a barred spiral galaxy that is a member of the "Leo Triplet" in the constellation of Leo. It has an apparent magnitude (v) of about 10.3 and is approximately 35 million light years in distance from planet Earth.

(M66)—is known to be a barred spiral galaxy that is a member of the "Leo Triplet" of galaxies. It has an apparent magnitude (v) of about 8.9 and is located approximately 35 million light years from planet Earth in the constellation of Leo.

(M67)—is known to be an open cluster in the constellation of Cancer. It has an apparent magnitude (v) of about 6.1 and is approximately 2250 light years in distance from planet Earth.

(M68)—is known to be a globular cluster in the constellation of Hydra. It has an apparent magnitude (v) of about 9.7 and lies approximately 32,000 light years from our solar system.

(M69)—is known to be a globular star cluster in the constellation of Sagittarius. It has an apparent magnitude (v) of about 8.3 and is approximately 25,000 light years from planet Earth.

(M70)—is known to be a globular cluster in the constellation of Sagittarius. It has an apparent magnitude (v) of about 9.0 and is approximately 28,000 light years from planet Earth.

(M71)—is noted to be a globular cluster in the constellation of Sagitta, also known as "the Arrow". It has an apparent magnitude (v) of about 6.1 and is approximately 12,000 light years from planet Earth.

(M72)—is known to be a globular cluster in the constellation of Aquarius. It has an apparent magnitude (v) of about 9.4 and is approximately 53,000 light years from our solar system.

(M73)—is known as an "asterism" in the constellation of Aquarius. It has an apparent magnitude (v) of about 9.0. No distance is given.

(M74)—is known as a spiral galaxy in the constellation of Pisces. It has an apparent magnitude (v) of about 10.0 and is approximately 35 million light years in distance from planet Earth.

(M75)—is known to be a globular cluster located in the constellation of Sagittarius. It has a an apparent magnitude (v) of about 9.2 and is approximately 58,000 light years in distance from planet Earth.

(M76)—is also known as the "Little Dumbbell Nebula". It is located about 3400 light years in distance in the constellation of Perseus. It is classified as a planetary nebula with an apparent magnitude (v) of about 10.1.

(M77)—is also known as "Cetus A". It is known to be a spiral galaxy located in the constellation of Cetus, often called "the Whale". It has an apparent magnitude (v) of about 9.6 and is approximately 60 million light years in distance from planet Earth.

(M78)—is known as a "diffuse nebula" in constellation of Orion. It has an apparent magnitude (v) of about 8.3 and is approximately 1600 light years from our solar system.

(M79)—is known as a globular cluster in the constellation of Lepus, also known as "the Hare". It has an apparent magnitude (v) of about 8.6 and is approximately 40,000 light years in distance from the planet Earth.

(M80)—is known to be a globular cluster in the constellation of Scorpius. It has an apparent magnitude (v) of about 7.9 and is approximately 27,000 light years from planet Earth.

(M81)—is also known as "Bode's Galaxy". It is a spiral galaxy located in the constellation of Ursa Major. It has an apparent magnitude (v) of about 6.9 and lies approximately 12 million light years from our solar system.

(M82)—is known as the famous "Cigar Galaxy" and is classified as a starburst galaxy located in the constellation of Ursa Major. It has a apparent magnitude (v) of about 8.4 and is approximately 11 million light years from the planet Earth.

(M83)—is commonly known as the "Southern Pinwheel Galaxy". It is a barred spiral galaxy in the constellation of Hydra. It has an apparent magnitude (v) of about 7.5 and is approximately 10 million light years from the planet Earth.

(M84)—is known as a "lenticular galaxy" in the constellation of Virgo. It has an apparent magnitude (v) of about 10.1 and is approximately 60 million light years in distance from our solar system.

(M85)—is known to be a "lenticular galaxy" in the Coma Berenices constellation. It has an apparent magnitude (v) of about 10.0 and is about 60 million light years in distance from the planet Earth.

(M86)—is also known to be a lenticular galaxy located in the constellation of Virgo. It has an apparent magnitude (v) of about 9.0 and is approximately 60 million light years in distance from our solar system..

(M87)—is also known as "Virgo A", an elliptical galaxy located in the constellation of Virgo. It has an apparent magnitude (v) of about 9.6 and is approximately 60 million light years from planet Earth.

(M88)—is known to be a spiral galaxy in Coma Berenices constellation. It has an apparent magnitude (v) of about 10.4 and is approximately 60 million light years from planet Earth.

(M89)—is known as an elliptical galaxy in the constellation of Virgo. It has an apparent magnitude (v) of about 10.7 and is approximately 60 million light years from planet Earth.

(M90)—is known as a spiral galaxy in the constellation of Virgo. It has a an apparent magnitude (v) of about 10.3 and is approximately 60 million light years from our solar system.

(M91)—is a barred spiral galaxy located in Coma Berenices. It has an apparent magnitude (v) of about 11.0 and is approximately 60 million light years distant.

(M92)—is a globular star cluster in the constellation of Hercules. It has an apparent magnitude (v) of about 6.3 and is about 26,000 light years from planet Earth.

(M93)—is an open cluster in Puppis constellation. It has an apparent magnitude (v) of 6.0 and is approximately 4500 light years from planet Earth.

(M94)—is also known as the "Cat's Eye Galaxy" or "Croc's Eye Galaxy". It is a spiral galaxy in Canes Venatici constellation. It has an apparent magnitude (v) of about 9.0 and is about 14.5 million light years from our solar system.

(M95)—is a barred spiral galaxy in the constellation of Leo. It has an apparent magnitude (v) of about 11.4 and is approximately 38 million light years from planet Earth.

(M96)—is a spiral galaxy in the constellation of Leo. It has an apparent magnitude (v) of about 10.1 and is about 38 million light years distance from planet Earth.

(M97)—is known as the "Owl Nebula". It is a planetary nebula located in the constellation of Ursa Major. It has an apparent magnitude (v) of about 9.9 and is approximately 2600 light years distant from planet Earth.

(M98)—is a spiral galaxy in Coma Berenices constellation. It has an apparent magnitude (v) of about 11.0 and is about 60 million light years from planet Earth.

(M99)—is a spiral galaxy in Coma Berenices. It has an apparent magnitude (v) of 10.4 and is about 60 million light years distant from Earth.

(M100)—is a spiral galaxy in Coma Berenices constellation. It has an apparent magnitude (v) of about 9.5 and is approximately 60 million light years from our solar system.

(M101)—is the famous "Pinwheel Galaxy", is a spiral galaxy in Ursa Major. It has an apparent magnitude (v) of about 7.9 and is about 27 million light years distance from planet Earth.

(M102)—listed as a galaxy, but the object has yet to be conclusively identified. The most likely candidate is the Spindle Galaxy (NGC 5866) in the constellation Draco. Est. Mag. 9.9

(M103)—is an open cluster in Cassiopeia constellation. It has an apparent magnitude (v) of 7.4 and is approximately 8000 light years distant from Earth.

(M104)—is known as the "Sombrero Galaxy". It is a spiral galaxy located in the constellation of Virgo. It has an apparent magnitude (v) of about 9.0 and is approximately 50 million light years from planet Earth.

(M105)—is an elliptical galaxy in the constellation of Leo. It has an apparent magnitude (v) of about 10.2 and is about 38 million light years from our solar system.

(M106)—is a spiral galaxy in the constellation of Canes Venatici. It has an apparent magnitude (v) of about 9.1 and is about 25 million light years distance from the planet Earth.

(M107)—is a globular cluster in the constellation of Ophiuchus It has an apparent magnitude (v) of about 8.9 and is approximately 20,000 light years from planet Earth.

(M108)—is a barred spiral galaxy in the constellation of Ursa Major. It has an apparent magnitude (v) of about 10.7 and is approximately 45 million light years distance from planet Earth

(M109)—is a barred spiral galaxy located in Ursa Major constellation. It has an apparent magnitude (v) of about 10.6 and is about 55 million light years from our solar system.

(M110)—is a dwarf elliptical galaxy in the constellation of Andromeda. It has an apparent magnitude (v) of about 8.92 and is approximately 2.2 million light years from planet Earth.

Selected NGC Objects

1. **(NGC 105)**—a spiral galaxy galaxy in Pisces Mag. (v) 14.1
2. **(NGC 110)**—an open star cluster in Cassiopeia. Mag. (v) 9.0
3. **(NGC 209)**—a lenticular galaxy in Cetus. Mag. (v) 14.7
4. **(NGC 221)**—a dwarf elliptical galaxy in Andromeda. Ma. (v) 8.08
5. **(NGC 225)**—an open cluster in Cassiopeia. Mag. (v) 7.0
6. **(NGC 253)**—a bright galaxy in the Scupltor. Mag. (v) 8.0
7. **(NGC 297)**—a galaxy in Cetus. Mag. (v) 17.3
8. **(NGC 381)**—an open cluster in Cassiopeia. Mag. (v) 9.3

9. **(NGC 598)**—a spiral galaxy in Triangulum. Mag. (v) 5.7
10. **(NGC 613)**—a barred spiral galaxy in the constellation Sculptor. Mag. (v) 10.0
11. **(NGC 660)**—a peculiar and polar-ring galaxy in Pisces. Mag. (v) 12.0
12. **(NGC 752)**—a bright cluster in Andromeda. Mag. (v) 5.7
13. **(NGC 869)**—h Persei, a double cluster with chi. Mag. (v) 3.7
14. **(NGC 884)**—chi Persei, a double cluster with h. Mag. (v) 3.8
15. **(NGC 891)**—an edge-on spiral galaxy in Andromeda. Mag. (v) 10.8
16. **(NGC 1055)**—an edge-on spiral galaxy in M77 group. Mag. (v) 11.4
17. **(NGC 1432)**—Maia Nebula in the Pleiades (M45). Mag. (v) 13.
18. **(NGC 1435/IC349/Merope)**—Tempel's Merope Nebula in the Pleiades (M45). Mag. (v) 13.0>IC 349/Mag. (v) 8.0>Merope
19. **(NGC 2023)**—is a bright reflection nebula located near the Horsehead Nebula.
20. **(NGC 2070)**—Tarantula Nebula in the Large Magellanic Cloud. Mag. (v) 8.0
21. **(NGC 2169)**—an open cluster in Orion. Mag. (v) 5.9
22. **(NGC 2175)**—an open cluster in Orion. Mag. (v) 6.8
23. **(NGC 2204)**—an open cluster in Canis Major. Mag. (v) 8.6
24. **(NGC 2237)**—is part of the Rosette Nebula. Mag. (v) 9.0
25. **(NGC 2238)**—is part of the Rosette Nebula.
26. **(NGC 2239)**—is part of the Rosette Nebula.
27. **(NGC 2244)**—a cluster in the Rosette Nebula. Mag. (v) 4.8
28. **(NGC 2246)**—is part of the Rosette Nebula.
29. **(NGC 2264)**—the Cone Nebula and associated cluster. Mag. (v) 3.9
30. **(NGC 2349)**—is an open cluster in Monoceros.
31. **(NGC 2360)**—is an open cluster in Canis Major. Mag. (v) 7.2
32. **(NGC 2362)**—is an open cluster in Canis Major. Mag. (v) 4.1
33. **(NGC 2403)**—is a Sc galaxy in the M81 group. Mag. (v) 8.2
34. **(NGC 2419)**—is an outlying globular cluster in Lynx. Mag. (v) 9.06
35. **(NGC 2438)**—is a planetary nebula in front of M46. Mag. (v) 10.8
36. **(NGC 2451)**—a bright cluster in Puppis. Mag. (v) 3.0
37. **(NGC 2477)**—a rich bright cluster in the constellation Puppis. Mag. (v) 5.8
38. **(NGC 2516)**—a bright cluster in Carina. Mag. (v) 3.8
39. **(NGC 2546)**—a considerable cluster in Puppis. Mag. (v) 6.3
40. **(NGC 2547)**—a very populated cluster in Vela. Mag. (v) 4.7
41. **(NGC 2903)**—a very bright spiral galaxy in Leo. Mag. (v) 9.7
42. **(NGC 2976)**—a faint companion of M81 and M82. Mag. (v) 10.8
43. **(NGC 3077)**—a companion of M81 and M82. Mag. (v) 9.6
44. **(NGC 3115)**—known as the Spindle Galaxy (Caldwell 53) in Sextans. Mag. (v) 9.9
45. **(NGC 3228)**—a considerable open cluster in Vela. Mag. (v) 6.0
46. **(NGC 3293)**—a bright open cluster in Carina. Mag. (v) 4.7
47. **(NGC 3372)**—also known as the Eta Carinae nebula. Mag. (v) 1.0
48. **(NGC 3532)**—a bright open cluster in Carina. Mag. (v) 3.0
49. **(NGC 3628)**—is the third of the Leo Triplet (with M65 and M66). Mag. (v) 10.2
50. **(NGC 3766)**—a concentrated cluster in Centaurus. Mag. (v) 5.3

51. **(NGC 3953)**—is a barred spiral galaxy near M109. Mag. (v) 10.8
52. **(NGC 4565)**—is a large bright edge-on spiral in Coma. Mag. (v) 10.4
53. **(NGC 4571)**—is a barred spiral in Virgo cluster. Mag. (v) 11.8
54. **(NGC 4631)**—is known as the Herring or Whale Galaxy. Mag. (v) 9.8
55. **(NGC 4656/57)**—an irregular looking spiral galaxy in Canes Venatici. Mag. 11.0
56. **(NGC 4755)**—is known as Kappa Cruxis, the Jewel Box cluster. Mag. (v) 4.2
57. **(NGC 4833)**—is a southern globular cluster in Musca. Mag. (v) 7.8
58. **(NGC 5128)**—is peculiar and a radio galaxy Centaurus A. Mag. (v) 6.84
59. **(NGC 5139)**—a globular cluster Omega Centauri. Mag. (v) 3.9
60. **(NGC 5195)**—a companion of **M51**. Mag. (v) 10.5
61. **(NGC 5281)**—a small compact open cluster in Centaurus. Mag. (v) 5.9
62. **(NGC 5662)**—a considerable southern cluster in Centaurus. Mag. (v) 5.5
63. **(NGC 5907)**—in the same group with **M102** candidate NGC 5866. Mag. (v) 11.1
64. **(NGC 6025)**—a considerable open cluster in Triangulum Australe. Mag. (v) 5.1
65. **(NGC 6124)**—a considerable open cluster in the constellation Scorpius. Mag. (v) 5.8
66. **(NGC 6231)**—a bright open cluster in the constellation Scorpius. Mag. (v) 2.6
67. **(NGC 6242)**—an open cluster in the constellation Scorpius. Mag. (v) 6.4
68. **(NGC 6397)**—a nearby globular cluster in Ara. Mag. (v) 6.7
69. **(NGC 6530)**—open cluster associated with the Lagoon Nebula **M8**. Mag. 4.6
70. **(NGC 6543)**—known as the Cat Eye nebula, a planetary near the N.E.P. Mag. 8.1
71. **(NGC 6603)**—an open cluster in the star cloud **M24.** Mag. 11.4
72. **(NGC 6633)**—a bright open cluster in the constellation Ophiuchus. Mag. 4.6
73. **(NGC 6712)**—a globular cluster in Scutum. Mag. 8.1
74. **(NGC 6819)**—an open cluster in the constellation Cygnus. Mag. 7.3
75. **(NGC 6822)**—known as Barnard's Galaxy, an irregular Local Group galaxy. Mag. 9.3
76. **(NGC 6866)**—an open cluster in the constellation Cygnus. Mag. 7.6
77. **(NGC 6946)**—a spiral galaxy in the constellation Cepheus. Mag. 8.9
78. **(NGC 7000)**—known as the North America Nebula. Apparent Mag. 4.0
79. **(NGC 7009)**—known as the Saturn Nebula. Mag. 8.0
80. **(NGC 7293)**—also known as the Helix Nebula. Mag. 7.3
81. **(NGC 7331)**—a conspicuous spiral galaxy in the constellation Pegasus. Mag. 9.5
82. **(NGC 7380)**—an open cluster with nebula in the constellation Cepheus. Mag. 7.2
83. **(NGC 7479)**—a barred spiral galaxy in the constellation Pegasus. Mag. 11.0
84. **(NGC 7789)**—a bright open cluster in the constellation Cassiopeia. Mag. 6.7

Selected IC Objects

1. **(IC 10)**—an outlying irregular dwarf member of the Local Group. Mag. (v) 10.4+−.2
2. **(IC 348)**—a star forming region in the constellation Perseus. Mag. (v) 7.3
3. **(IC 349)**—known as Barnard's Merope Nebula in the Pleiades **(M45)**. Mag. (v) 13.0
4. **(IC 434)**—the emission nebula behind the Horsehead Nebula. App. Mag. (v) 7.3
5. **(IC 1434)**—an open cluster in Lacerta. Mag. (v) 9.0
6. **(IC 2391)**—Open cluster known as Omicron Velorum in Vela. Mag. (v) 2.5
7. **(IC 2395)**—vdB-Ha 47, an open scattered cluster in Vela. Mag. (v) 4.6
8. **(IC 2488)**—an inconspicuous cluster in the constellation Vela. Mag. (v) 7.4p
9. **(IC 2602)**—known as the Theta Carinae cluster, (the Southern Pleiades). Mag. (v) 1.9
10. **(IC 5148)**—a planetary nebula located in the constellation Grus. Mag. (v) 16.5
11. **(IC 4665)**—a coarse bright cluster in the constellation Ophiuchus. Mag. (v) 4.2
12. **(IC 5152)**—an irregular dwarf member of the Local Group. Mag. (v) 10.5

Some of the Author's Favorite Intragalactic Objects

1. **(Barnard 33)**—known as the Horsehead Nebula.
2. **(Barnard 92)**—a dark nebula associated with M24.
3. **(Barnard 93)**—a dark nebula associated with M24.
4. **(Barnard's Loop)**—(Sharpless 276, Sh2-276).
5. **(Brocchi's Cluster)**—(Collinder 399, Cr 399).
6. **(Canis Major Dwarf)**—a disputed nearby irregular galaxy in the local group.
7. **(Coalsack Dark Nebula)**—that appears in the southern Milky Way.
8. **(Collinder 140)**—(Cr 140)—an open cluster in southern Canis Major.
9. **(Collinder 228)**—(Cr 228)—an open cluster within the Great Carina Nebula.
10. **(The Hyades)**—(Meylotte 25, Mel 25)—a beautiful open cluster in Taurus.
11. **(Eta Carinae)**—one of the most massive and luminous, and spectacular stars.
12. **(G1 (Mayall II)**—the brightest globular in **M31**.
13. **(Coma Star Cluster)**—(Melotte 111, Mel 111).
14. **(The Large Magellanic Cloud)**—(LMC).
15. **(Leo I—The Regulus Galaxy)**—a Dwarf Elliptical Galaxy in Local Group.
16. **(M-42)**—The Great Nebula in Orion.
17. **(The Maffei 1 Group of galaxies)**—thought to be once part of our Local Group.

Selected Objects in the Milky Way: Our Galaxy

1. **(M8)**—the Lagoon Nebula in Sagittarius.
2. **(SagDEG)**—nearby Sagittarius Dwarf Elliptical Galaxy containing globular **M54**
3. **(Sculptor Group of Galaxies)**—also South Polar Group
4. **(The Small Magellanic Cloud)**—**(SMC)**
5. **(Trumpler 10)**—(Tr 10), open cluster, presumably Lacaille II.6
6. **(The Ursa Major Moving Cluster)**—(Collinder 285, Cr 285)
7. **(Van den Bergh-Hagen 47)**—(vdB-Ha 47, BH 47)—an open cluster, presumably **IC I IC IC2395**
8. **(Wolf-Lundmark-Melotte)**—**(WLM)**—a remote Local Group galaxy

Note: NGC, IC objects…etc. listed above are just some of the author's favorite celestial objects…and all of them are easily found on most sky atlases and star maps. Some of them can present a challenge for any keen observer even with a large binoscope. *Lists above are adapted from : en.wikipedia.org.

Astronomical Societies

1. Amateur Astronomers Association of Pittsburgh
2. American Association of Variable Star Observers
3. American Astronomical Society
4. American Meteor Society, an amateur organization specializing in meteor observations.
5. Association of Lunar and Planetary Observers, an amateur organization
6. Astronomical League, an umbrella organization of U.S. amateur astronomy societies.
7. Astronomical Society of Australia (ASA)
8. Astronomical Society of Glasgow
9. Astronomical Society of New South Wales, based in Sydney, Australia.
10. Astronomical Society of South Australia
11. Astronomical Society of Southern Africa
12. Astronomical Society of the Pacific
13. Astronomical Society of Victoria, based in Melbourne, Australia.
14. Astronomical Society Ruder Boskovic, from Belgrade Serbia
15. Astronomische Gesellschaft (German Astronomical Society)
16. Birmingham Astronomical Society, in Birmingham, Alabama
17. British Astronomical Association (BAA)
18. Confederation of Indian Amateur Astronomer Association (India)
19. Cornell Astronomical Society.
20. Crayford Manor House Astronomical Society
21. Escambia Amateur Astronomers Association, Northwest Florida.

22. EAAE—European Association for Astronomy Education, a European Association that promotes activities for schools, teachers and students.
23. Federation of Astronomical Societies (UK)
24. International Meteor Organization, an international organization dealing in meteor observations.
25. Jyotirvidya Parisanstha (Pune, India)
26. Kauai Educational Association for Science and Astronomy (Kauai, Hawaii)
27. Khagol vishwa (India)
28. Khagol Mandal (Mumbai, India)
29. Kopernik Astronomical Society (Vestal, New York)
30. Louisville Astronomical Society
31. Macarthur Astronomical Society, based in south-western Sydney, Australia
32. Mohawk Valley Astronomical Society, Central New York, USA
33. Mornington Peninsula Astronomical Society, based in Victoria, Australia.
34. Network for Astronomy School Education (NASE)
35. Northern Virginia Astronomy Club
36. Northumberland Astronomical Society
37. Nottingham Astronomical Society, Nottingham, England, UK
38. Pakistan Amateur Astronomers Society
39. The Planetary Society
40. Polish Astronomical Society
41. Royal Astronomical Society in London
42. Royal Astronomical Society of Canada
43. Royal Astronomical Society of New Zealand
44. SETI Institute
45. Shreveport-Bossier Astronomical Society
46. Society astronomique de France, the French astronomical society
47. Society for the History of Astronomy
48. Society for Popular Astronomy based in the United Kingdom for beginners to amateur astronomy
49. Southern Cross Astronomical Society, based in Miami, Florida, US.
50. Sutherland Astronomical Society based in the southern suburbs of Sydney, Australia.
51. Spaceturk, an amateur organization specializing in planetary, lunar, and other solar system observations in Turkey
52. Whakatane Astronomical Society

Note: For those who are looking for an astronomy club or astronomical society to join, the list above will help you locate a club or society near you. Their websites are easily found on the internet along with their contact information. (Adapted from en.wikipedia.org) (Figs. 10.42, 10.43, 10.44, 10.45, and 10.46).

Fig. 10.42 Jan Pavlacka a Virginia amateur astronomer is putting his 70's Jason 7×50 binoc's to good use scanning the sky just before sunset for any mysterious cosmic interlopers. Lucky for Jan on this particular fine evening that a Venus/Jupiter conjunction has caught his observing attention. (Image credit: Jan and Carolyn Pavlacka)

Fig. 10.43 (Cartoon credit: Jack Kramer)

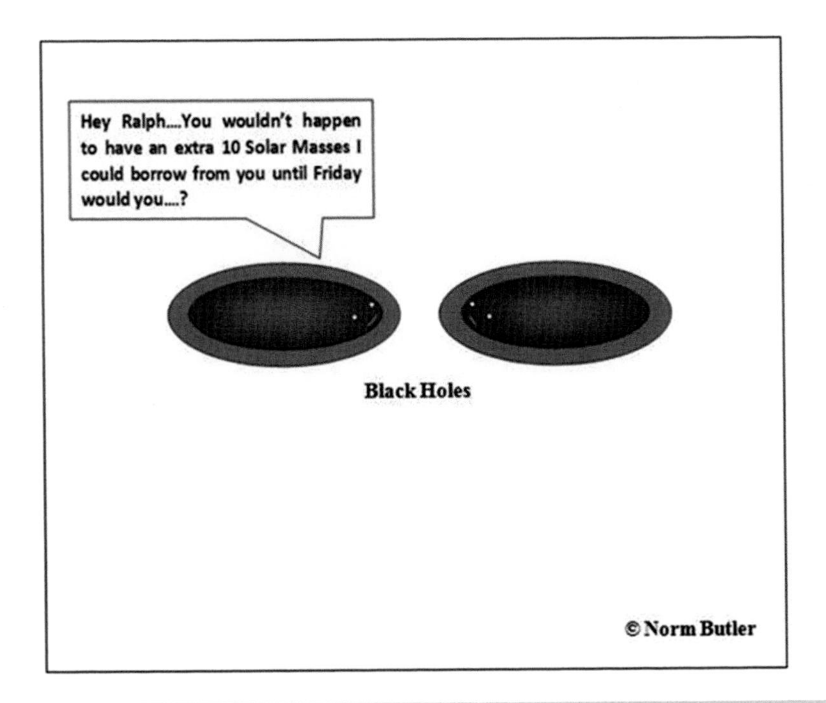

Fig. 10.44 (Cartoon credit: the Author)

Fig. 10.45 (Cartoon credit: Les Lamb)

Fig. 10.46 IC434 More commonly known as the "Horse Head Nebula" in the constellation Orion (Image credit: Sara Wager - www.swagastro.com)

Further Reading

Alex Tat-Sang Choy. A asymmetric binocular telescopes. http://www.telescopereviews.com, www.astronomics.com

Astrogizmos. E Z Gazer. http://www.astrogizmos.com

Liming, G. The Bino Chair. http://www.liming.org

Schell, Jochen. http://www.binoscopes.de

The Bolton Group. http://www.deep-sky.co.uk

Wabash Valley Astronomical Society. Eight inch binocular telescope. http://www.stargazing.net

Walton, G. Easy To build binocular chair. http://www.mpas.asn.au

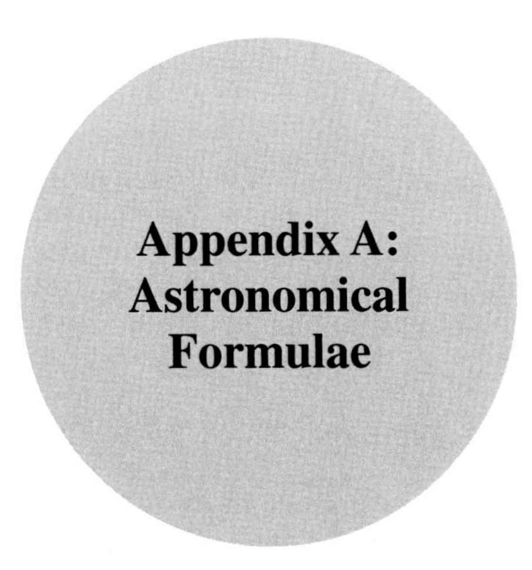

Appendix A: Astronomical Formulae

Introduction

Magnification

The magnification of an astronomical telescope changes with the eyepiece used. It is calculated by dividing the focal length of the telescope (usually marked on the optical tube) by the focal length of the eyepiece (both in millimeters). Thus:

$$\frac{\text{Telescope focal length}}{\text{Eyepiece focal length}} = \text{Magnification}$$

For example, a telescope with a 1200-mm focal length using a 12-mm ocular is operating at 100× magnification (1200/12 = 100).

Focal Ratio (f/stop)

The focal ratio, or f/stop, of any lens system (including telescopes) is computed by dividing the focal length by the clear aperture (usually expressed in millimeters). In other words, the focal ratio is the ratio of the focal length and clear aperture. Thus,

$$\frac{\text{Telescope focal length}}{\text{Clear aperture}} = \text{Focal ratio}$$

N. Butler, *Building and Using Binoscopes*, The Patrick Moore Practical Astronomy Series, DOI 10.1007/978-3-319-46789-4

For example, a telescope with a focal length of 1200 mm and a 150 mm (6″) clear aperture has a focal ratio of f/8 (1200/150 = 8).

True Field of View

There are two ways to calculate the true field of view (FOV) in degrees of a telescope and eyepiece combination. An easy method to use is to divide the apparent field of view (AFOV) of the ocular by the magnification of the system. The AFOV for almost all eyepieces is provided by the manufacturer and is easy to derive the magnification of any telescope/ocular combination. Thus,

$$\frac{\text{AFOV}}{\text{Magnification}} = \text{FOV}\left(\text{field of view}\right)$$

For example, a 30-mm Plossl eyepiece generally has an AFOV of 52°. Used in a telescope with a 1200-mm prime focal length, the magnification is 40×. The true field of view is therefore 1.30° (52/40 = 1.3).

Power per Inch

This is good to know because it is a truism among amateur astronomers that the power per inch (PPI) figure of a telescope and ocular should not exceed 50 PPI in excellent seeing conditions. In average seeing conditions, I figure about 30 PPI as a practical maximum.

PPI can be calculated by dividing the magnification of the telescope and eyepiece combination by the telescope's clear aperture in inches (1 in. = approximately 25 mm). Thus,

$$\frac{\text{Magnification}}{\text{Aperture}}\left(\text{IN}\right) = \text{PPI}\left(\text{power per inch}\right)$$

For example, a 150-mm clear aperture is approximately 6″, so such a telescope operating at 120× magnification is at 25 PPI (120/6 = 20).

Exit Pupil

The exit pupil is the diameter of the "light pencil" that emerges from the eye-piece. The pupil of fully dark-adapted human eye can dilate to about 7-mm diameter, so an exit pupil in excess of 7 mm is passing more light than the eye can accept. On the other hand, as the exit pupil decreases below 7 mm, lack of light becomes the basic limiting factor to what you can see at night. Exit pupils of less than about 0.5 mm are so small and pass so little light to the eye that they are functionally useless. Actually, I like exit pupils of at least 1.0 mm for decent viewing.

Exit pupil can be calculated by dividing the telescope's clear aperture (in millimeters) by the magnification produced by the ocular in use. Thus,

$$\frac{\text{Aperture}}{\text{Magnification}} = \text{Exit pupil}$$

For example, our 150-mm (f/8) clear aperture telescope with a 10-mm ocular is operating at 120× magnification and therefore has a 1.25 mm exit pupil (150/120 = 1.25).

Another way to calculate exit pupil is to divide the eyepiece focal length in millimeters by the telescope's focal ratio (f/stop).

$$\frac{\text{Ocular focal length}}{\text{Telescope focal ratio}} = \text{Exit pupil}$$

Thus, a 10-mm ocular in our f/8 (150-mm clear aperture and 1200-mm focal length) telescope has a 1.25-mm exit pupil (10/8 = 1.25). Either formula results in the same answer.

Resolution

The resolution of a telescope must be considered. In terms of ideal "seeing" conditions, the following formula applies:

$$\text{Resolution} = \frac{116}{D}$$

$$D = \text{diameter of telescope in mm}$$

Example: a 254-mm telescope (10″)

$$\text{Resolution} = \frac{116}{\text{mm}\,254}$$

$$\text{Resolution} = 0.457(0.5)\text{seconds}$$

How faint an object can your telescope see:

$$m = 2.7 + 5\log D$$

$$D = \text{Objective lens diameter}$$

where m is the limiting magnitude, for example, a 10″ telescope:

$$m = 2.7 + 5\log(254\,\text{mm})$$

$$m = 14.7$$

The faintest object a 10″ telescope can see (depending on seeing conditions) is a visual magnitude of 14.7 (Pluto has a magnitude of around 13.8).

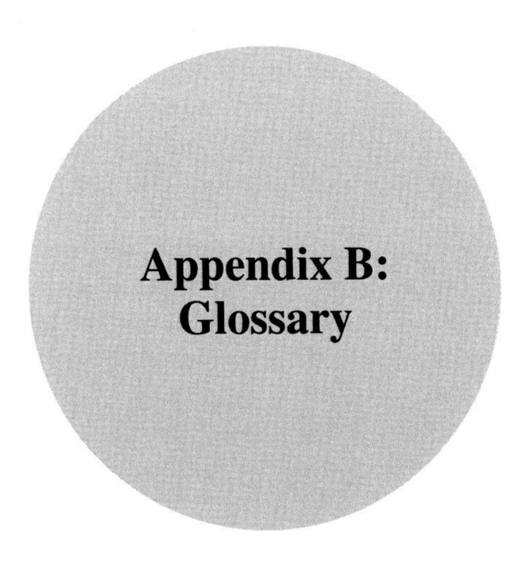

Airy Disk The best focused spot of light that can be created by a perfect lens system, assuming a circular aperture and limited by light diffraction.

Altazimuth A telescope mounting that allows motion of the telescope about a vertical axis (in azimuth) and a horizontal axis (in altitude).

Antireflection Coating (AR) Coating is a type of optical coating applied to the surface of lenses and other optical devices to reduce reflection.

Astigmatism An optical system with astigmatism is one where rays that propagate in two perpendicular planes have different foci.

Autoguider An electronic device that makes use of a CCD camera to detect guiding errors and makes automatic corrections to the telescope's drive system in order for it to track accurately.

Autotracking The ability of a Dobsonian mount to track celestial objects as they move across the night sky.

Barlow Lens It is named after Peter Barlow and is a diverging lens which, used in series with other optics in an optical system, increases the effective focal length of an optical system as perceived by all components after it is in the system.

Central Obstruction It is usually defined as a telescope's secondary mirror, for example, in a Newtonian telescope's optical system that acts as a central obstruction in the telescope's light path.

Collimation Refers to all the optical elements in an instrument being on their designed optical axis. It also refers to the process of adjusting an optical instrument so that all its elements are on that designed axes (in line and parallel). With regard to a telescope, the term refers to the fact that the optical axes of each optical component should all be centered and paralleled, so that collimated light emerges from the eyepiece.

N. Butler, *Building and Using Binoscopes*, The Patrick Moore Practical Astronomy Series, DOI 10.1007/978-3-319-46789-4

Coma In optics (especially telescopes), the *coma* (aka comatic aberration) in an optical system refers to aberration inherent to certain optical designs or due to imperfection in the lens or other components which results in off-axis point sources such as stars appearing distorted, appearing to have a tail (coma) like a comet.

Conical Mirror A cone-shaped mirror that can be mounted more easily and cools down faster than conventional "cylindrical" mirrors.

Dawes Limit A formula to express the maximum resolving power of a microscope or telescope.

Depth of Focus A lens optics concept that measures the tolerance of placement of the image plane (the film plane in a camera) in relation to the lens.

Declination (Dec) An axis of rotation on an Altazimuth telescope mount that is perpendicular to the polar axis and allows the telescope to be pointed at objects of different declinations.

Dielectric Coatings These consist of layers of dielectric materials (i.e., not metals, which are used in mirrors for consumer goods) and are designed to achieve the highest possible reflectance, usually at specific wavelengths.

Diffraction The bending of light waves around the edge of an obstacle. When light strikes an opaque body, for instance, a shadow forms on the side of the body that is shielded from the light source. Ordinarily light travels in straight lines through a uniform, transparent medium, but those light waves that just pass the edges of the opaque body are bent, or deflected.

Diffraction Limited The minimum angular separation of two sources that can be distinguished by a telescope depends on the wavelength of the light being observed and the diameter of the telescope. This angle is called the diffraction limit.

Dispersion For a refracting, transparent substance, such as a prism of glass, the dispersion is characterized by the variation of refractive index with change in wavelength of the radiation.

ED Short for extra low dispersion, usually referring to glass that focuses red, green, and blue light more tightly than a regular crown flint objective and resulting in better color correction.

Equatorial Mount A mount for instruments that follows the rotation of the sky (celestial sphere) by having one rotational axis parallel to the Earth's axis of rotation. This type of mount is used for astronomical telescopes.

Extrafocal The focus is outside of the focal plane.

Eye Relief For an optical instrument (such as a telescope, a microscope, or binoculars), it is the distance from the last surface of an eyepiece at which the user's eye can obtain the full viewing angle.

Focal Length The focal length of an optical system is a measure of how strongly the system converges or diverges light. For an optical system in air, it is the distance over which initially collimated rays are brought to a focus.

Focal Point The point where a light cone converges is called the primary mirror's focal point.

Focal Ratio The numerical value of the relative aperture. If the relative aperture is f8, 8 is the f-number and it indicates that the focal length of the lens is 8 times the size of the lens aperture.

Fresnel Lens Series of concentric rings, each consisting of a thin part of a simple lens, assembled on a flat surface.

Gamma Stretch Gamma refers to the degree of contrast between the midlevel gray values of a raster dataset. Gamma does not affect the black or white values in a raster dataset, only the middle values. By applying a gamma correction, you can control the overall brightness of a raster dataset. Additionally, gamma changes not only the brightness but also the ratios of red to green to blue.

GOTO Represents a type of electronic telescope drive that can automatically go to a selected celestial object on demand.

Interpupillary Distance (IPD) The distance between the center of the pupils of the human eye. Average interpupillary distance is 65 mm.

Intrafocal The focus is inside the focal plane.

Lazy Susan Bearing A circular revolving frame that rotates with the use of ball bearings; sometimes used in Altazimuth telescope mounts.

Magnification A measure of the ability of a lens or other optical instrument to magnify, expressed as the ratio of the size of the image to that of the object.

Multicoated Refers to lens elements of optical system with many layers of lens coating (of transparent dielectric material) to increase transmission of light and reduce reflection.

Parabolic Mirror A cone-shaped concave mirror with a rounded-off tip, whose cross section is shaped like the tip of a parabola.

Peak-to-Valley Wavelength A term describing how sounds and light waves are measured. The peak of a wave is known as the peak, while the low dip of the wave is known as the valley. Wavelength is determined by measuring the distance between each crest.

Rayleigh Criterion or Resolving Power A criterion for how finely a set of optics may be able to distinguish the location of objects which are near each other.

Right Ascension (RA) The arc of the celestial equator measured eastward from the vernal equinox to the foot of the great circle passing through the celestial poles and a given point on the celestial sphere, expressed in degrees or hours. Right ascension is the celestial equivalent of terrestrial longitude. Both right ascension and longitude measure an angle that increases toward the east as measured from a zero point on an equator, which, by convention, is the first point of Aries.

Ronchi Test In optical testing, a Ronchi test is a method of determining the surface shape (figure) of a mirror used in telescopes and other optical devices.

Sagitta In optics is where it is used to find the depth of a spherical mirror or lens.

Secondary Mirror A secondary mirror (or secondary) is the second deflecting or focusing mirror element in a reflecting telescope.

Seeing Disk Diameter The point spread function diameter (seeing disk diameter or “seeing”) is a reference to the best possible angular resolution which can be achieved by an optical telescope in a long photographic exposure and corresponds to the diameter of the fuzzy blob seen when observing a point-like star through the atmosphere.

Sled Focuser The sled-type focuser enables the whole secondary mirror together with eyepiece to move closer or further away from the primary to achieve focus instead of the usual way of racking the eyepiece in and out.

Spherical Aberration An optical effect observed in an optical device (lens, mirror, etc.) that occurs due to the increased refraction of light rays when they strike a lens or a reflection of light rays when they strike a mirror near its edge, in comparison with those that strike nearer the center.

Spherical Mirror A mirror whose surface is curved liked a sphere and is usually found in smaller telescopes with and f/10 focal ratio or higher.

Strehl Ratio A measure of optical quality that measures how much an optic deviates from perfection. A Strehl ratio of 1.0 is the best one can attain.

Truss Tube A type of round tube design used primarily on Dobsonian telescopes that uses trusses (lightweight poles) to secure and maintain the optical components in precise alignment.

Turned Down Edge An aberration that occurs when the edge does not end abruptly but curls over gradually, starting from about 80% of the way out from the center of the mirror.

Vignetting A condition caused by any portion of an incoming light column (or the resulting light cone) failing to reach the focal plane. The primary causes of vignetting are obstructions blocking the light's path or light missing a reflecting surface.

Zonal Errors Localized defects that arise during the initial figuring and polishing of optical mirrors.

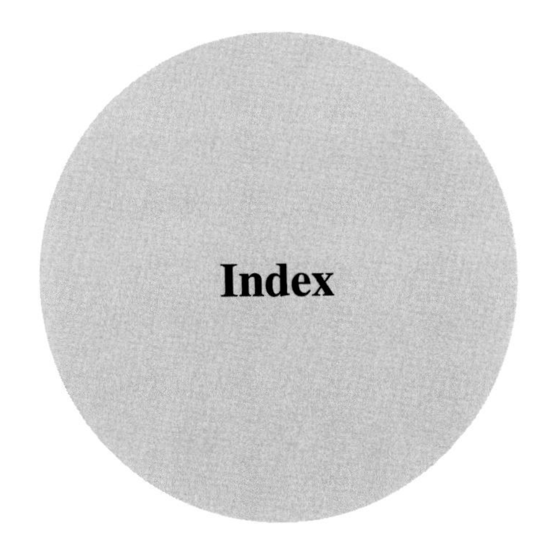

Index

N. Butler, *Building and Using Binoscopes*, The Patrick Moore Practical Astronomy Series, DOI 10.1007/978-3-319-46789-4